DETAILING FOR ACOUSTICS

DETAILING FOR ACOUSTICS

Duncan Templeton and Peter Lord
with graphics by Aidan Potter

The Architectural Press: London
Nichols Publishing Company: New York

First published in 1983 by
The Architectural Press Ltd, 9 Queen Anne's Gate,
London SW1H 9BY

ISBN 0 85139 794 8

Published in the USA by Nichols Publishing Company,
PO Box 96, New York, NY10024

Library of Congress Cataloging in Publication Data

Templeton, Duncan.
Detailing for acoustics.

1. Building—Details—Drawings.
2. Soundproofing.
3. Architectural acoustics. I. Lord, P. (Peter)
II. Title.
TH2031.T46 1983 693.8′34 83–8067
ISBN 0–89397–161–8

No guarantee is given that any detail shown herein will be suitable for a particular use, and any detail similar to or based upon those shown should not be used without expert advice as to suitability. Although some details are referenced to a particular project, there may be slight amendments from the details as built or originally designed.

Printed in Great Britain by
Biddles Ltd, Guildford, Surrey
using photosetting by Phoenix Photosetting, Chatham, Kent

CONTENTS

ACKNOWLEDGEMENTS

We are particularly grateful to the following:

Richard Cowell, Dr Judith Lang, Nils Jordan, Hadyn Bodycombe, and colleagues at BDP and University of Salford, Department of Applied Acoustics.

Material from CIRIA Report 88, 'Sound Insulation of House Separating and External Walls (with lightweight masonry for thermal insulation)' is reproduced by permission of the Director of CIRIA.

FOREWORD

Effective acoustic design—unfortunately still a Cinderella element of our architecture—depends heavily on detail. There are many references for guidance on the principles of good acoustic design, not enough tackling the details.

Here is a book which emphasizes details, and the authors set out to make it useful. Details, whether for sound separation, absorbent linings or impact isolation need to take due account of non-acoustic influences, and favourite solutions differ widely. Whether taken from product technical literature or prised from private collections these details represent a useful reference for all those involved with building.

I expect to refer to them often and hope they will help stimulate debate and development in a field where they are badly needed.

Richard Cowell
Arup Acoustics

1 OBJECTIVES

There is paucity of practical material relating to architectural acoustics, compared with academic material dealing with the mathematics and physics of sound. This book is composed largely of guidance details intended for reference by architects, students, structural and services engineers and interior designers. Some knowledge of the theory of sound is assumed. Exhaustive coverage is not claimed, nor does the content replace the need for specialist aid. Component assembly and choice of materials are 'moving targets' and sheets could be added ad infinitum to this selection of standard and built examples.

Sources
Wherever possible laboratory test results have been obtained for the standard assemblies of materials: often the basic 'favourites' could be traced to a number of sources, with minor variations on a theme. Inevitably, for light partitions and ceilings in particular, manufacturers' information has been an important source of data. Such data should be treated with some caution as results may be given in the best light. The material presented has been checked as far as possible. Exact trade name specifications and references are given only where critical, as some products are subject to frequent change and development.

Format
The material comprises details of standard elements of building construction, assemblies of elements from projects and diagrams for assessing basics. Standard elements consider both sound insulation and sound absorption aspects. In the case of the former, a single value is normally given; for sound reduction, values are in the range 100 – 3150 Hz (with values listed for octaves, ie 125, 250, 500, 1000, 2000, and 4000 Hz). The scale of details as originally drawn is generally 1:5 unless noted otherwise, but all drawings are reproduced at reduced scale—use the drawn scale on each sheet for reference. Details of specialist components—eg suspended discs—illustrate specific usage and are not for copying on another project: we want to encourage dialogue with acoustic consultants, not replace it.

Content
The stated performance of the sound insulation of component assemblies is not intended to assign each a definite value. A particular assembly does not have a 'magic number' because its performance will depend on the context of use. Taking a partition type as an example, the sound insulation performance depends not only on its sectional constituents but on its size, degree of restraint at all sides, flanking effects and receiving room absorption. The values stated are only an average and two elements with the same average may have widely varying sound transmission at a particular frequency. For this reason, some important examples have their performance illustrated throughout the frequency range.

The single-figure values are to be used to provide a choice of comparable elements of construction, consistent in acoustical terms. A 'kit of parts' of elements can be put together where no part of the whole assembly is significantly weaker than the rest in performance. Doors and windows are problematic in that they are inherently lighter and gap-prone, but all is not lost when doors and windows are not up to the performance of the surrounding walls, floors and ceiling, and composite performance of assemblies can be assessed. A philosophy that the whole is only as good as the weakest link is analogous to components in a sound-reproduction system. Care in detailing junctions is particularly important where discontinuous construction details are used to achieve good sound isolation. Cost-cutting exercises through the design stage should not allow part of the 'kit' to be downrated to meet budget.

In the case of sound-absorption values, the figures should be used together with the tabulated values to sum the total effect of absorption in a particular space. In a large-area space, for example, an open-plan office, the characteristics of the ceiling and floor will be of particular concern.

Information can only reflect the current 'state of the art'. There are exciting developments in materials technology, for instance in the development of thin but very effective isolating membranes. In a future of lighter buildings and higher external noise levels, such materials could be a valuable asset to the designer.

Design process
For basic knowledge, use a primer text: eg the classic 'Acoustics Noise and Buildings', 4th Edition, Parkin, Humphreys and Cowell. The details we give here are intended to help at the practical rather than the crucial concept end of the acoustical design process. Stages of acoustical design input are summarized to show the total picture through a project. Acoustic consultants should be approached where a close check of characteristics is required.

1 Briefing
The client may state required standards in the finished building or, more likely, the functions stated dictate the fabric performance. Target criteria should be set at this stage.

2 Site selection
If a choice of sites is available noise sources may be included with other environmental criteria in a scoring matrix to determine the most suitable.

3 Noise survey
Background noise levels may be taken to assist brief-making studies or to determine the best position of a building on site. Quantification of adjacent noise sources may influence the design of the external envelope, in order to avoid noise intrusion.

4 Outline design
The form of the building may allow 'buffer' screening of sensitive spaces, and the arrangement of rooms may allow those requiring quiet conditions to be remote from noisy rooms, with fenestration limited and not facing any external noise source.

5 Detail design
In order that rooms can be sized to suit function and good aural reception, proportion, volume/occupant, reverberation time and surface geometry may be considered and adjusted. The quantity, shape and specification of finishes may be made optimal to required sound reflection, diffusion or absorption properties. The required separation between rooms or to outside noise is designed-in by each element having adequate sound insulating properties. Noise-control standards for services should be compatible with fabric-design standards: air-conditioning systems, plumbing, lighting, generators and switchgear plant are frequent noise culprits in large buildings. Design of specialist sound systems to suit the acoustics of the spaces is finalized at this stage.

6 Supervision
Care must be taken in getting many of the assemblies shown built properly, as some principles go against the grain of normal building practice—discontinuous construction at junctions and cavity walls with no ties will tempt the average contractor used to tying the whole lot together for stability.

7 Commissioning
Testing and tuning of services systems and sound systems and checking that sound insulation standards are provided in practice are advisable as the logical follow-up to care in detailing.

2 DETAILS

ROOFS

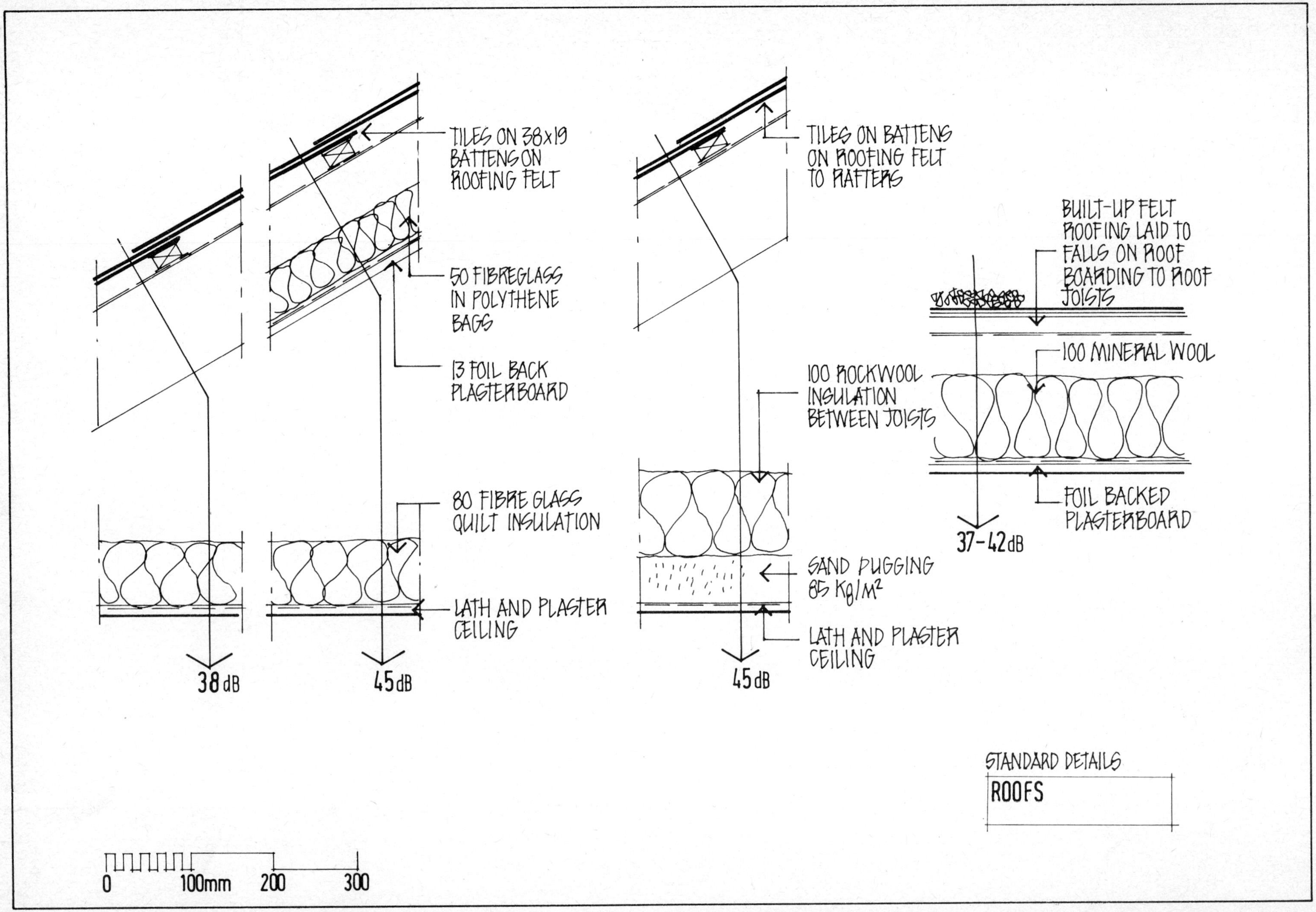
TILES ON 38x19 BATTENS ON ROOFING FELT
50 FIBREGLASS IN POLYTHENE BAGS
13 FOIL BACK PLASTERBOARD
80 FIBRE GLASS QUILT INSULATION
LATH AND PLASTER CEILING
38 dB
45 dB
TILES ON BATTENS ON ROOFING FELT TO RAFTERS
100 ROCKWOOL INSULATION BETWEEN JOISTS
SAND PUGGING 85 Kg/M²
LATH AND PLASTER CEILING
45 dB
BUILT-UP FELT ROOFING LAID TO FALLS ON ROOF BOARDING TO ROOF JOISTS
100 MINERAL WOOL
FOIL BACKED PLASTERBOARD
37-42 dB
STANDARD DETAILS
ROOFS
0
100mm
200
300

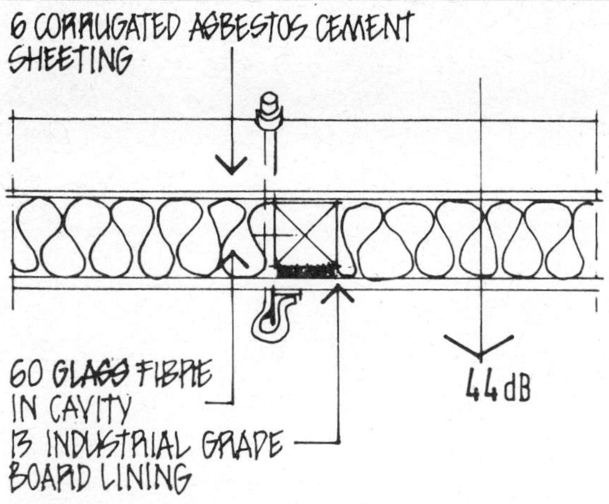

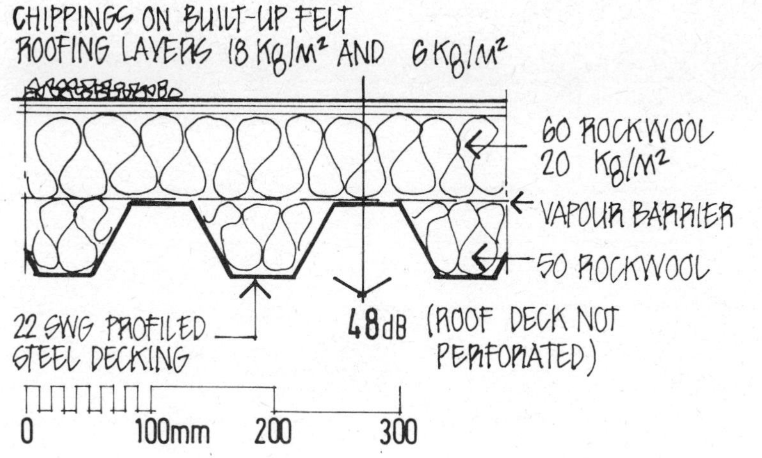

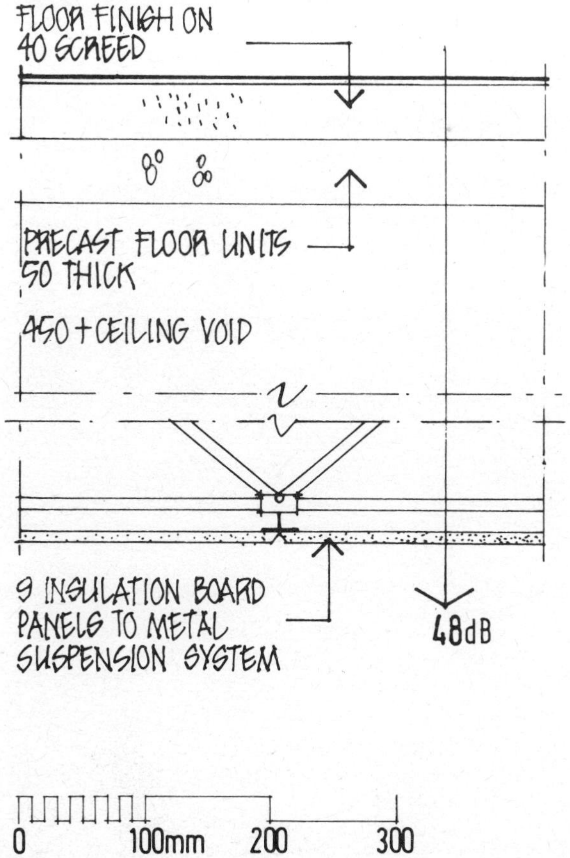

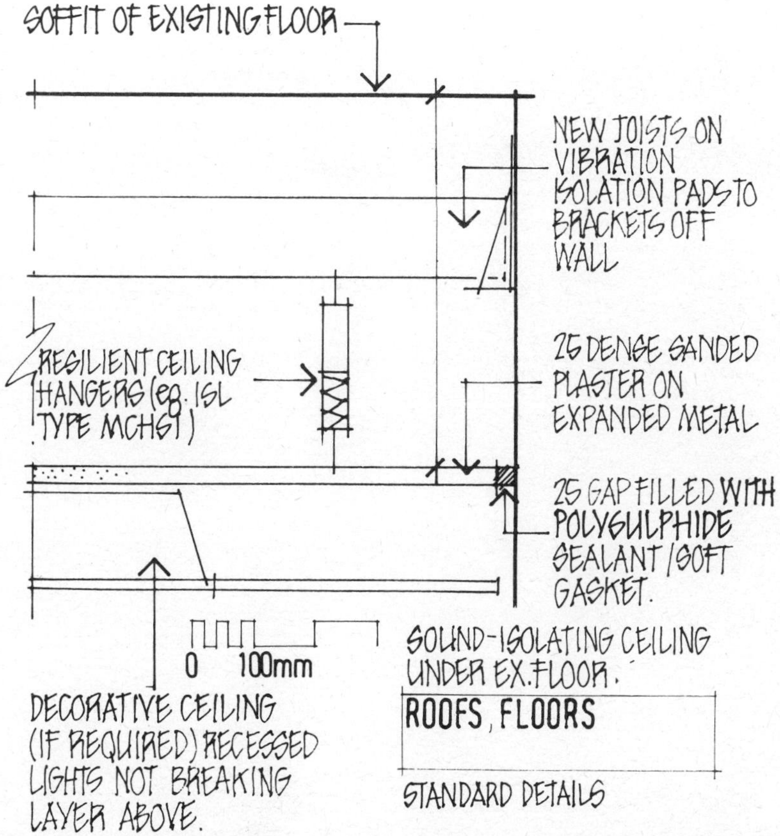

SOUND-ISOLATING CEILING UNDER EX.FLOOR.

ROOFS, FLOORS

STANDARD DETAILS

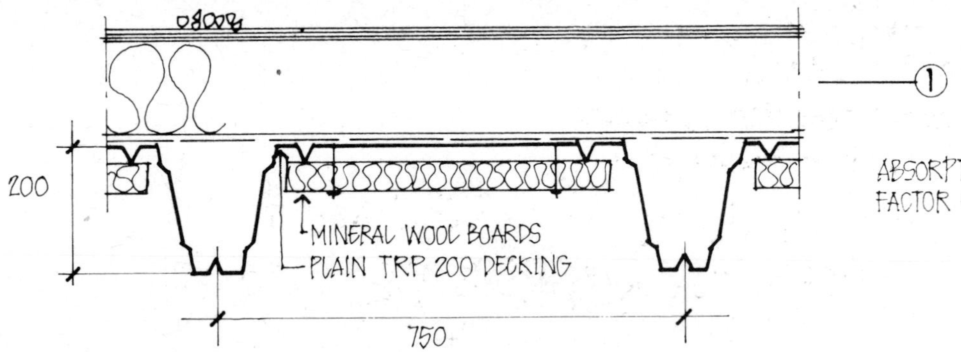

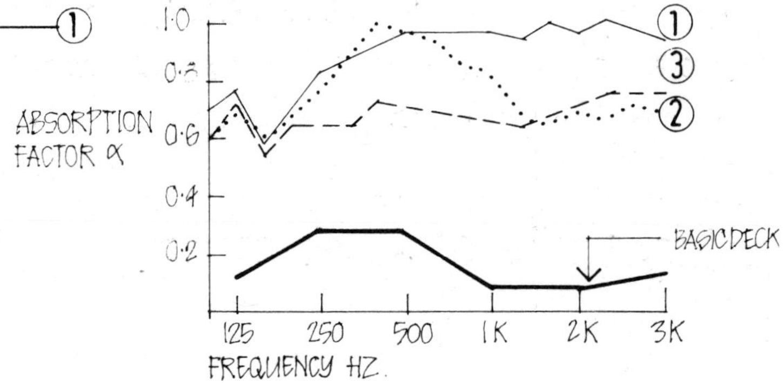

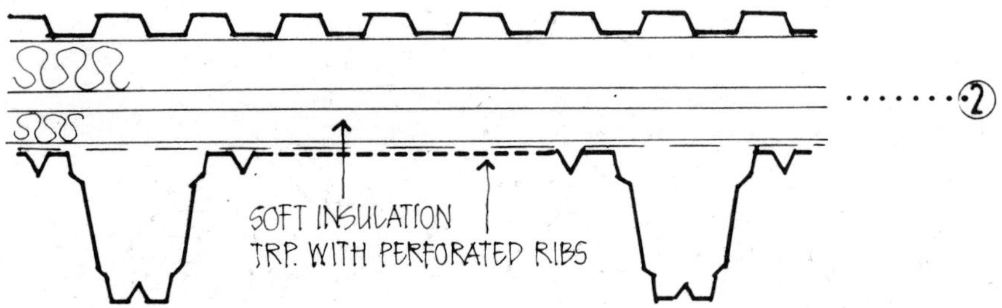

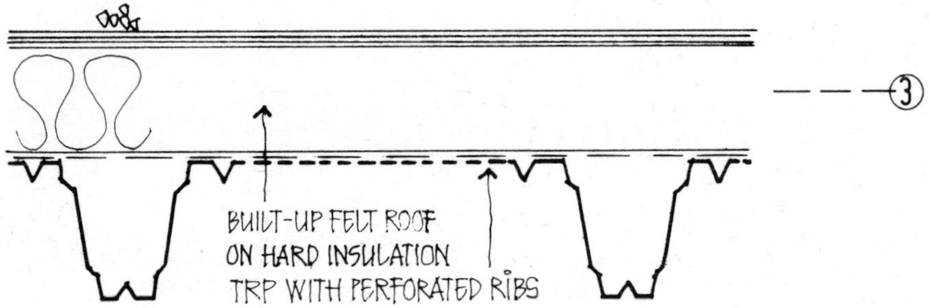

STEEL DECKING CAN BE PROBLEMATIC WHEN EXPOSED INTERNALLY IN INDUSTRIAL OR EVEN OFFICE BUILDINGS. THIS FORM OF ROOFING HAS DEEP RIBS WHICH DIFFUSE REFLECTED SOUND AT THE HIGHER [SPEECH] FREQUENCIES, AND A GOOD PROPORTION OF PERFORATIONS LOCATED ON PART OF THE DECKING NOT CRITICAL FOR STRENGTH.

IN ASSESSING PERFORATED STEEL DECK ROOFS, CHECK GAUGE OF STEEL AND THICKNESS OF MEMBRANE IMMEDIATELY ABOVE THE PERFORATIONS – BOTH CAN AFFECT α.

HEAVY LINE IS AVERAGE VALUE OF STEEL 'PLANNJA' DECK WITH NO PERFORATIONS OR ABSORPTION

ROOFS
SOURCE: PLANNJA

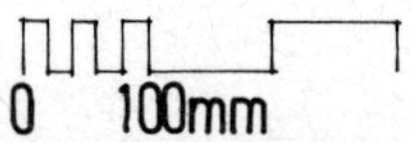

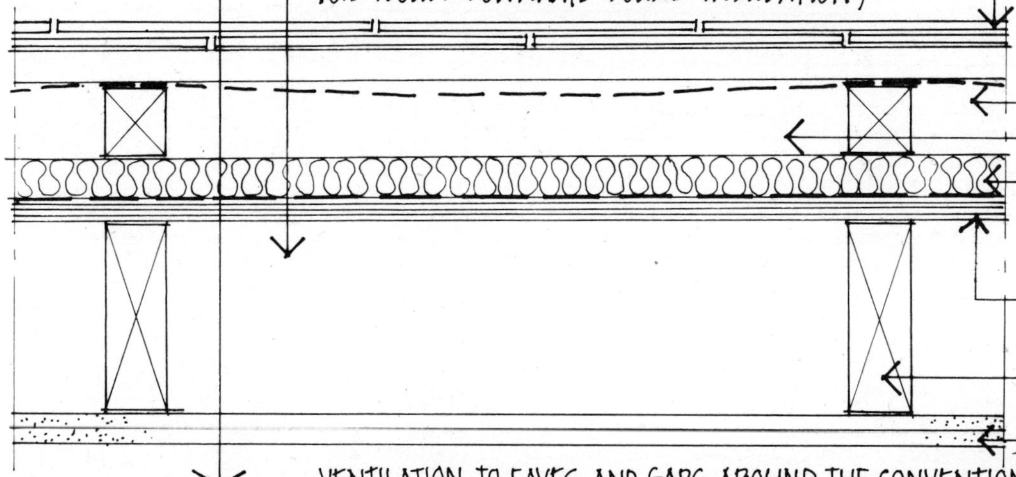

VENTILATION TO EAVES AND GAPS AROUND THE CONVENTIONAL ROOF FINISH OF SLATES OR TILES/BATTENS/ROOFING FELT/JOISTS SEVERELY LIMIT ITS SOUND INSULATION PROPERTIES. IN THIS EXAMPLE, A TRADITIONAL SLATED ROOF FINISH WAS REQUIRED FOR THE REBUILDING OF A CONCERT HALL. LONG SPANS AND EXISTING SUPPORT MASONRY PREVENTED USE OF A 'HEAVY ROOF,' BUT A REASONABLE PERFORMANCE WAS REQUIRED TO MEET NR=20 WITHIN THE HALL. OLDER BUILDINGS HAVE INHERENTLY MORE MASSIVE CONSTRUCTION AND THE MOST SHOULD BE MADE OF THIS. IN A CONVERSION SCHEME TO PUT TV/FILM STUDIOS IN A LISTED WAREHOUSE BUILDING, THE EXISTING HEAVY ROOF BOARDING WAS RETAINED AND INSULATION BOARD/VAPOUR BARRIER, COUNTER BATTENS, ROOF FELT & SLATES ADDED TO GIVE A BUILD UP OF LAYERS VERY SIMILAR TO THIS DETAIL. INTERLOCKING CONCRETE ROOF TILES ARE HEAVIER AND MORE CLOSELY FITTING THAN NATURAL OR ASBESTOS CEMENT SLATES, AND SO WOULD GIVE MARGINALLY BETTER SOUND INSULATION.

ROOFS
SOURCE: BDP.

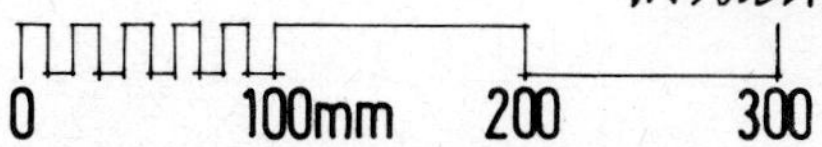

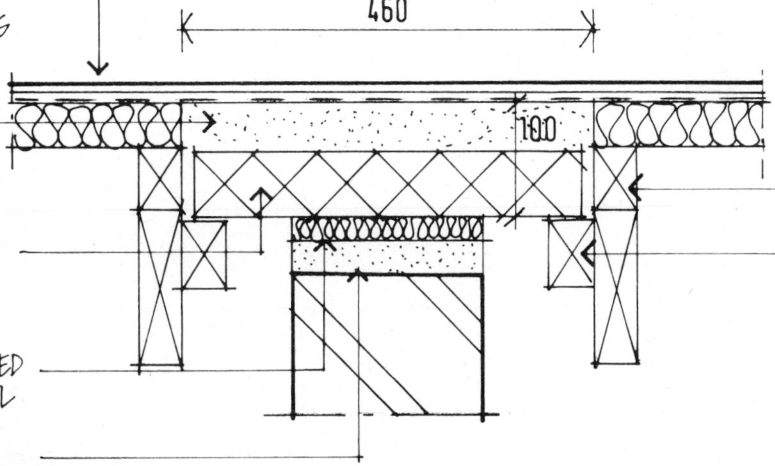

PURPOSE OF DETAIL IS PRIMARILY TO MEET THE LONDON BUILDING BYE LAWS WITH RESPECT TO FIRE SEPARATION. THE CONSTRUCTION SERVES EXCELLENTLY, HOWEVER, TO PREVENT NOISE FLANKING AT THE PARTY WALL–ROOF SOFFIT JUNCTION.

AVOIDANCE OF NOISE FLANKING OVER PARTY WALL
NOT TO SCALE.

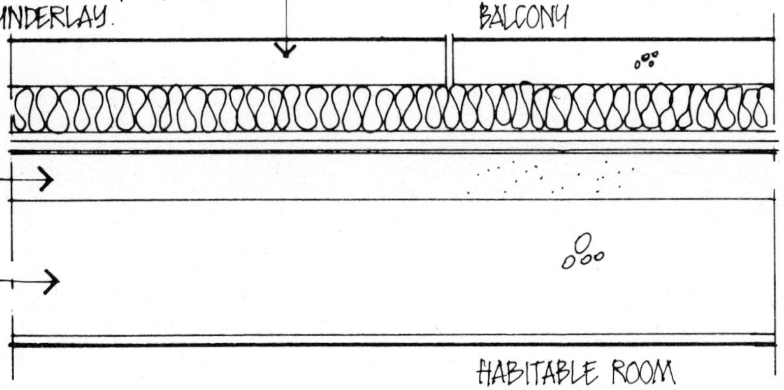

THE INCORPORATION OF THERMAL INSULATION TO MEET PART F OR FF GIVES THE OPPORTUNITY TO ISOLATE STRUCTURE FROM IMPACT SOUND. SIMILARLY ROOF PARKING CAN HAVE IMPACT CAR NOISE ISOLATED BY WEARING DECK SEPARATED FROM STRUCTURAL SLAB BY WATERPROOF MEMBRANE AND RIGID INSULATION BOARDS – THE 'UPSIDE DOWN' ROOF PRINCIPLE.

ISOLATION FROM IMPACT SOUND.
NOT TO SCALE

ROOFS
SOURCE: GLC

not to scale

CEILINGS

SOLID
SOFFIT.

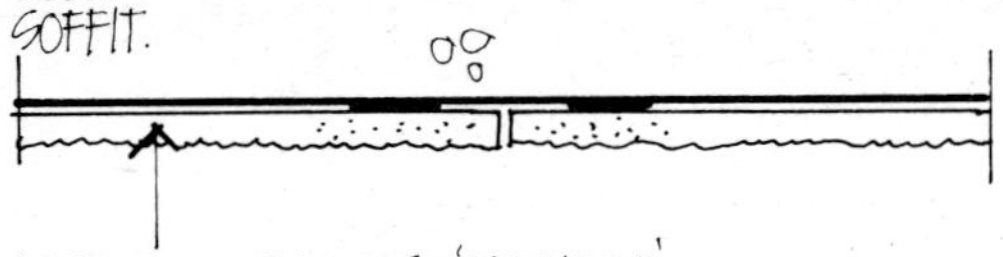

15 HEAVILY FISSURED 'KOMFORT' CEILING TILES GLUED DIRECT TO SOFFIT

0·09 0·30 0·82 0·95 0·76 0·61

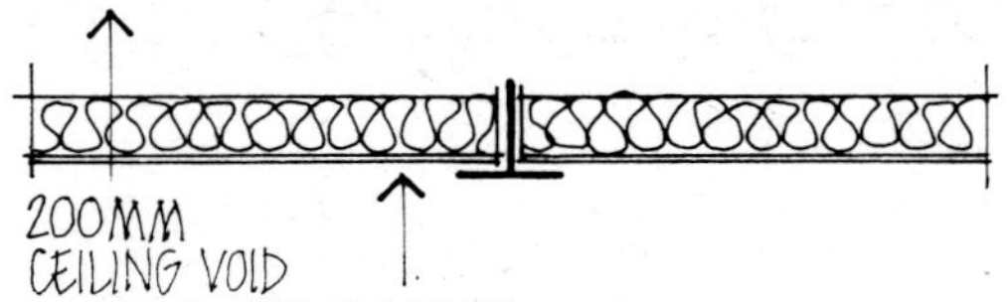

200MM
CEILING VOID

25 LIGHTWEIGHT LAY-IN GLASSFIBRE SUSPENDED CEILING TILES, MATT WHITE COAT FINISH TO UNDERSIDE (TYPE: ROCLAINE DIAPAISON 'P')

0.33 0.84 0.84 0.88 0.89 0.92

40 FIBREGLASS TILES OVER 400 VOID
0.70 0.75 0.65 0.75 0.60 0.35

300 VOID

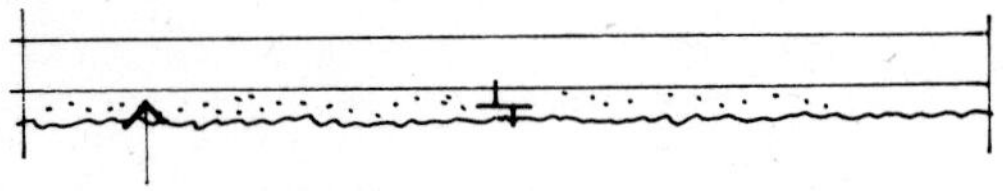

15 HEAVILY FISSURED CONCEALED GRID PLASTER 'GYPTONE' CEILING TILES. METAL SUSPENSION SYSTEM

0.30 0.35 0.40 0.55 0.80 0.70

ABSORPTION FIGURES RELATE TO FREQUENCIES.
125Hz. 250 500 1K 2K 4K

300 CEILING VOID

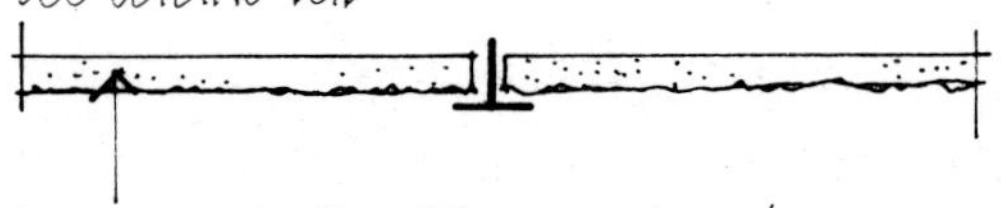

15 FISSURED-FINISH ARMSTRONG 'CERAMGUARD' KILN-FIRED MINERAL FIBRE TILES (FOR USE IN CORROSIVE OR HUMID SITUATION EG. SWIMMING POOLS)

0.25 0.25 0.45 0.70 0.08 1.10

STANDARD DETAILS
CEILINGS
SOUND ABSORPTION

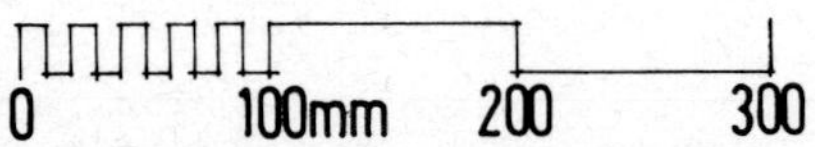

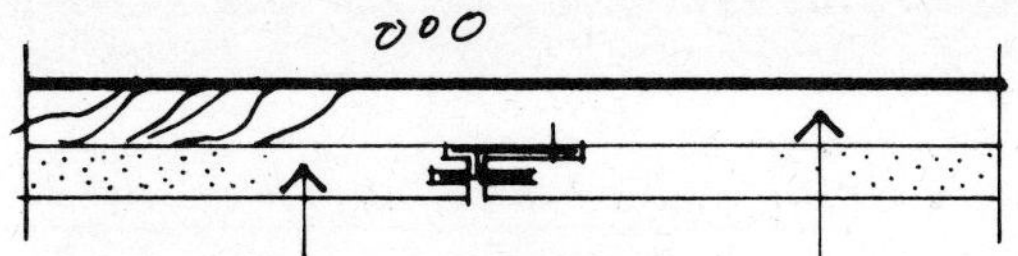

FABRIC COVER TO PROPRIETARY
PANELS 600 WIDE. PANELS
19 PERFORATED MINERAL
FIBREBOARD BY ARMSTRONG.
50×25 BATTENS ON MASONRY

0.22 0.58 0.56 0.72 0.76 0.81

50×25 BATTENS

25 AIR GAP

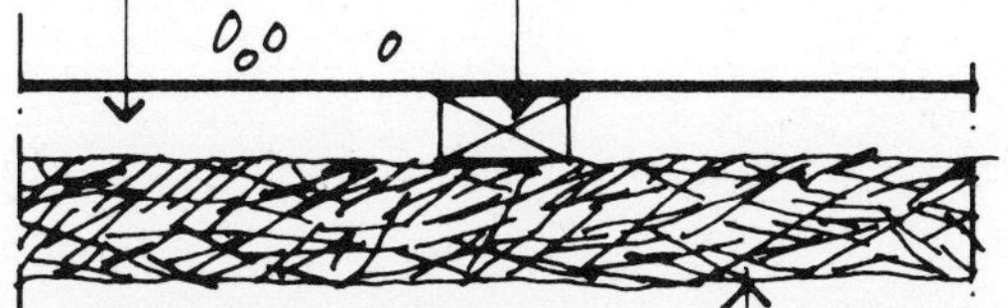

50 WOODWOOL / CEMENT
BUILDING SLABS (PRESCREEDED WOODCEMAIRE)

0.30 0.40 0.50 0.85 0.50 0.65

0.15 0.20 0.55 0.75 0.65 0.85 (NO AIR GAP)

ABSORPTION FIGURES RELATE TO
FREQUENCIES
125 250 500 1K 2K 4K.

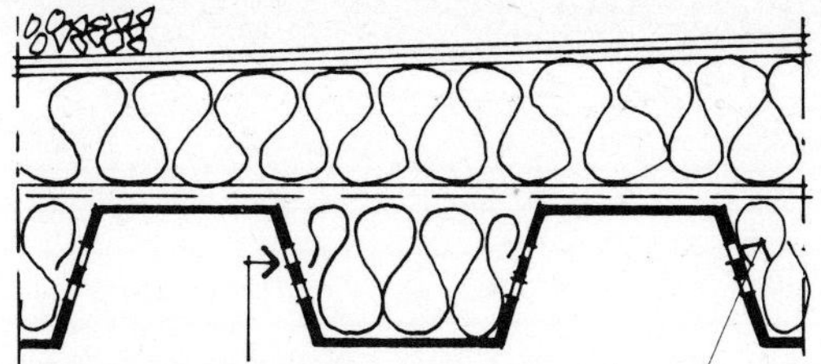

5 PERFORATIONS TO PLANNJA
PROFILED DECKING
EXPOSED TO UNDERSIDE

MINERAL WOOL PADS
NOTE: PERFORATIONS MEAN THAT ROOF
HAS POOR SOUND INSULATION PROPERTIES

0.20 0.50 0.80 0.85 0.60 0.50

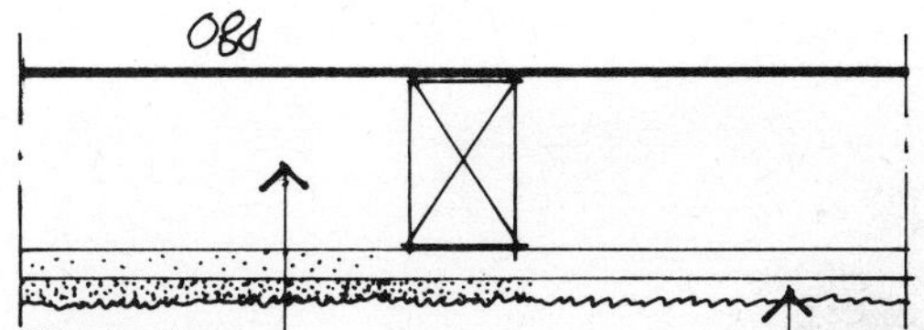

75 AIR GAP
13 ACOUSTIC PLASTER (SPRAY
APPLIED EG. 'AUDEX') ON 13
PLASTERBOARD BACKING
0.30 0.35 0.55 0.70 0.85 0.95

STANDARD DETAILS.

CEILINGS / WALL FINISHES
SOUND ABSORPTION

CEILING VOID 370

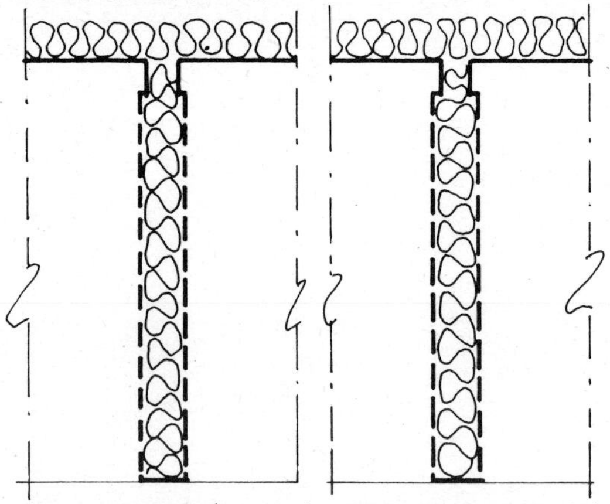

PROPRIETARY 'DAMPA' BAFFLES OF PERFORATED ALUMINIUM ON 600 X 600 MODULE PACKED WITH 25 MINERAL WOOL PACKED IN POLYTHENE FOIL. 22% PERFORATIONS.

0·22 0·27 0·61 0·71 0·71 0·80

CEILING VOID 125

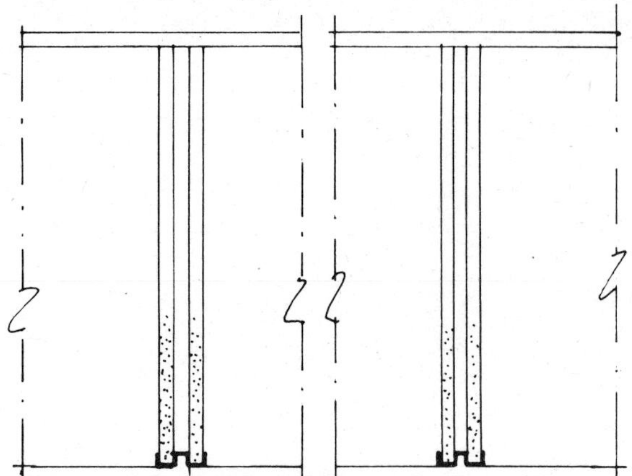

ARMSTRONG FISSURED 15 THICK TWIN PANELS OF MINERAL WOOL

0·10 0·30 0·55 1·35 1·15 0·90

CEILING VOID

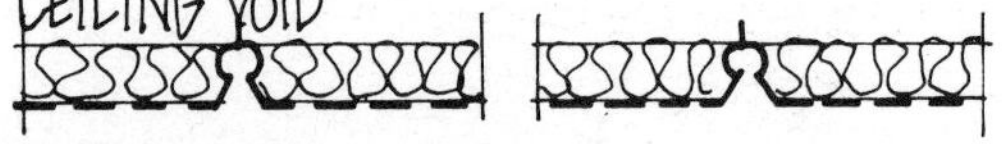

PERFORATED 'DAMPA' TRAYS WITH 25 MINERAL WOOL PADS ON POLYTHENE FOIL. 600 X 600 MODULE IN ALUMINIUM OR STEEL

0·68 0·66 0·57 0·58 0·52 0·62

CEILING VOID.

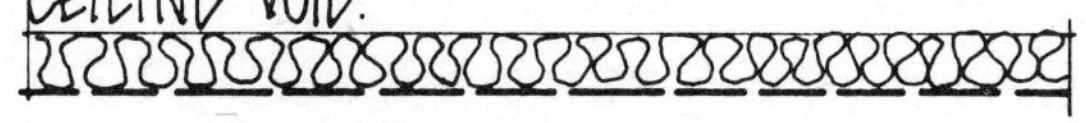

'ROCLAINE' METAL PANELS 5% PERFORATED WITH 20 GLASS FIBRE FACED IN BLACK FIBRE FELT.

0·13 0·27 0·55 0·79 0·90 1·00

ABSORPTION FIGURES RELATE TO FREQUENCIES:

125Hz 250 500 1K 2K 4K

STANDARD DETAILS

CEILINGS
SOUND ABSORPTION.

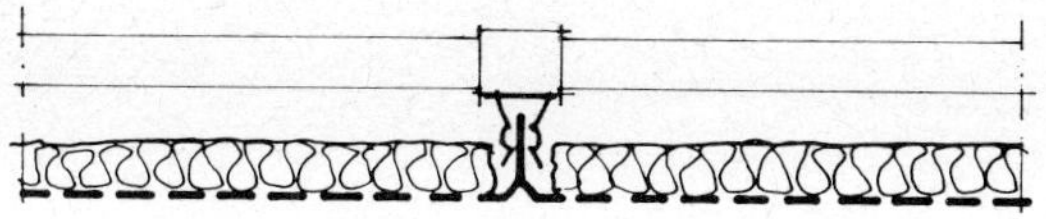

PERFORATED METAL TILES CLIPPED
TO SUPPORT FRAMING, 25 PROPRIETARY
PADS - DENSITY 64 KG/M2.

SOUND ABSORPTION:
0·20 0·55 0·80 0·80 0·80 0·75

SOUND ATTENUATION (BS 2750 ROOM TO ROOM)
32 dB (40 dB FOR SOLID FACE TILE)

LINEAR AIR DIFFUSER

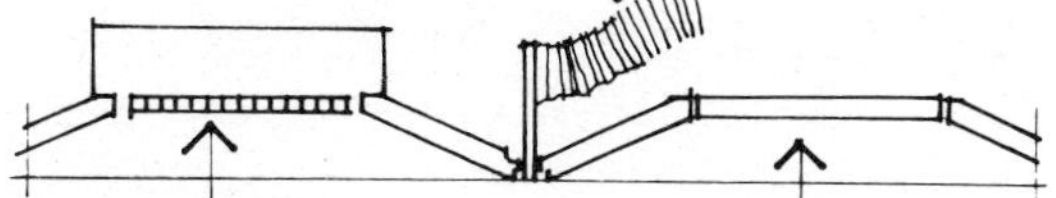

LIGHT FITTINGS IN
900 MODULAR COFFERED
CEILING PANELS

PERFORATED METAL SKIN WITH
MINERAL WOOL BACKING
NOT TO SCALE
0·45 0·50 0·75 0·90 0·90 0·90

METAL DECK CLADDING
OR ROOF FINISH, SPACER

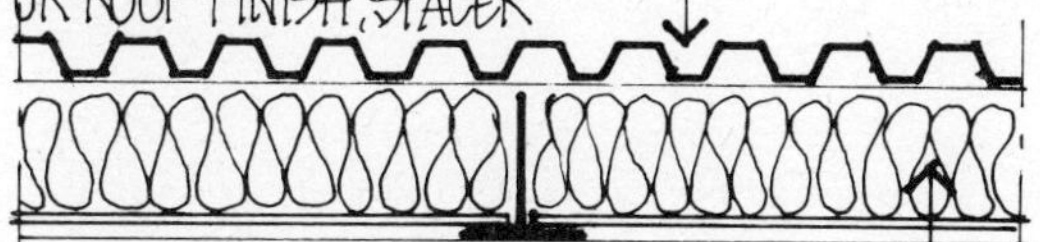

FACTORY LINER' FIBREGLASS LTD.
PANELS 50 THICK, IN METAL
TEE-SECTION FRAMING TO PURLINS
0·45 1·00 0·90 0·70 0·50 0·35
STANDARD DETAILS.

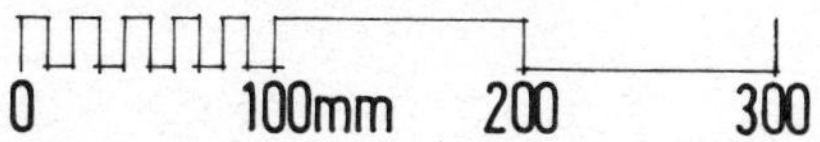

VARYING THICKNESSES OF
MINERAL WOOL ON
LIGHTWEIGHT CEILING

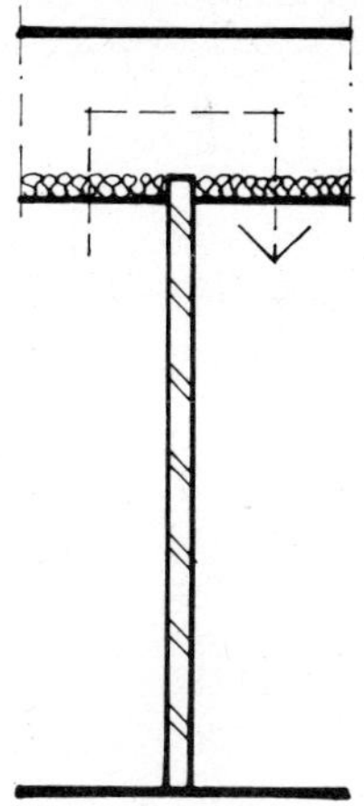

SECTION. NTS

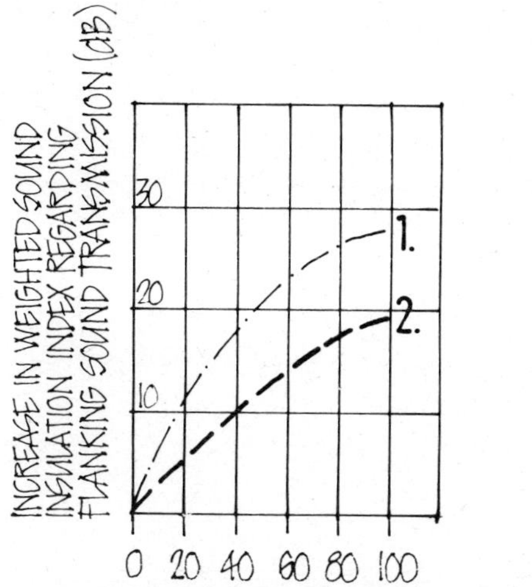

THICKNESS OF MINERAL WOOL
1. AIRTIGHT CEILINGS
2. SUSPENDED PROPRIETARY
CEILINGS - NO AIRTIGHT
JOINTS
SOURCE: DR. LANG.
TECHNOLOGISCHES
GEWERBE MUSEUM
VIENNA.

CEILINGS

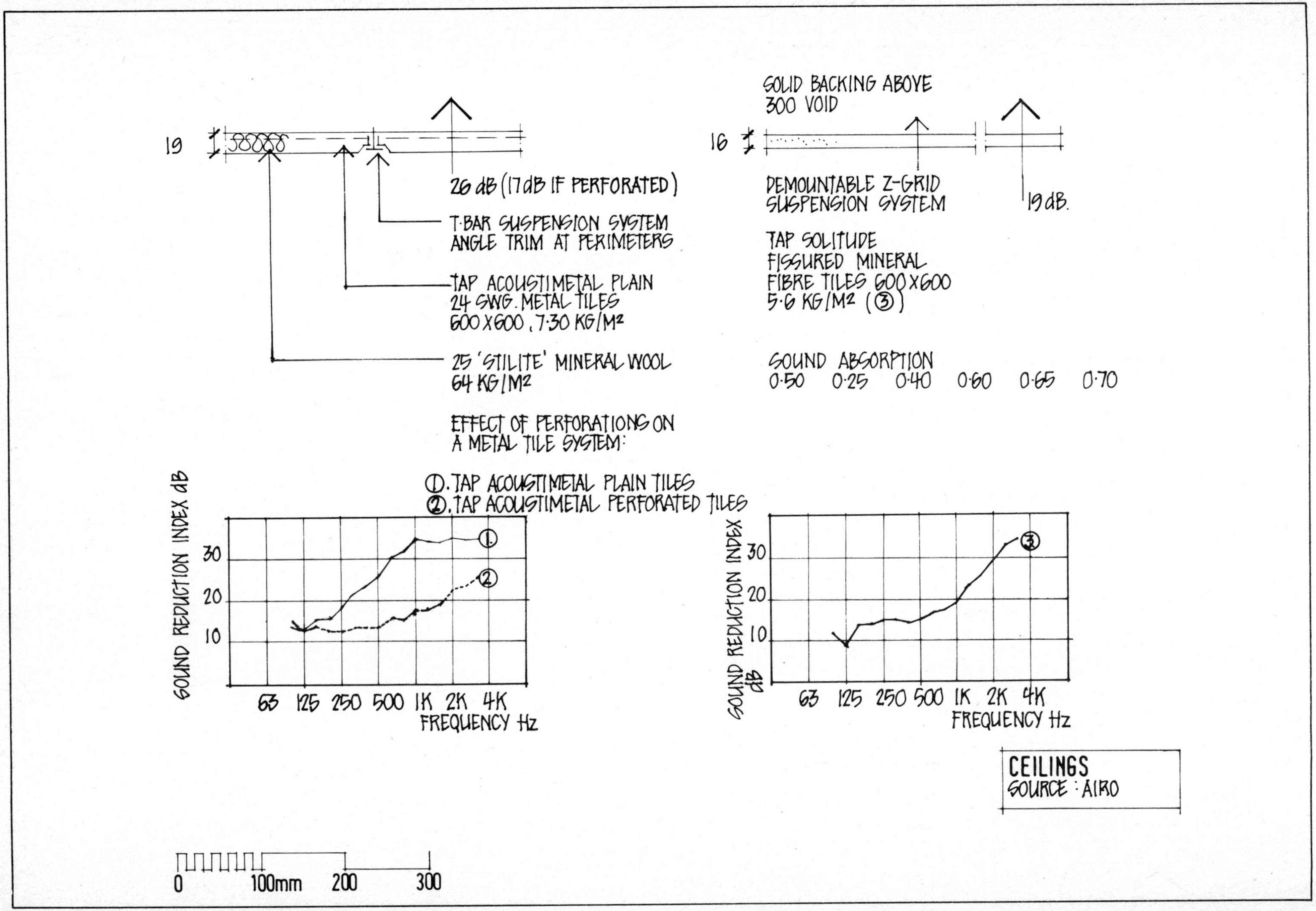
19
26 dB (17dB IF PERFORATED)
T·BAR SUSPENSION SYSTEM
ANGLE TRIM AT PERIMETERS
TAP ACOUSTIMETAL PLAIN
24 SWG. METAL TILES
600 X 600, 7·30 KG/M²
25 'STILITE' MINERAL WOOL
64 KG/M²
EFFECT OF PERFORATIONS ON
A METAL TILE SYSTEM:
SOLID BACKING ABOVE
300 VOID
16
DEMOUNTABLE Z-GRID
SUSPENSION SYSTEM
19 dB.
TAP SOLITUDE
FISSURED MINERAL
FIBRE TILES 600 X 600
5·6 KG/M² (③)
SOUND ABSORPTION
0·50 0·25 0·40 0·60 0·65 0·70
①. TAP ACOUSTIMETAL PLAIN TILES
②. TAP ACOUSTIMETAL PERFORATED TILES
SOUND REDUCTION INDEX dB
30
20
10
63 125 250 500 1K 2K 4K
FREQUENCY Hz
CEILINGS
SOURCE : AIRO
0 100mm 200 300

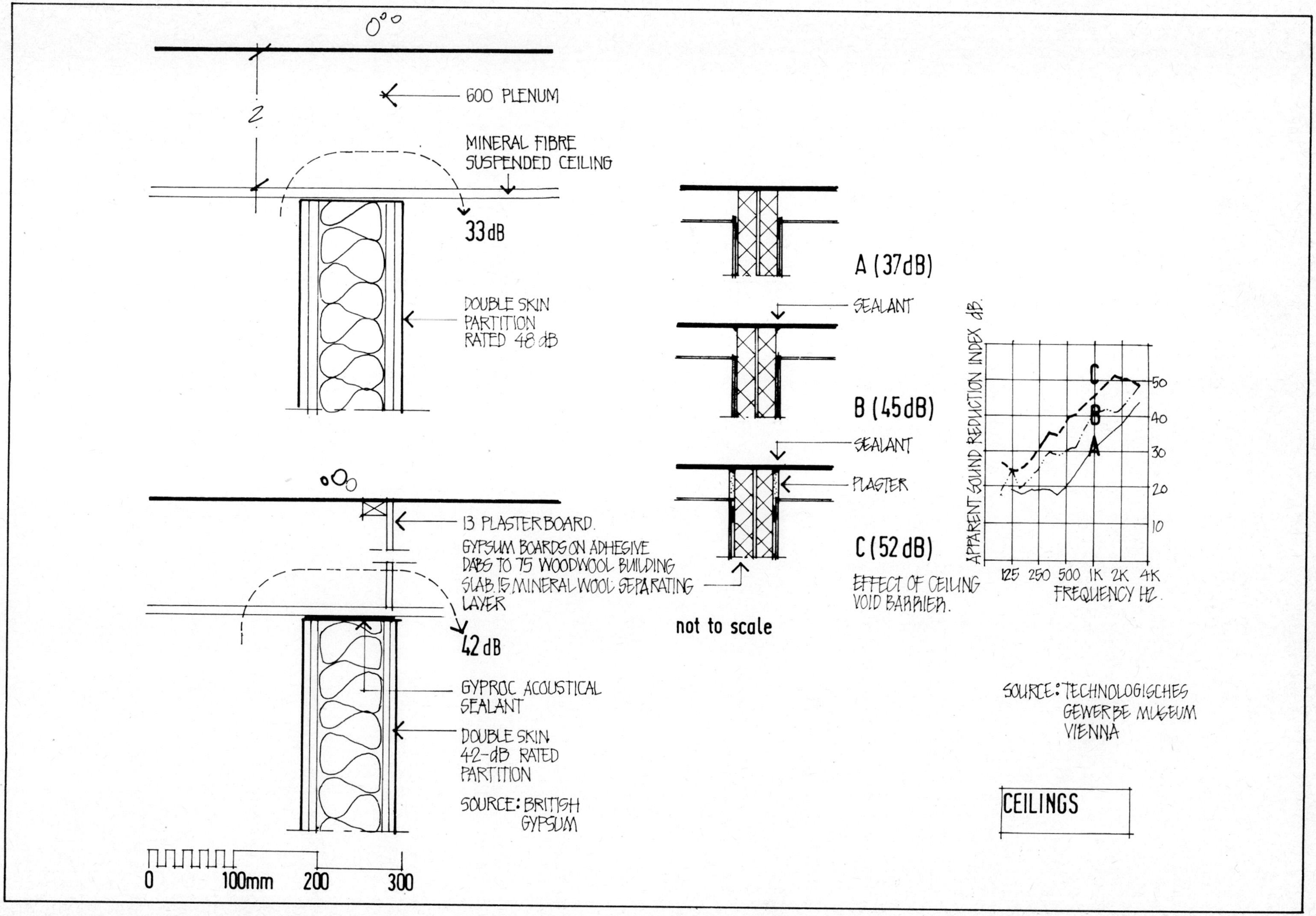
600 PLENUM
MINERAL FIBRE SUSPENDED CEILING
33dB
DOUBLE SKIN PARTITION RATED 48 dB
13 PLASTERBOARD.
GYPSUM BOARDS ON ADHESIVE DABS TO 75 WOODWOOL BUILDING SLAB. 15 MINERAL WOOL SEPARATING LAYER
42 dB
GYPROC ACOUSTICAL SEALANT
DOUBLE SKIN 42-dB RATED PARTITION
SOURCE: BRITISH GYPSUM
0
100mm
200
300
A (37dB)
SEALANT
B (45dB)
SEALANT
PLASTER
C (52 dB)
not to scale
EFFECT OF CEILING VOID BARRIER.
APPARENT SOUND REDUCTION INDEX dB.
50
40
30
20
10
C
B
A
125
250
500
1K
2K
4K
FREQUENCY HZ.
SOURCE: TECHNOLOGISCHES GEWERBE MUSEUM VIENNA
CEILINGS

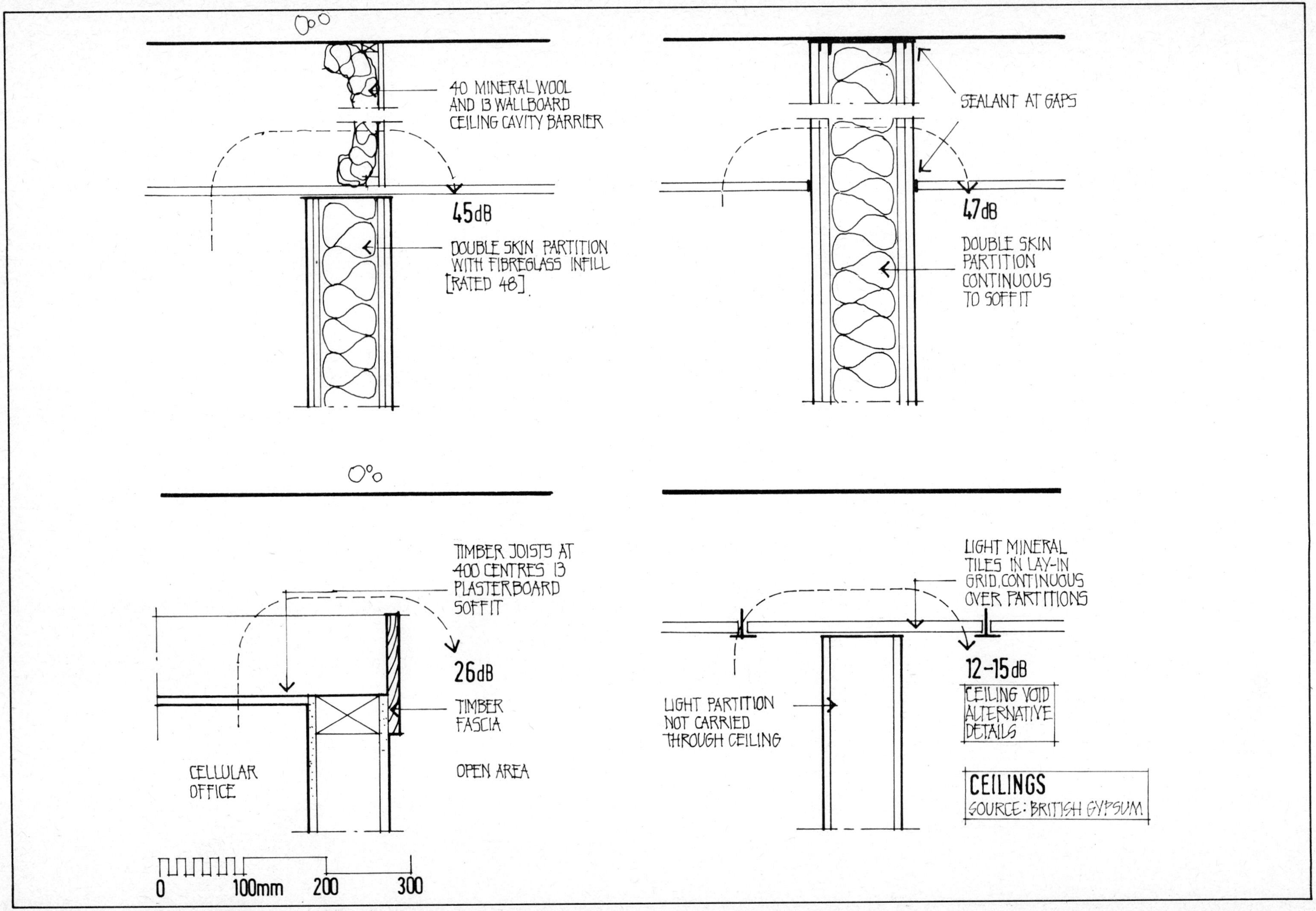

40 MINERAL WOOL AND 13 WALLBOARD CEILING CAVITY BARRIER
45dB
DOUBLE SKIN PARTITION WITH FIBREGLASS INFILL [RATED 48]
SEALANT AT GAPS
47dB
DOUBLE SKIN PARTITION CONTINUOUS TO SOFFIT
TIMBER JOISTS AT 400 CENTRES 13 PLASTERBOARD SOFFIT
26dB
TIMBER FASCIA
CELLULAR OFFICE
OPEN AREA
LIGHT MINERAL TILES IN LAY-IN GRID, CONTINUOUS OVER PARTITIONS
12–15dB
CEILING VOID ALTERNATIVE DETAILS
LIGHT PARTITION NOT CARRIED THROUGH CEILING
CEILINGS
SOURCE: BRITISH GYPSUM
0
100mm
200
300

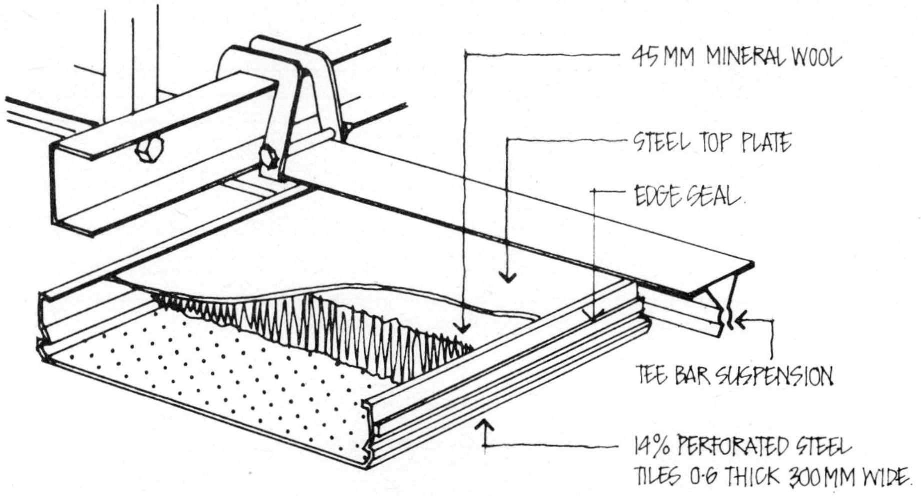

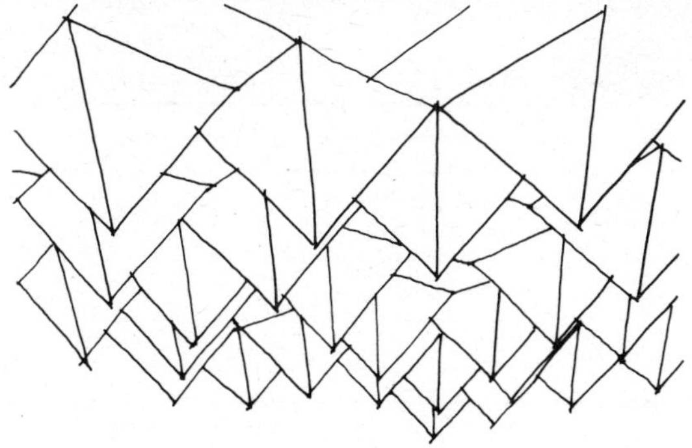

SPECIAL METAL CEILING UNITS 600 X 600 X 0·6 THICK PYRAMIDS FORMING A SOUND DIFFUSING CEILING

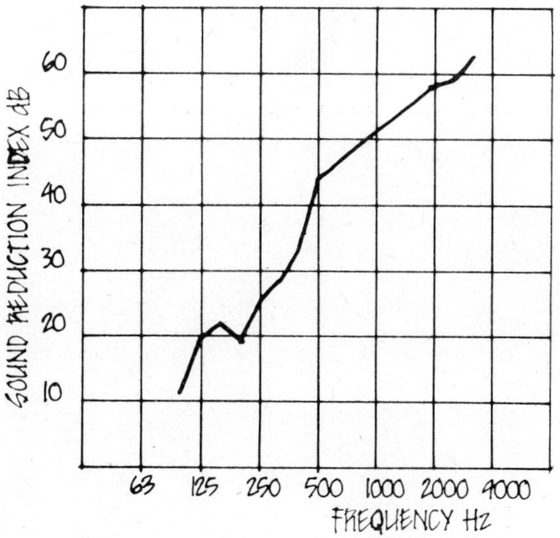

SOUND ABSORPTION
0.50 0.70 0.80 1·00 1·00 1·00

UNPERFORATED METAL TILES CAN GIVE ROOM-TO-ROOM (BS 2750) AVERAGE NORMALIZED S.L. DIFFERENCE 41 dB.

PROPRIETARY CEILING TILE FOR SOUND INSULATION AND SOUND ABSORBTION

CEILINGS
SOURCE: BURGESS ARCHITECTURAL PROD LTD.

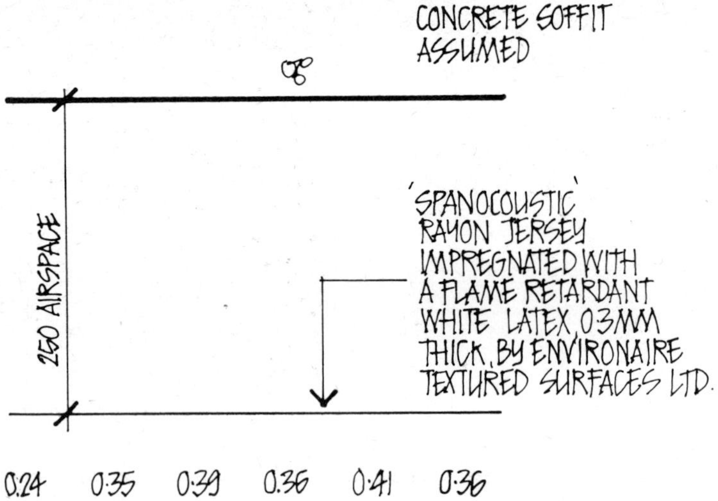

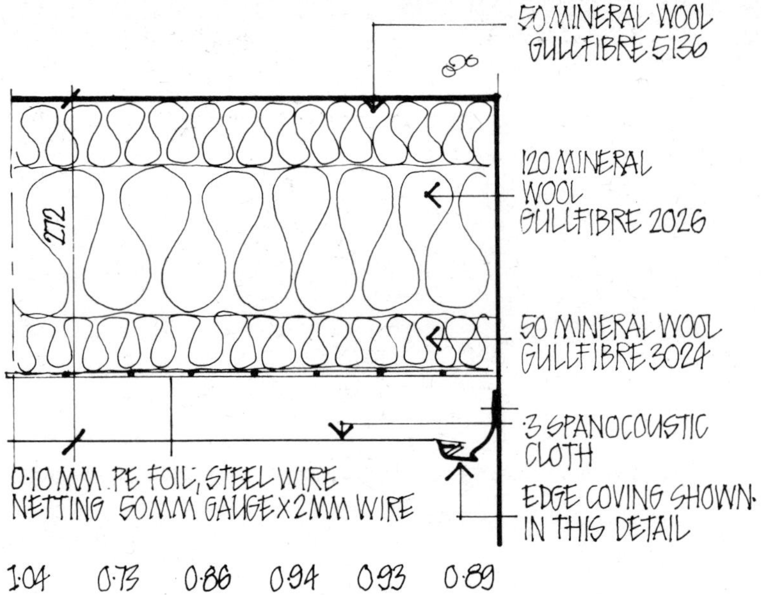

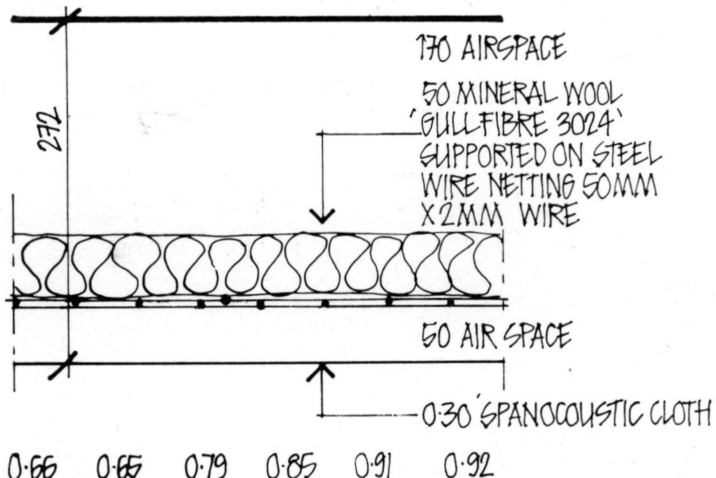

(TEST STANDARD: ISO/R354, SIMILAR TO BS 3638/1963)

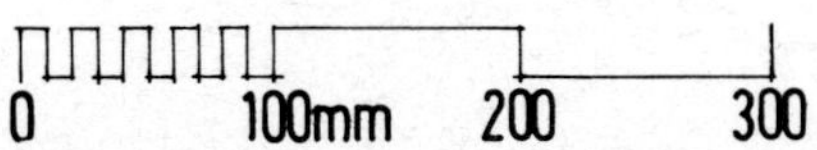

THIS IS A RECENT INNOVATION TO THE U.K.: 'MOIST TEXTILE MATERIAL IS STRETCHED TO EDGE BATTENS AND A ONE-OFF EVAPORATION RESULTS IN A TAUT CEILING MEMBRANE UNAFFECTED BY SUBSEQUENT CHANGES IN HUMIDITY AND TEMPERATURE. CLAIMED LIFE EXPECTANCY IS 20 YRS. NO FIRE PROTECTION GIVEN BY THIS CEILING, WHICH IS TESTED TO BS 476: PART 7: CLASS 1. A LIGHT CEILING OF THIS FORM MAY BE CONSIDERED IN REFURBISHMENT WHERE WEIGHT OF SUSPENDED CEILING WOULD BE PROBLEMATIC.

CEILINGS

SOURCE: UNIVERSITY OF TECHNOLOGY, LUND, SWEDEN

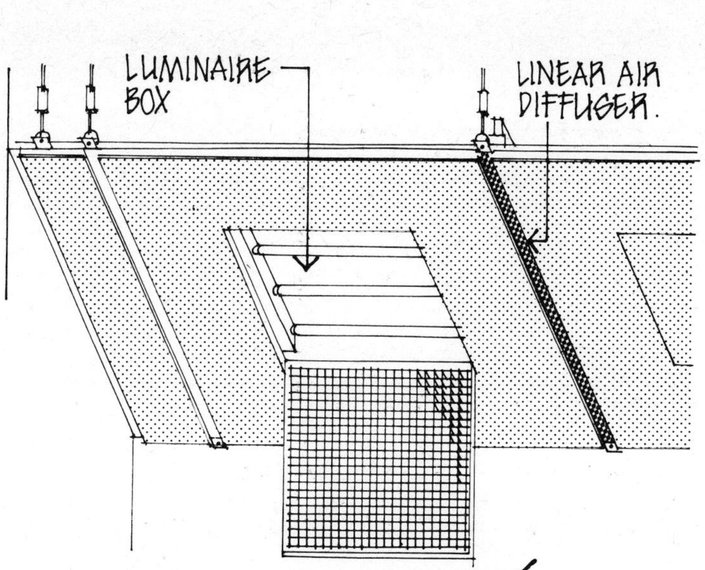

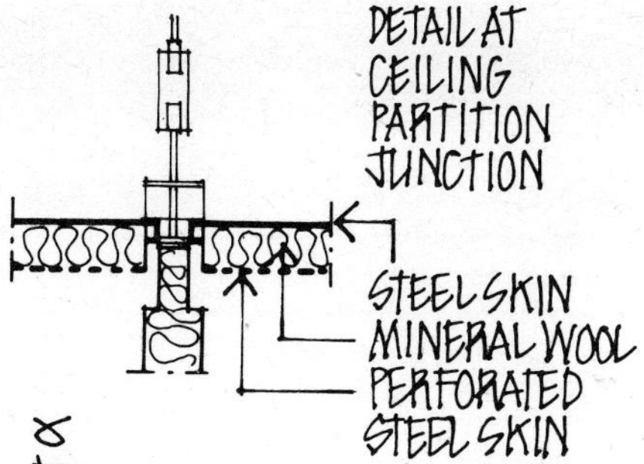

WHEN SOUND REDUCTION AS WELL AS ABSORPTION IS REQUIRED OF A CEILING, CARE SHOULD BE TAKEN THAT CEILING FITTINGS DO NOT DE-RATE. IN THIS EXAMPLE, LIGHT BOXES ARE FULLY ENCLOSED TO MAINTAIN SEPARATION, PARTITION JUNCTIONS ARE SEALED, AND AIR DIFFUSERS CAN BE LINEAR, SQUARE OR ROUND, ALL WITH SEALED PLENUM BOXES BEHIND.

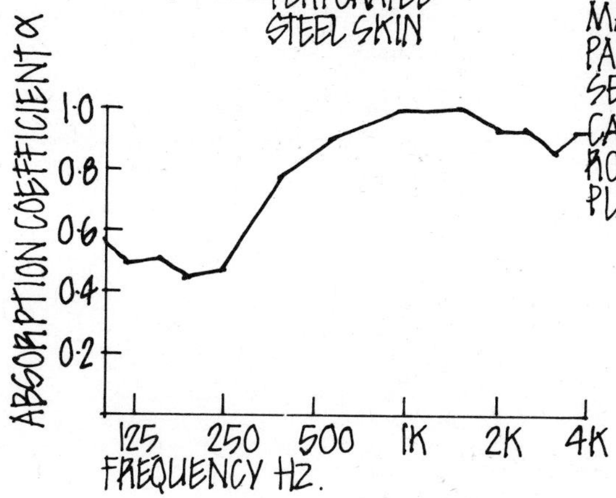

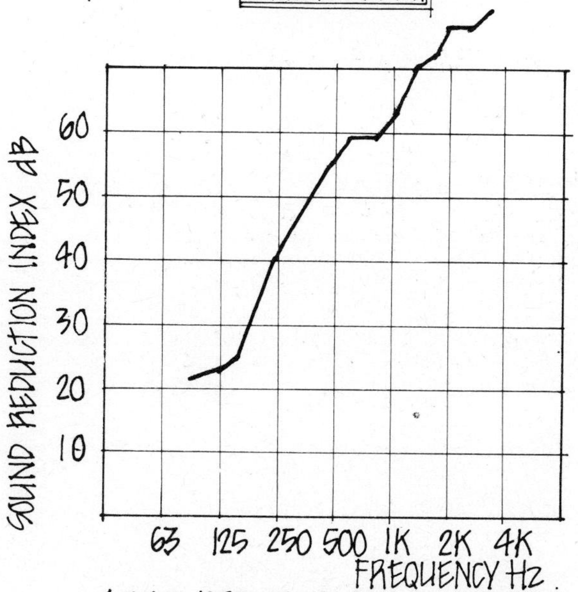

SOUND REDUCTION FOR REINFORCED PANEL OPTION (TYPE 'COMPAC REGULAR')

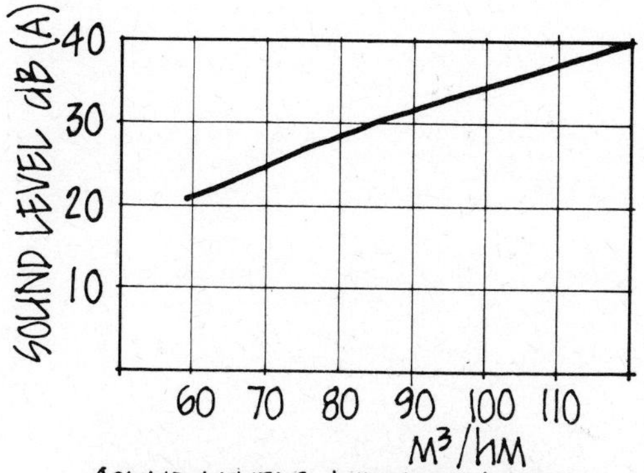

SOUND LEVELS THROUGH LINEAR 3-DIRECTIONS DIFFUSER
(INDICATION ONLY: DEPENDS ON ROOM CONDITIONS)

CEILINGS
SOURCE:
HAUSERMAN LTD

PARTITIONS

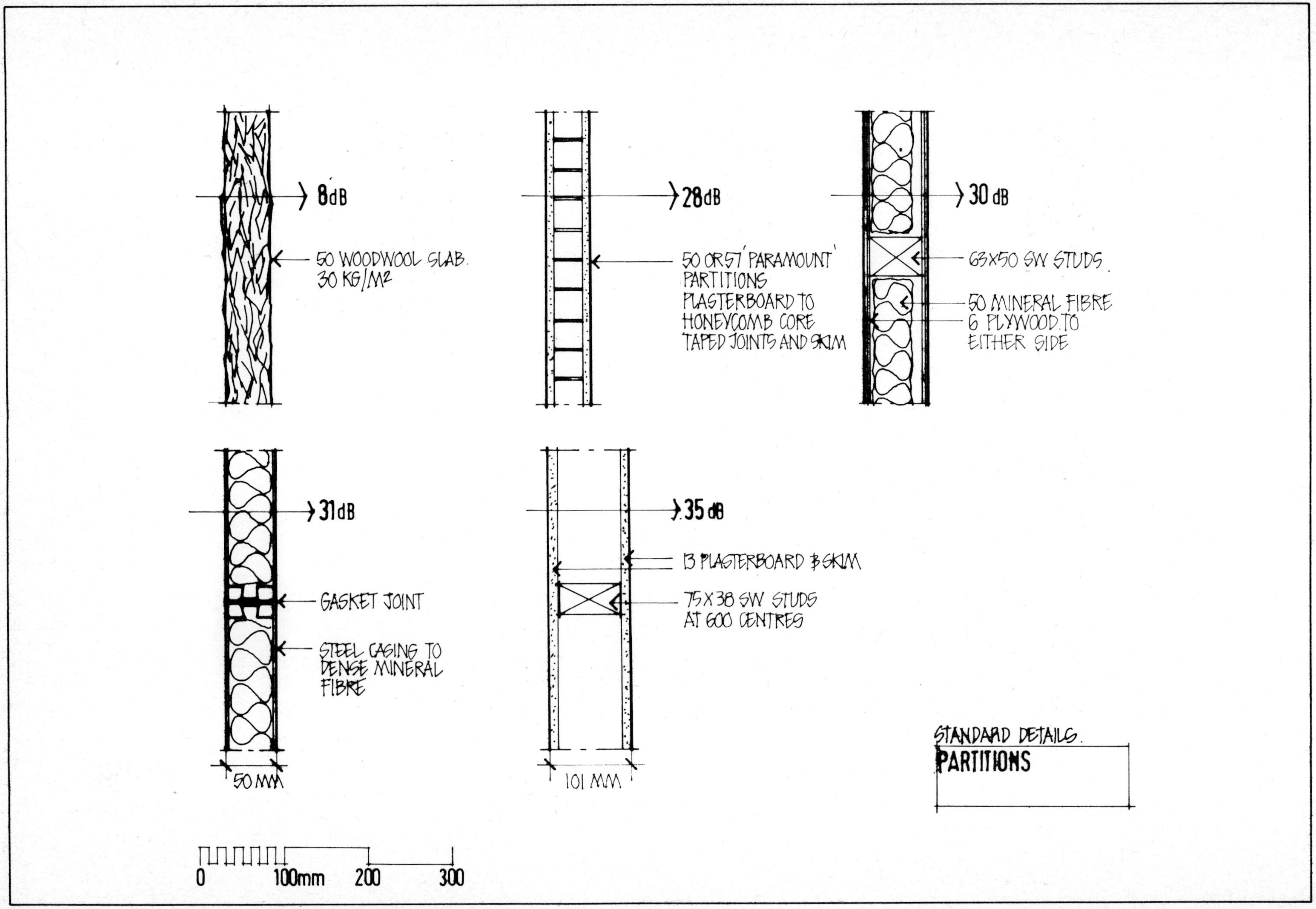
8dB
50 WOODWOOL SLAB.
30 KG/M2
28dB
50 OR 57 'PARAMOUNT'
PARTITIONS
PLASTERBOARD TO
HONEYCOMB CORE
TAPED JOINTS AND SKIM
30 dB
63x50 SW STUDS
50 MINERAL FIBRE
6 PLYWOOD TO
EITHER SIDE
31dB
GASKET JOINT
STEEL CASING TO
DENSE MINERAL
FIBRE
50 MM
35 dB
13 PLASTERBOARD & SKIM
75 X 38 SW STUDS
AT 600 CENTRES
101 MM
STANDARD DETAILS.
PARTITIONS
0
100mm
200
300

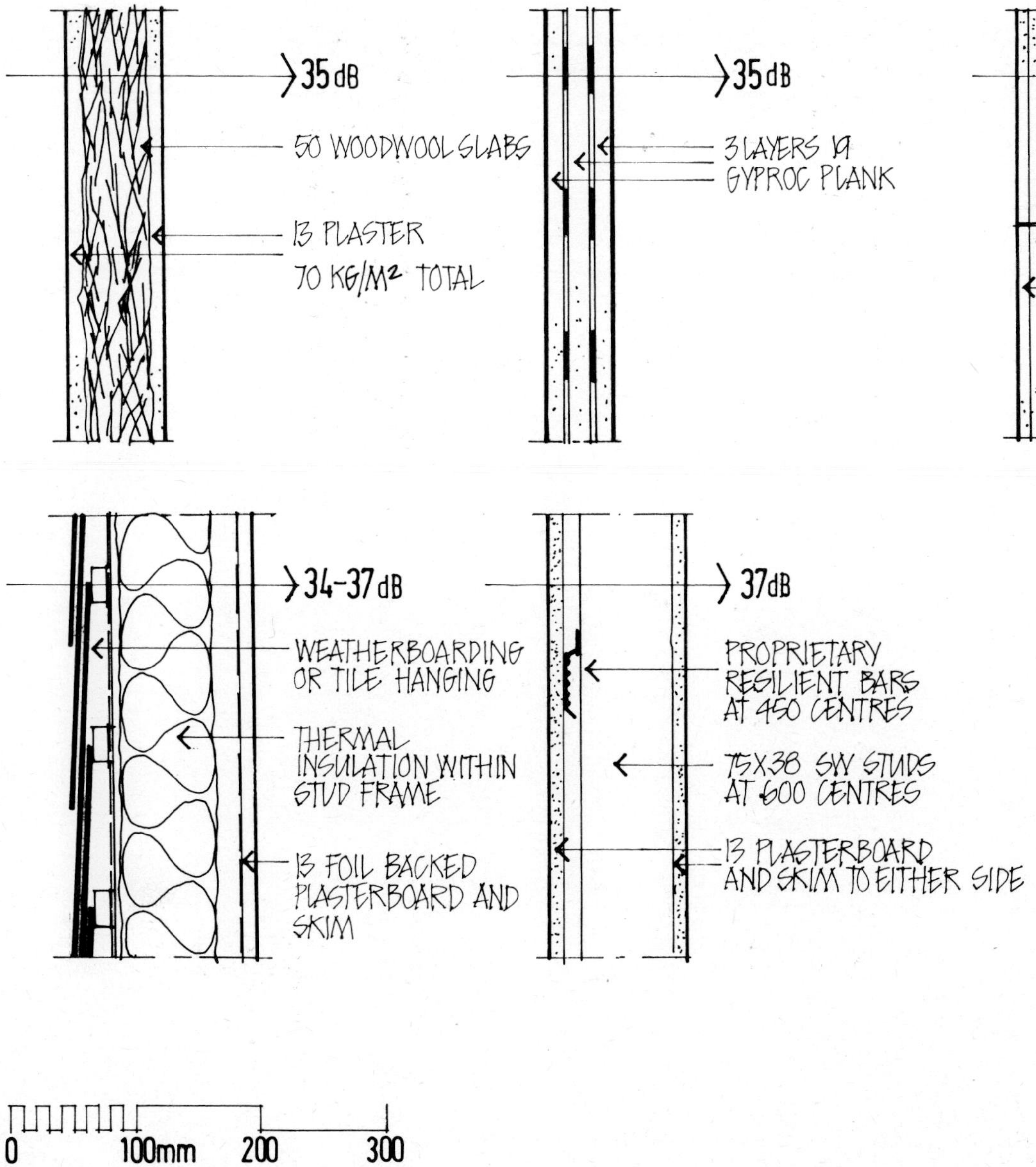

STANDARD DETAILS

PARTITIONS

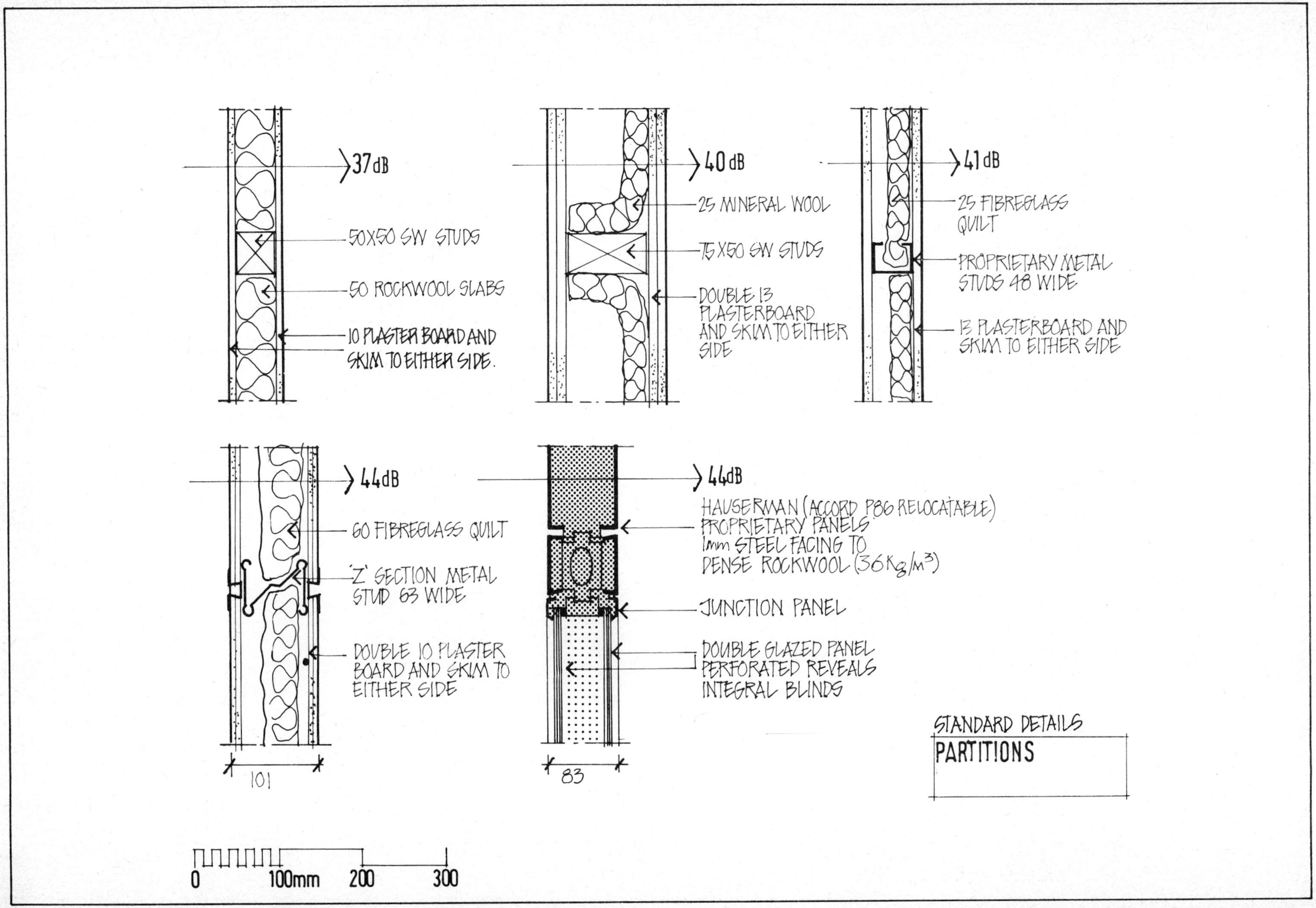

37dB
50X50 SW STUDS
50 ROCKWOOL SLABS
10 PLASTER BOARD AND SKIM TO EITHER SIDE.
40 dB
25 MINERAL WOOL
75 X50 SW STUDS
DOUBLE 13 PLASTERBOARD AND SKIM TO EITHER SIDE
41 dB
25 FIBREGLASS QUILT
PROPRIETARY METAL STUDS 48 WIDE
13 PLASTERBOARD AND SKIM TO EITHER SIDE
44dB
60 FIBREGLASS QUILT
'Z' SECTION METAL STUD 63 WIDE
DOUBLE 10 PLASTER BOARD AND SKIM TO EITHER SIDE
101
44dB
HAUSERMAN (ACCORD P86 RELOCATABLE) PROPRIETARY PANELS 1mm STEEL FACING TO DENSE ROCKWOOL (36 Kg/m³)
JUNCTION PANEL
DOUBLE GLAZED PANEL PERFORATED REVEALS INTEGRAL BLINDS
83
STANDARD DETAILS
PARTITIONS
0
100mm
200
300

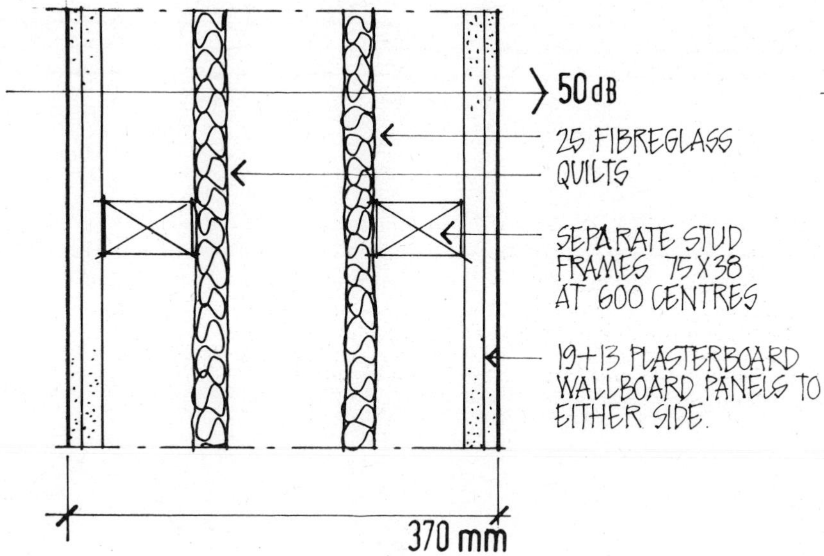

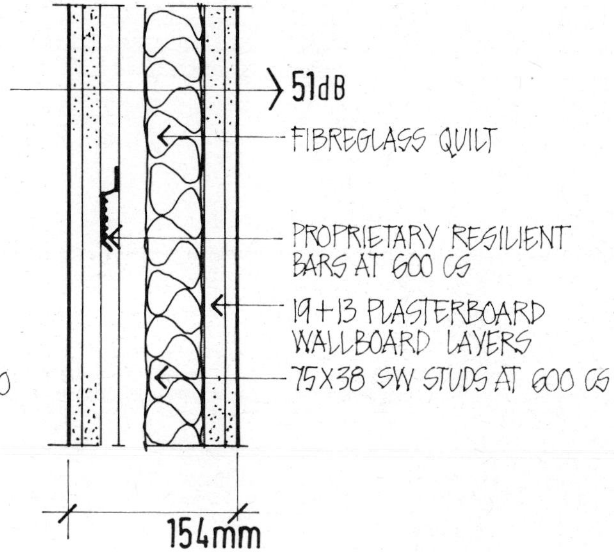

STANDARD DETAILS.

PARTITIONS

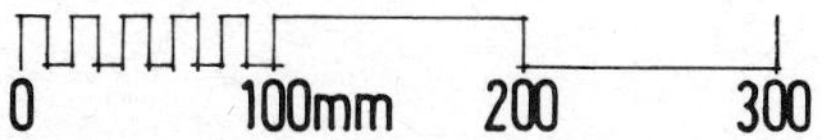

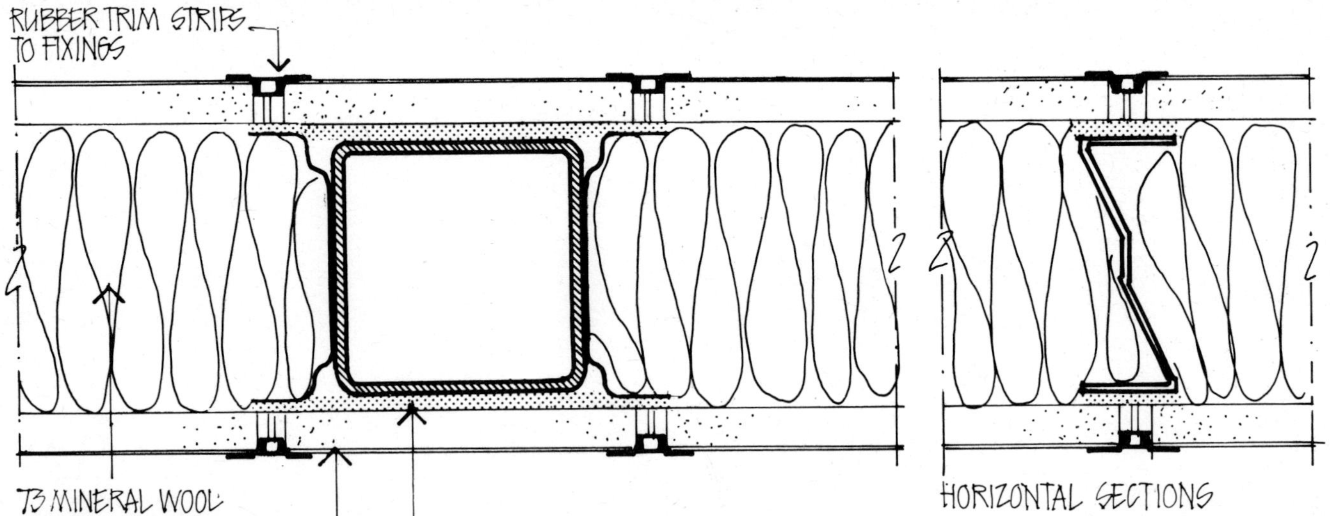

STEEL SUPPORTS AND STUDDING

1. FASTENING OF GYPSUM PANELS AS DETAILED (R=47 dB)

2. GYPSUM PANELS FIXED DIRECTLY TO STEEL SUPPORTS (R=45 dB)

RESILIENT MOUNTING TO FRAMING GIVES LIGHTWEIGHT PARTITIONS GOOD PERFORMANCE AT HIGH FREQUENCIES. BENEFIT FOR SPEECH PRIVACY (GOOD PERFORMANCE IN 1-2K RANGE) BUT SMALL IMPROVEMENT FOR LOW FREQUENCY PERFORMANCE: MASS REQUIRED.

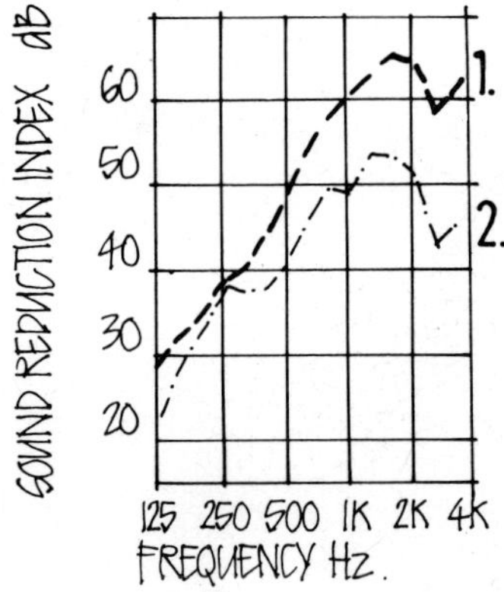

PARTITIONS
SOURCE: DR LANG
TECHNOLOGISCHES
GEWERBE MUSEUM VIENNA

0 50mm

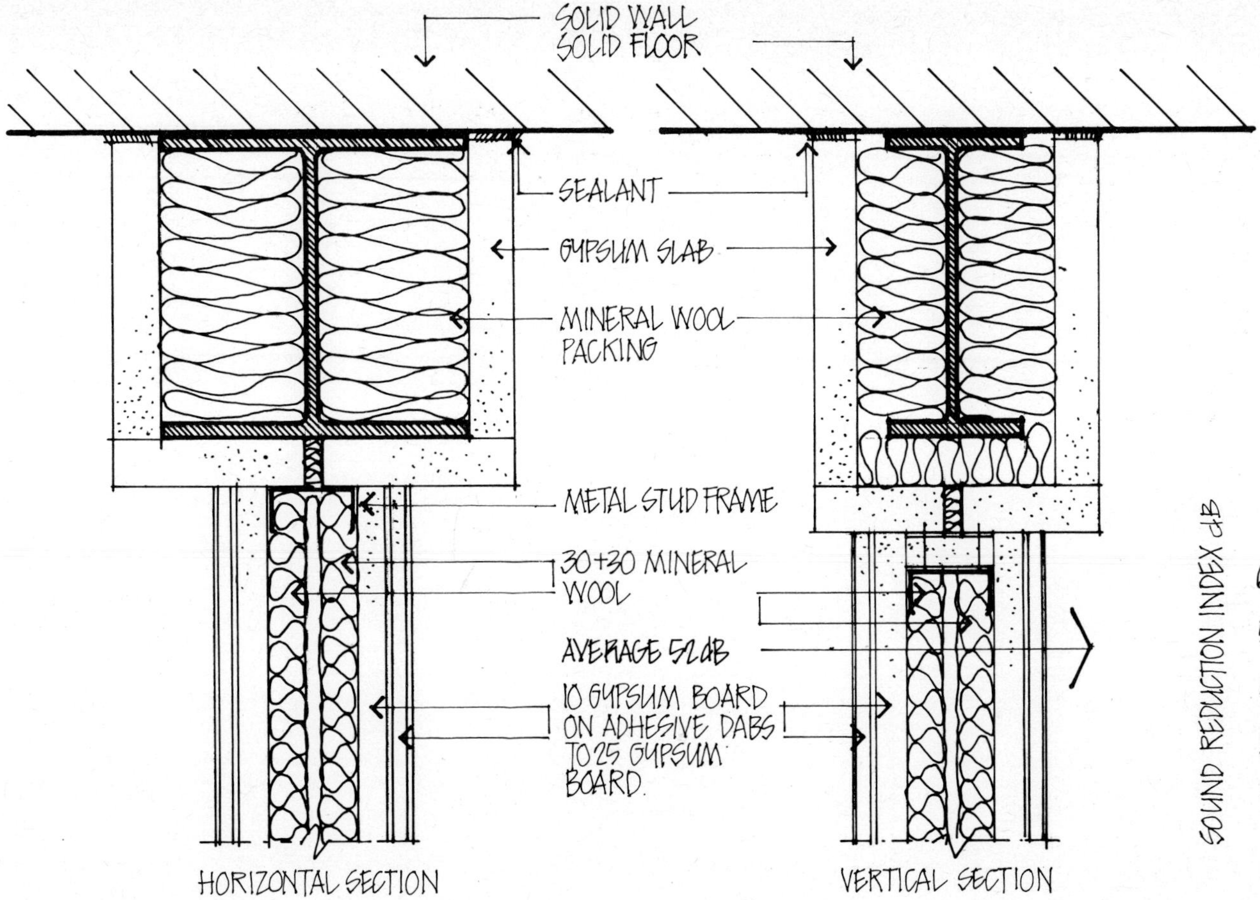

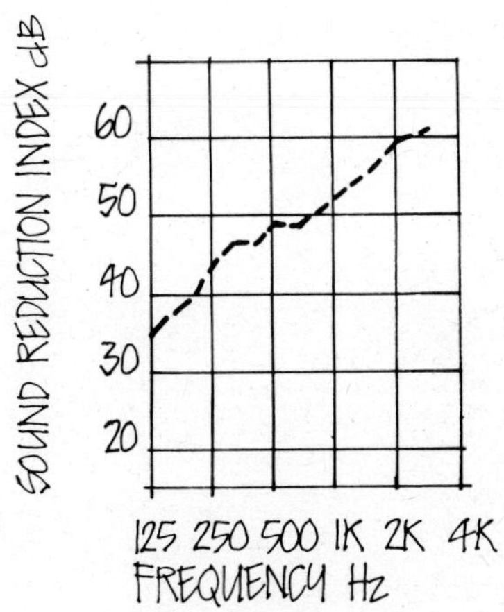

JUNCTIONS TO STRUCTURAL STEELWORK

PARTITIONS
SOURCE: DR LANG
TECHNOLOGISCHES
GEWERBE MUSEUM, VIENNA

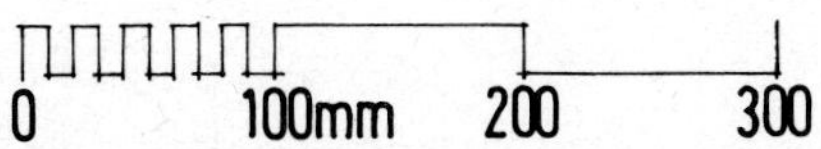

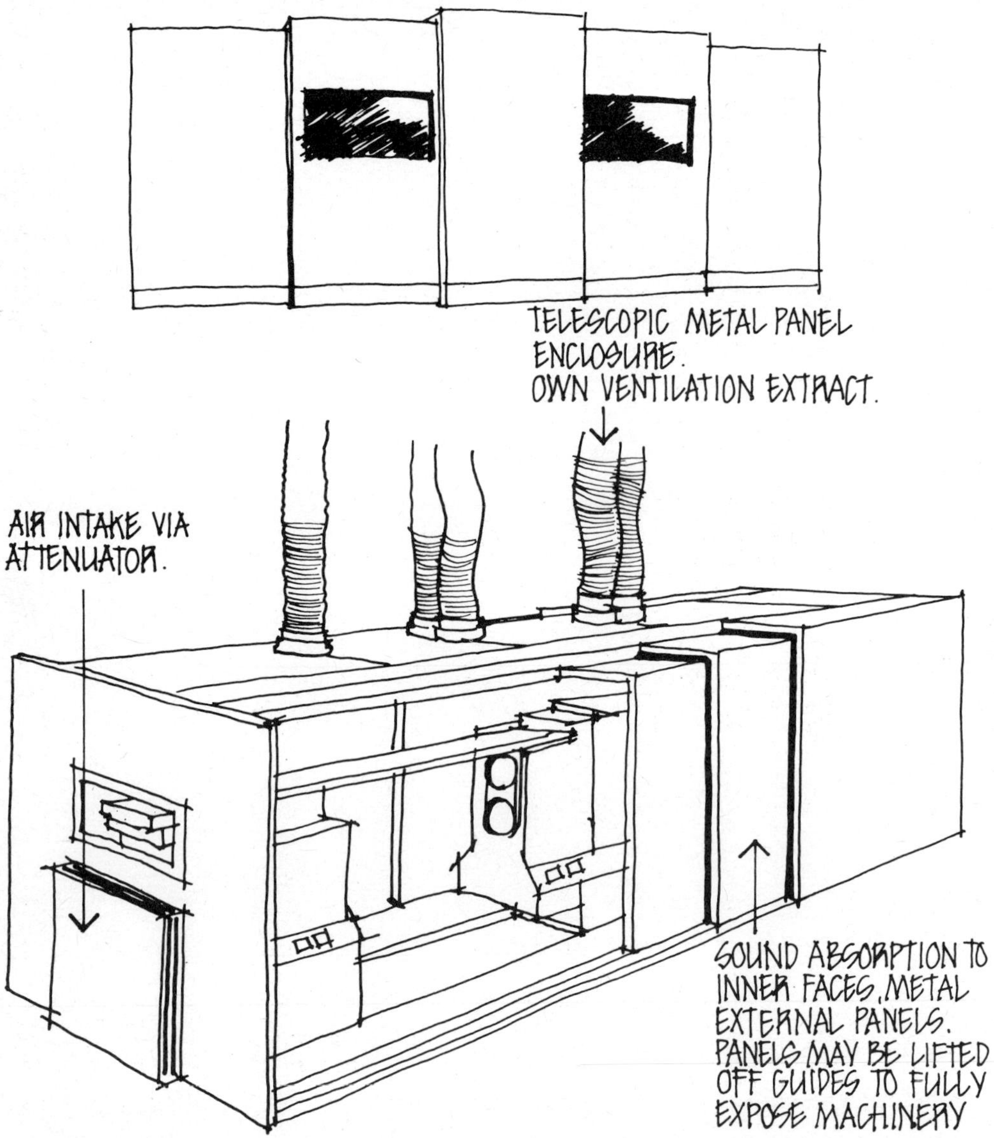

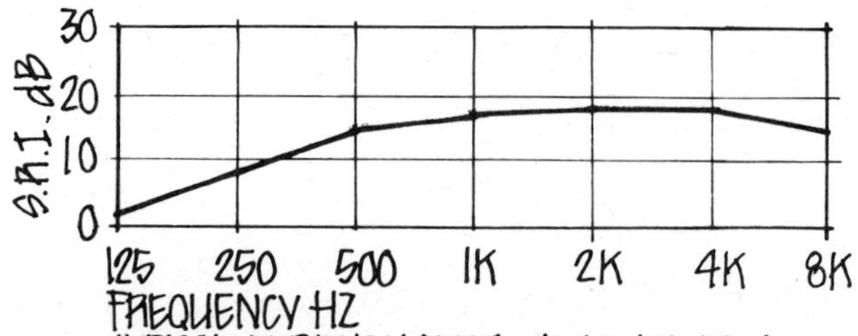

TYPICAL PERFORMANCE OF METAL PANEL ENCLOSURE.
(OVERALL: 15–20dBA)
SOURCE: THE NOISE CONTROL CENTRE LTD.

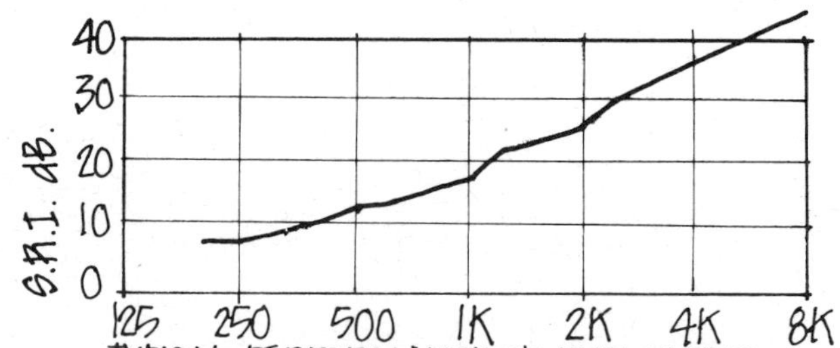

TYPICAL PERFORMANCE OF ENCLOSURE BY PROPRIETARY LIMP LOW-STIFFNESS ACOUSTIC CURTAINS
SOURCE: BESTOBELL ACOUSTICS.
SUCH SYSTEMS SHOULD BE CONSIDERED TO SCREEN PRESSES, MACHINE TOOLS, PUMPS, COMPRESSORS ETC, MOUNTED WITHIN WORKING AREAS. NEITHER SYSTEM EFFECTIVE TO SCREEN LOW FREQUENCY NOISE, BUT USEFUL IMPROVEMENT IN SPEECH FREQUENCIES.

MACHINERY ENCLOSURES

WALLS

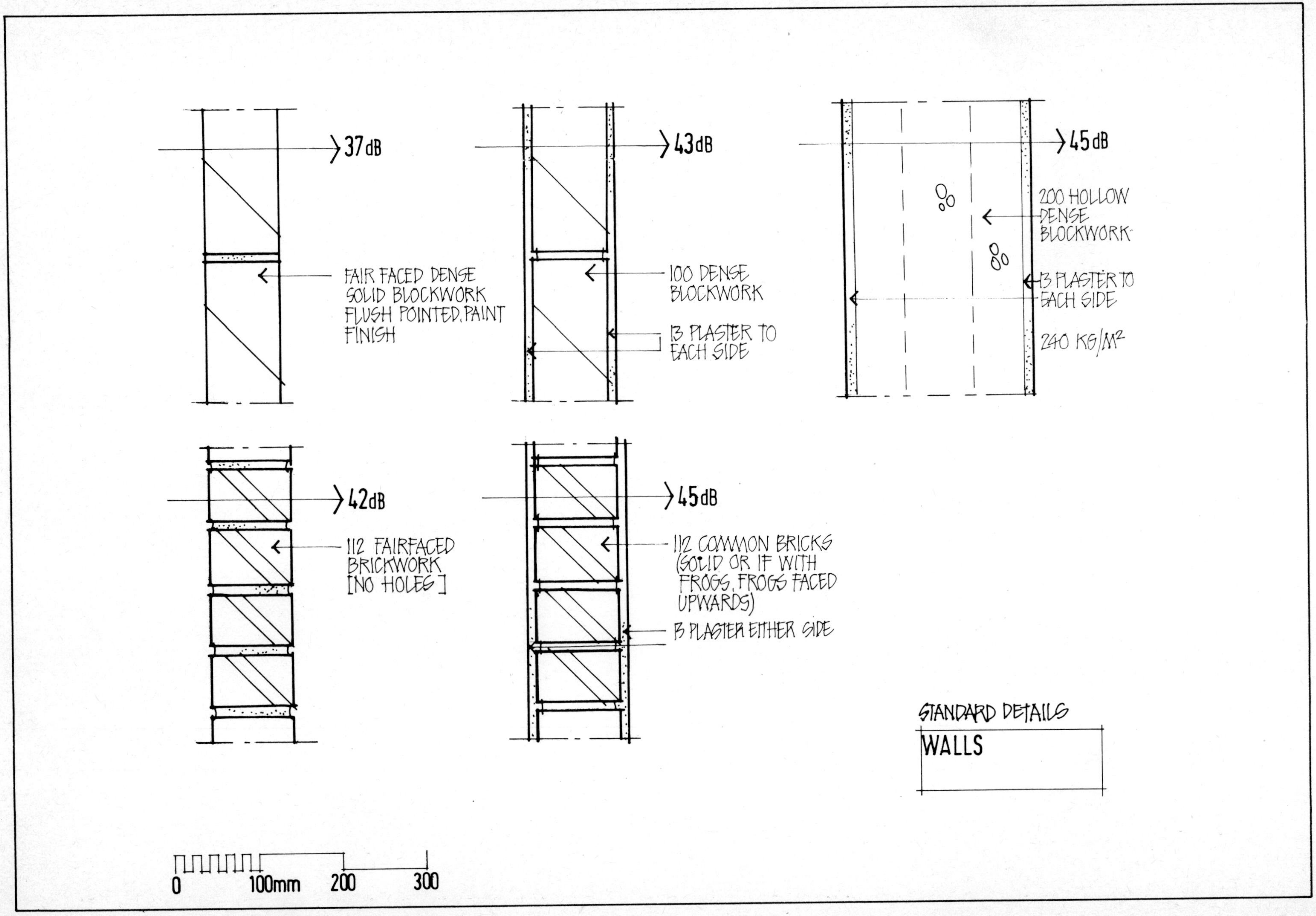
37 dB
FAIR FACED DENSE SOLID BLOCKWORK FLUSH POINTED. PAINT FINISH
43 dB
100 DENSE BLOCKWORK
13 PLASTER TO EACH SIDE
45 dB
200 HOLLOW DENSE BLOCKWORK
13 PLASTER TO EACH SIDE
240 KG/M²
42 dB
112 FAIRFACED BRICKWORK [NO HOLES]
45 dB
112 COMMON BRICKS (SOLID OR IF WITH FROGS, FROGS FACED UPWARDS)
13 PLASTER EITHER SIDE
STANDARD DETAILS
WALLS
0
100mm
200
300

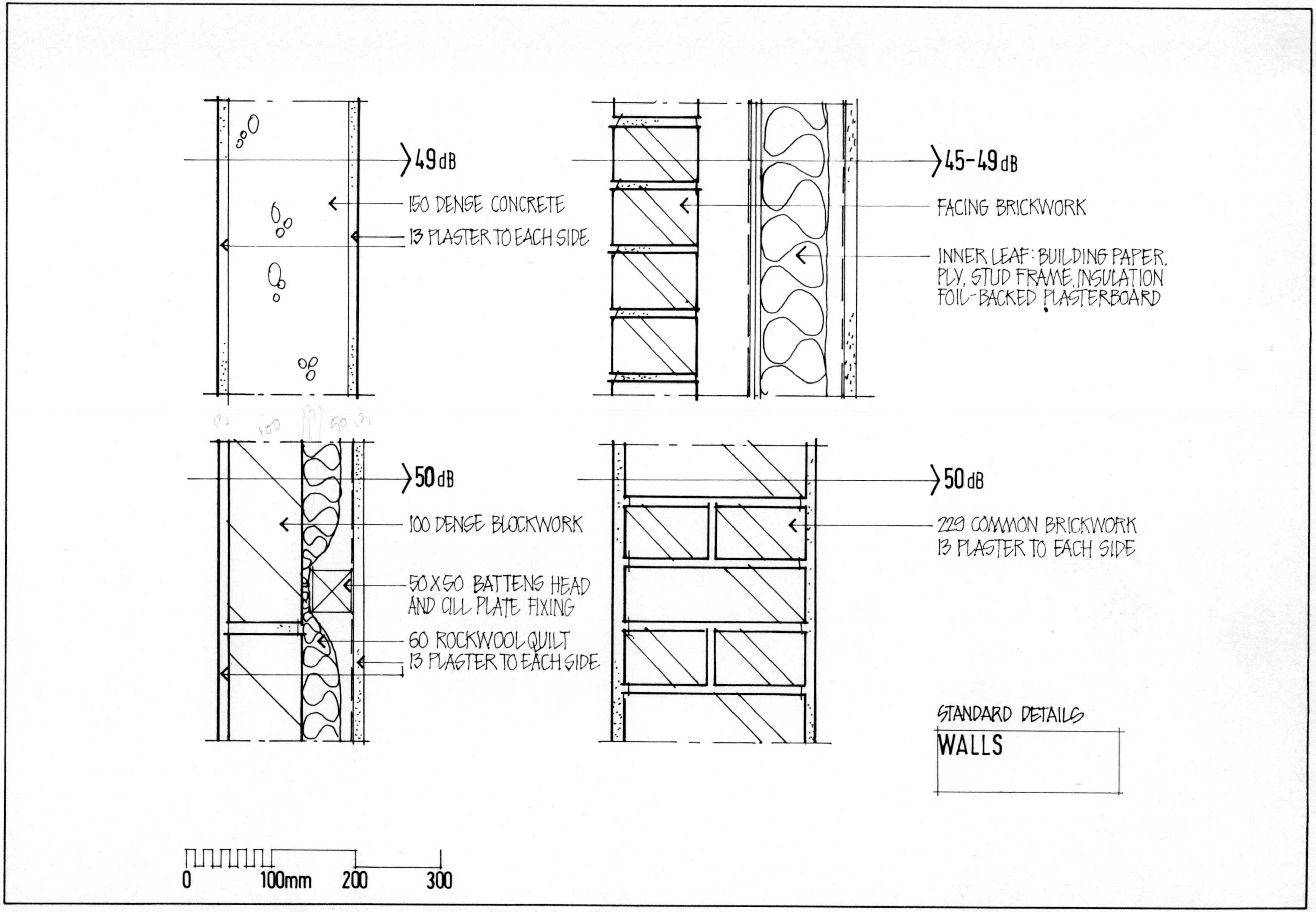
49 dB
150 DENSE CONCRETE
13 PLASTER TO EACH SIDE
45-49 dB
FACING BRICKWORK
INNER LEAF: BUILDING PAPER, PLY, STUD FRAME, INSULATION FOIL-BACKED PLASTERBOARD
50 dB
100 DENSE BLOCKWORK
50 X 50 BATTENS HEAD AND CILL PLATE FIXING
60 ROCKWOOL QUILT
13 PLASTER TO EACH SIDE
50 dB
229 COMMON BRICKWORK
13 PLASTER TO EACH SIDE
STANDARD DETAILS
WALLS
0
100mm
200
300

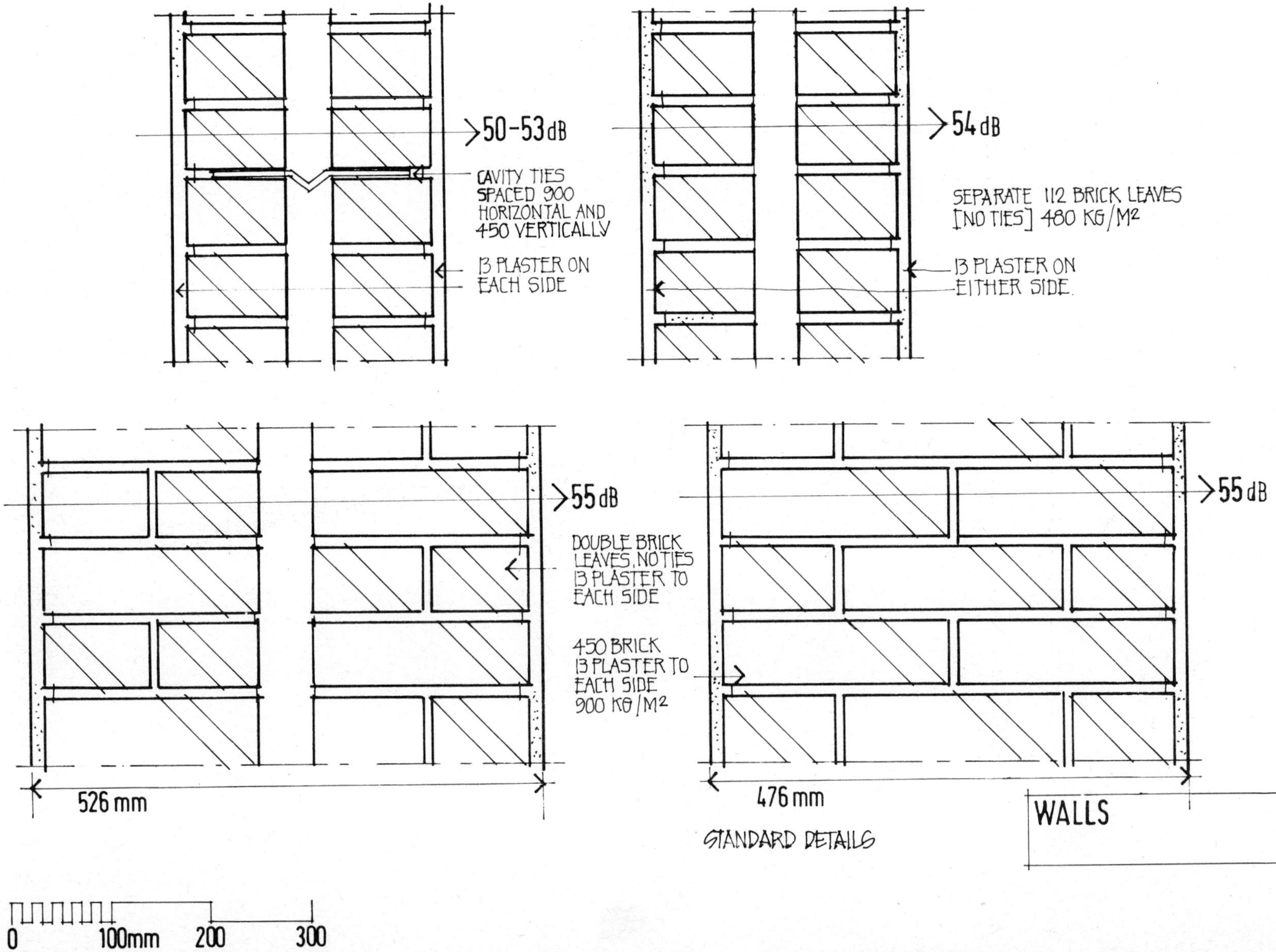

WALLS

STANDARD DETAILS

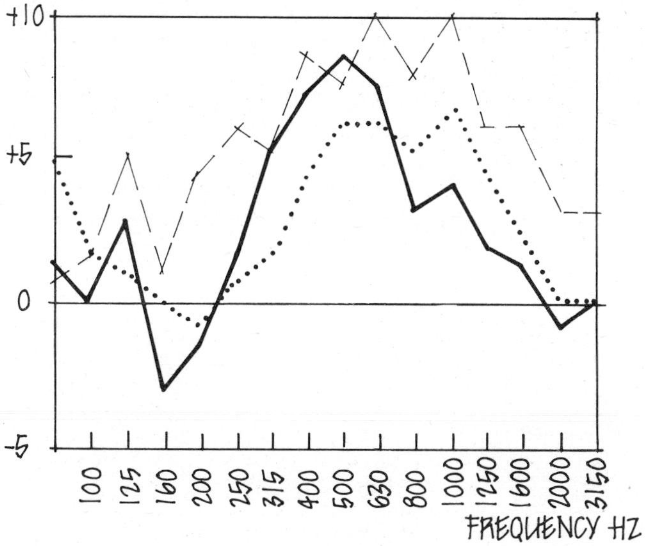

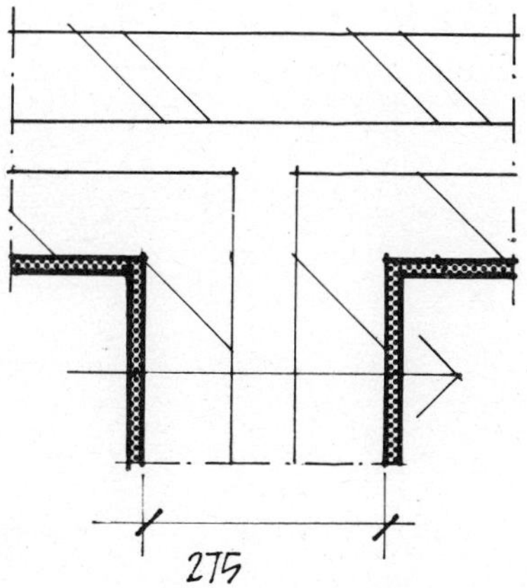

EXTERNAL AND SEPARATING WALL LINING EITHER:

——— 1 LAYER 9·5 PLASTERBOARD

– – – – 2 LAYER 9·5 PLASTERBOARD (1 LAYER TO EXTERNAL WALL)

·········· 12·5 LIGHTWEIGHT PLASTER.

SOURCE: CIRIA REPORT 1980

PARTY WALLS.

AS WELL AS THE MASS OF SEPARATING AND EXTERNAL WALLS THE FOLLOWING ASPECTS AFFECT THE SOUND INSULATION THROUGH THE OPERATING WALLS

(1.) TIES. ABSENCE OF TIES CONSIDERABLY IMPROVES THE SOUND INSULATION. BUTTERFLY TIES PREFERABLE TO STRIP TIES.

(2.) CAVITY INSULATION. 60–70 KG./M^3 MINERAL WOOL LINING TO SEPARATING WALL CAVITY IMPROVES SOUND INSULATION BY IMPEDING SOUND TRACKING ACROSS THE CAVITY. (MUST BE LOOSELY PLACED IN CAVITY.)

(3.) JOIST DIRECTION. JOISTS PARALLEL TO SEPARATING WALLS HAVE IN TESTS GIVEN CONSISTANLY HIGHER SOUND INSULATION THAN JOISTS PERPENDICULAR TO SEPARATING WALL

(4.) JOIST TIGHTNESS. RATHER THAN 'LOOSE' POCKETS AT JOIST ENDS, CONSIDER USE OF JOIST HANGERS TO ALLOW THE WALL INTEGRITY TO BE MAINTAINED.

(5.) DRY LINING. THE EFFECT OF WALL FINISHES IS SHOWN AT LHS. 2 LAYERS TO SEPARATING WALL ARE NECESSARY TO IMPROVE ON PLASTER. 1 LAYER CAN ACTUALLY REDUCE SOUND INSULATION AT SOME FREQUENCIES BECAUSE OF RESONANCE.

WALLS

not to scale

PARTY WALLS

THE BUILDING REGULATIONS REQUIRE THAT SEPARATING WALLS IN DWELLINGS SHOULD BE OF PARTY WALL GRADE. THIS IS STATED AS DECIBEL REDUCTION WITH A TOTAL MINIMUM ADVERSE DEVIATION OF 23 dB MEASURED OVER A SPECIFIED RANGE OF FREQUENCIES. THE ONLY WAY OF ENSURING CONDITIONS ARE MET IS BY MEASUREMENT RATHER THAN DEEMED TO SATISFY PROVISIONS.

FREQUENCY Hz	100	125	160	200	250	315	400	500	630	800	1000	1250	1600	2000	2500	3150
SOUND REDUCTION INDEX dB.	40	41	43	44	45	47	48	49	57	52	53	55	56	56	56	56

BLOCKWORK

FOR SOUND INSULATION PURPOSES BLOCKWORK IS BEST USED PLASTERED (AS THIS CLOSES FISSURES IN THE BLOCK FACES) USING NO-VOID SOLID D.C.M. BLOCKS BEWARE SHRINKAGE CRACKS, PARTICULARLY AT CONCRETE COLUMN OR BEAM JUNCTIONS

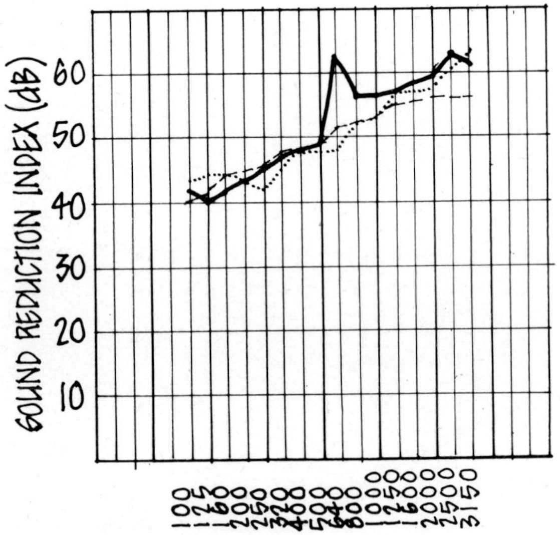

SOUND REDUCTION FREQUENCY Hz

SOUND ABSORPTION (FOR UNPAINTED BLOCKWORK)
0·2 0·2 0·2 0·15 0·2 0·25

IN TESTS BLOCKWORK WAS LAID IN 1:1:6 CEMENT: LIME: SAND MORTAR AND PAINTED 2 COATS (UNPAINTED PERFORMANCE 1-2 dB LESS THROUGHOUT FREQUENCY RANGE.)

······· SOLID 175 BLOCKWORK
—— SOLID 190 BLOCKWORK
– – – BUILDING REGULATIONS PARTY WALL GRADE
SOURCE: AIRO

SRI dB	THICKNESS MM	DENSITY KG/M2	TYPE OF FORTICRETE DCM
41	90 100	150	SOLID HOLLOW
43	90 140	180	NO VOID SOLID HOLLOW
46	190	240	HOLLOW
48	140	240	SOLID
50	190	340	SOLID

WALLS

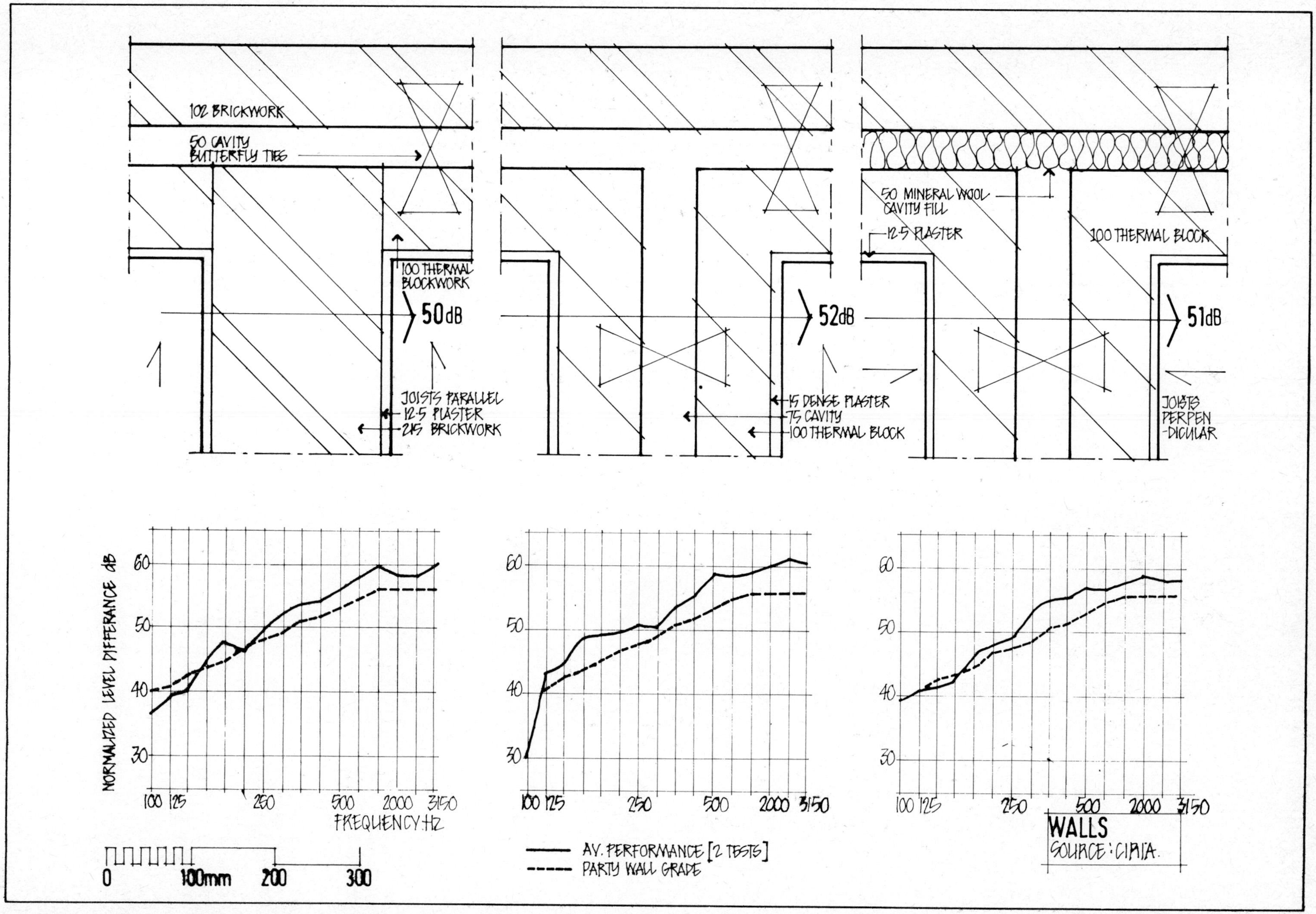
102 BRICKWORK
50 CAVITY
BUTTERFLY TIES
100 THERMAL BLOCKWORK
50dB
JOISTS PARALLEL
12·5 PLASTER
215 BRICKWORK
52dB
15 DENSE PLASTER
75 CAVITY
100 THERMAL BLOCK
50 MINERAL WOOL CAVITY FILL
12·5 PLASTER
100 THERMAL BLOCK
51dB
JOISTS PERPEN-DICULAR
NORMALIZED LEVEL DIFFERANCE dB
60
50
40
30
100
125
250
500
2000
3150
FREQUENCY Hz
0
100mm
200
300
AV. PERFORMANCE [2 TESTS]
PARTY WALL GRADE
WALLS
SOURCE: CIRIA.

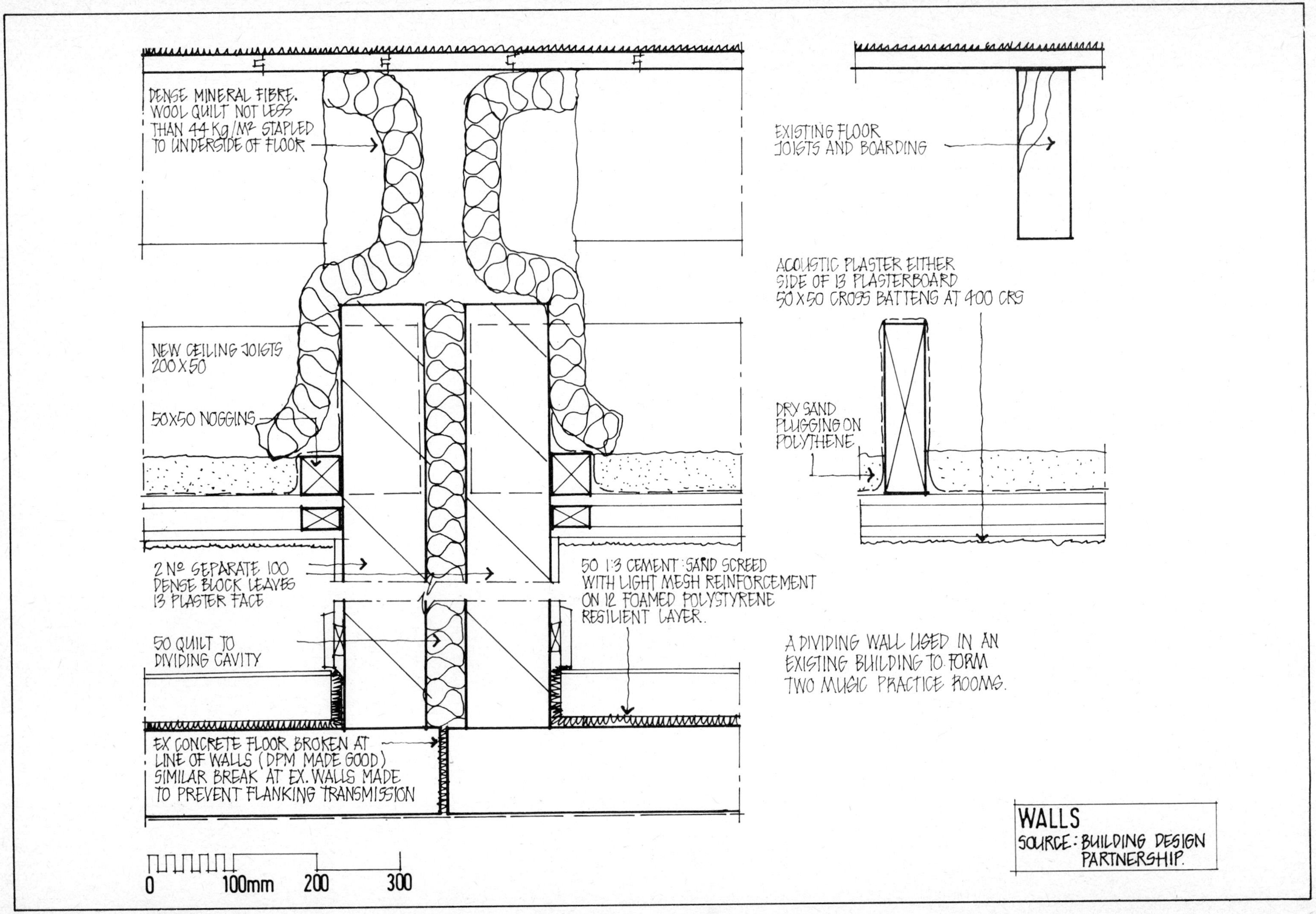
DENSE MINERAL FIBRE. WOOL QUILT NOT LESS THAN 44 Kg/M² STAPLED TO UNDERSIDE OF FLOOR
EXISTING FLOOR JOISTS AND BOARDING
ACOUSTIC PLASTER EITHER SIDE OF 13 PLASTERBOARD 50 X 50 CROSS BATTENS AT 400 CRS
NEW CEILING JOISTS 200 X 50
50X50 NOGGINS
DRY SAND PLUGGING ON POLYTHENE
2 Nº SEPARATE 100 DENSE BLOCK LEAVES 13 PLASTER FACE
50 1:3 CEMENT : SAND SCREED WITH LIGHT MESH REINFORCEMENT ON 12 FOAMED POLYSTYRENE RESILIENT LAYER.
50 QUILT TO DIVIDING CAVITY
A DIVIDING WALL USED IN AN EXISTING BUILDING TO FORM TWO MUSIC PRACTICE ROOMS.
EX CONCRETE FLOOR BROKEN AT LINE OF WALLS (DPM MADE GOOD) SIMILAR BREAK AT EX. WALLS MADE TO PREVENT FLANKING TRANSMISSION
0
100mm
200
300
WALLS
SOURCE: BUILDING DESIGN PARTNERSHIP.

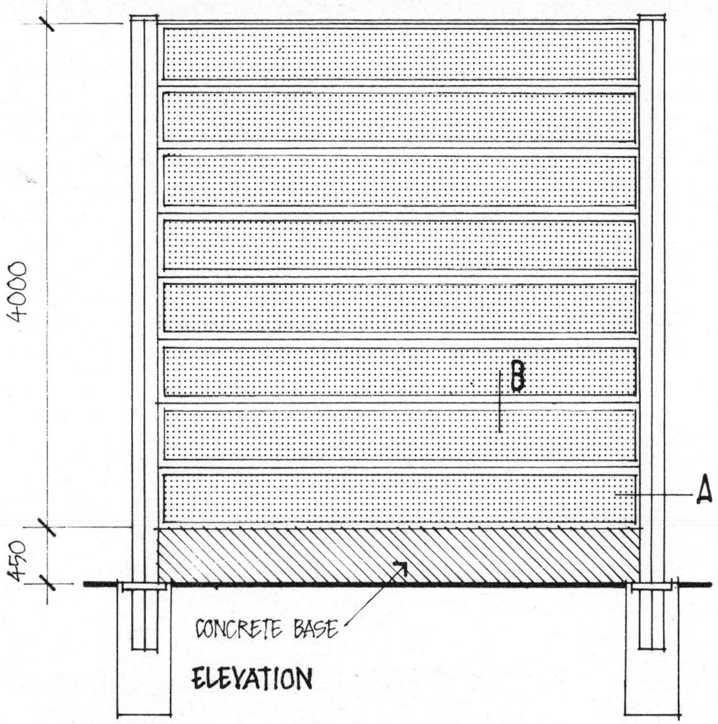

ELEVATION

not to scale

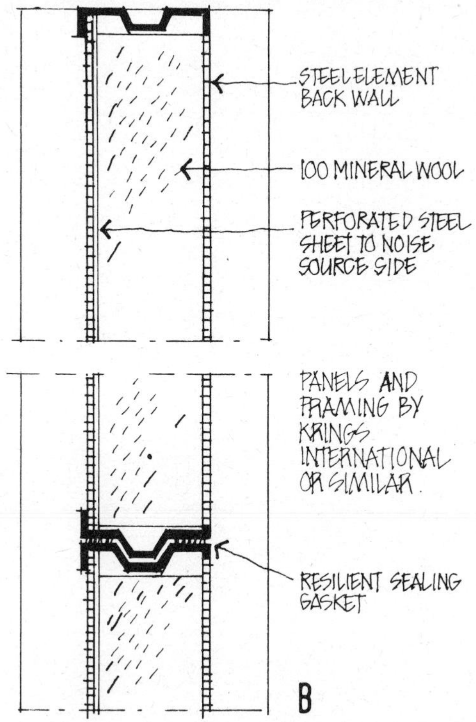

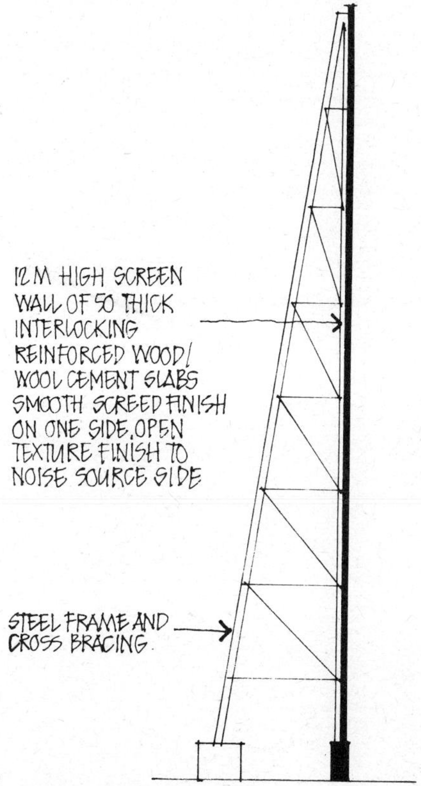

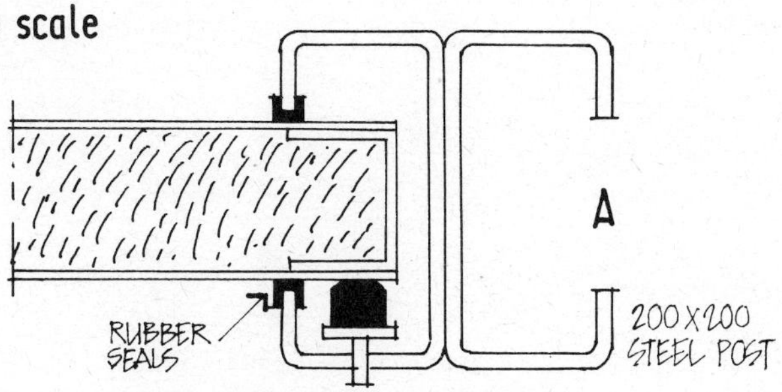

DETAIL OF RESILIENT MOUNTING OF PANELS

PROPRIETARY SOUND ABSORBING SCREEN PANELS. SECTION 1:5

USED ALONG SIDE OF ROADS TO PROTECT ADJACENT HOUSING. RESILIENT MOUNTING ALLOWS 'LIMP' ACTION OF PANEL TO ABSORB NOISE.
DAMPING EFFECT (DIN 52210) AVERAGE 27 dB.

BAFFLE WALL TO INDUSTRIAL PLANT. NTS

HIGH SCREENS AROUND PLANT NECESSARY WHERE A LARGE AREA OF PROCESS PLANT HAS TO BE SCREENED.

NOISE BARRIERS

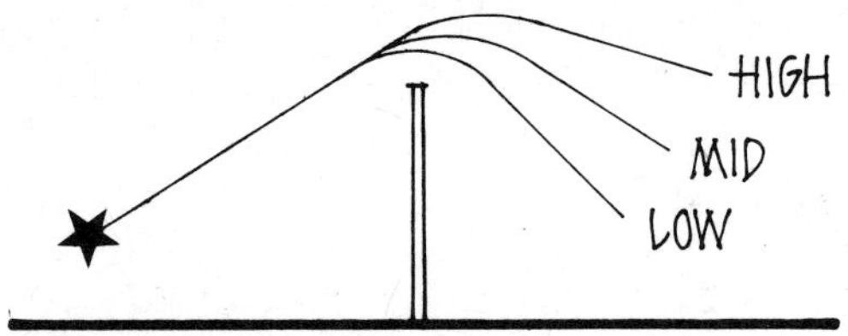

VARYING DIFFRACTION OVER A BARRIER AT DIFFERENT FREQUENCIES

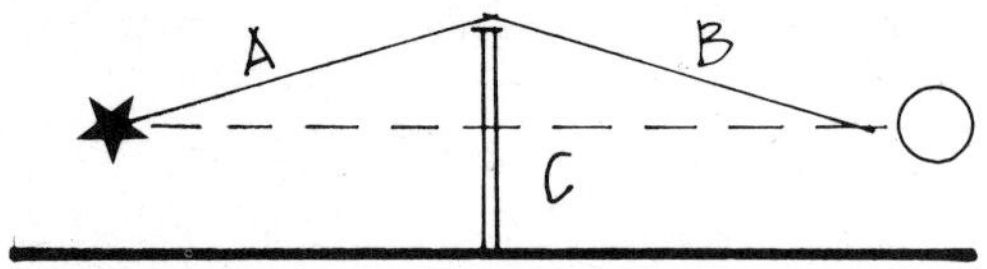

DIFFERENCE IN SOUND PATHS, d = A + B − C

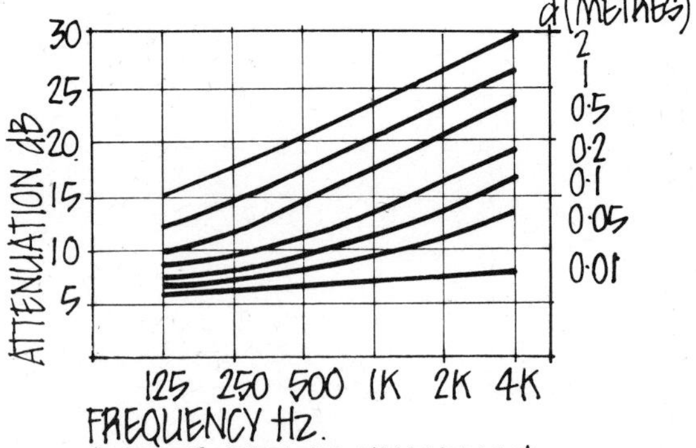

SOURCE: PARKIN HUMPREYS & COWELL

ATTENUATION DUE TO SCREENS IN OPEN AIR IS RELATED TO HOW MUCH FURTHER THE SOUND HAS TO TRAVEL AROUND THE SCREEN. FOR ATTENUATION DUE TO A FINITE SCREEN LENGTH, CARRY OUT PATH DIFFERENCE CALCULATIONS IN PLAN AS WELL AS SECTION AND COMBINE ATTENUATIONS. WEIGHT OF SCREEN MATERIAL IS IMMATERIAL, BUT NEEDS TO BE SOLID FACED TO NOISE SIDE.

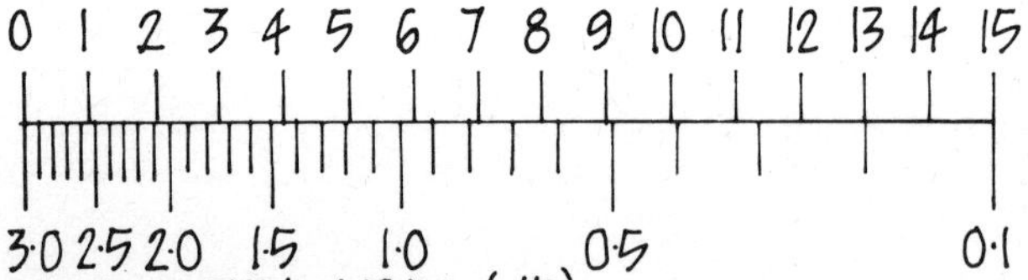

ADD TO HIGHER LEVEL (dB)
SOURCE: T.E. MOORE.

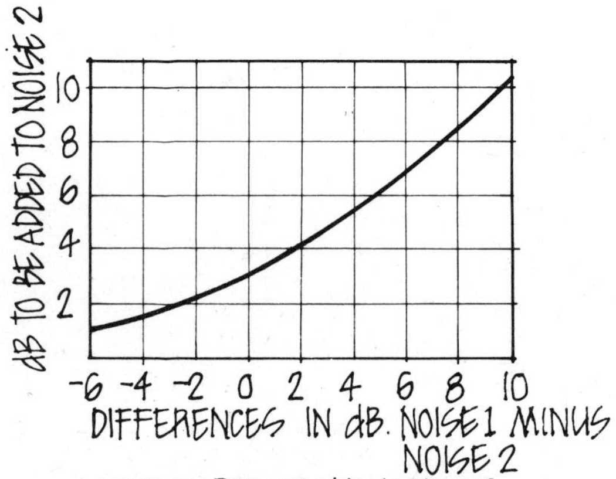

SOURCE: PARKIN HUMPREYS & COWELL

COMBINING SOUND LEVELS, FOR EXAMPLE IN PATH DIFFERENCE CALCULATIONS ABOVE, CAN BE CARRIED OUT USING EITHER OF THE FIGURES AT LHS. FOR INSTANCE, A TELEPHONE BELL 3M. AWAY AT 73dB ADDS TO THE BACKGROUND 70 dB. IN A TYPISTS POOL TO GIVE A RESULTANT SOUND OF 74·75 dB (i.e. 75)

BARRIERS AND SCREENS

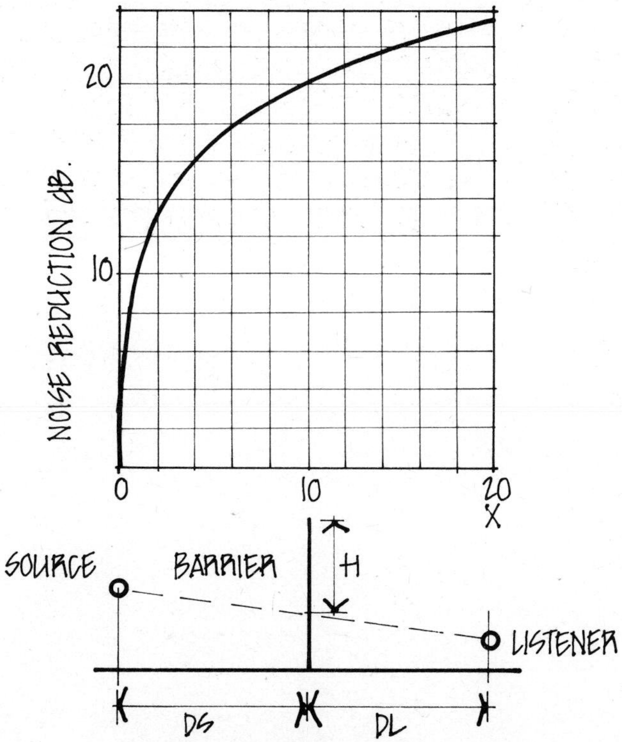

X = H²/λDS PROVIDED DL>>DS AND DS>H

HEIGHT OF BARRIER CAN BE DETERMINED IF WAVELENGTH OF NOISE SOURCE AND DS ARE KNOWN AND REQUIRED dB. REDUCTION SPECIFIED.

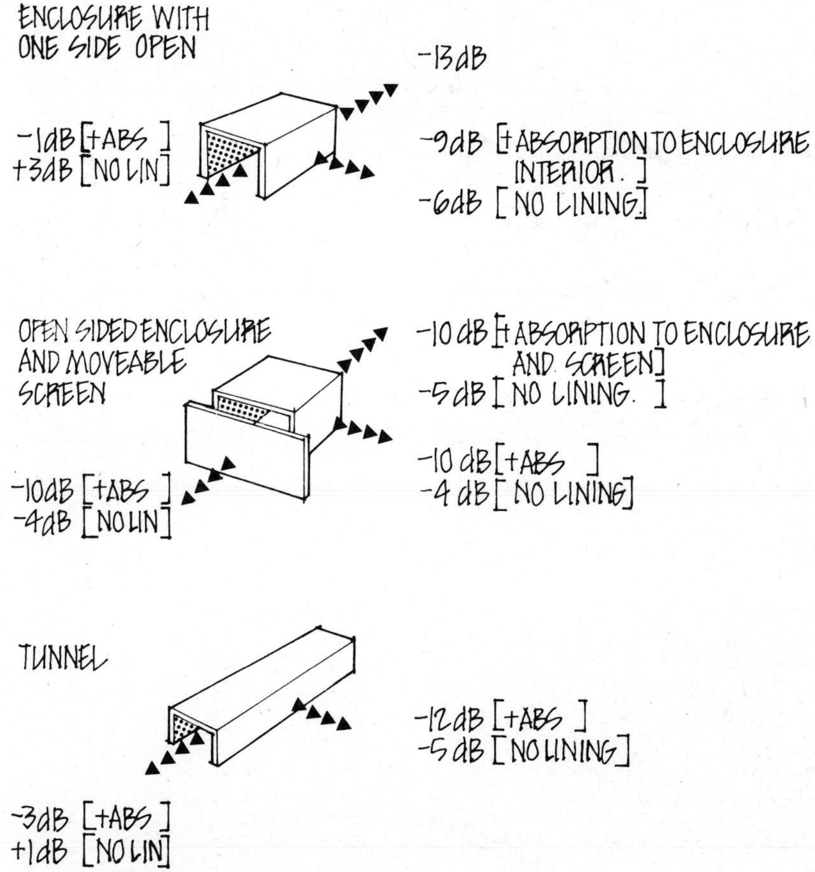

EFFECT ON DIRECT SOUND PRESSURE LEVEL FREQUENCY RANGE 500-4K Hz OF SOME PARTIAL ENCLOSURES.
SOURCE: WOODS OF COLCHESTER.
(NOTE: INDICATION ONLY FOR POINTS NEAR ENCLOSURES.)

NOISE BARRIERS

FLOORS

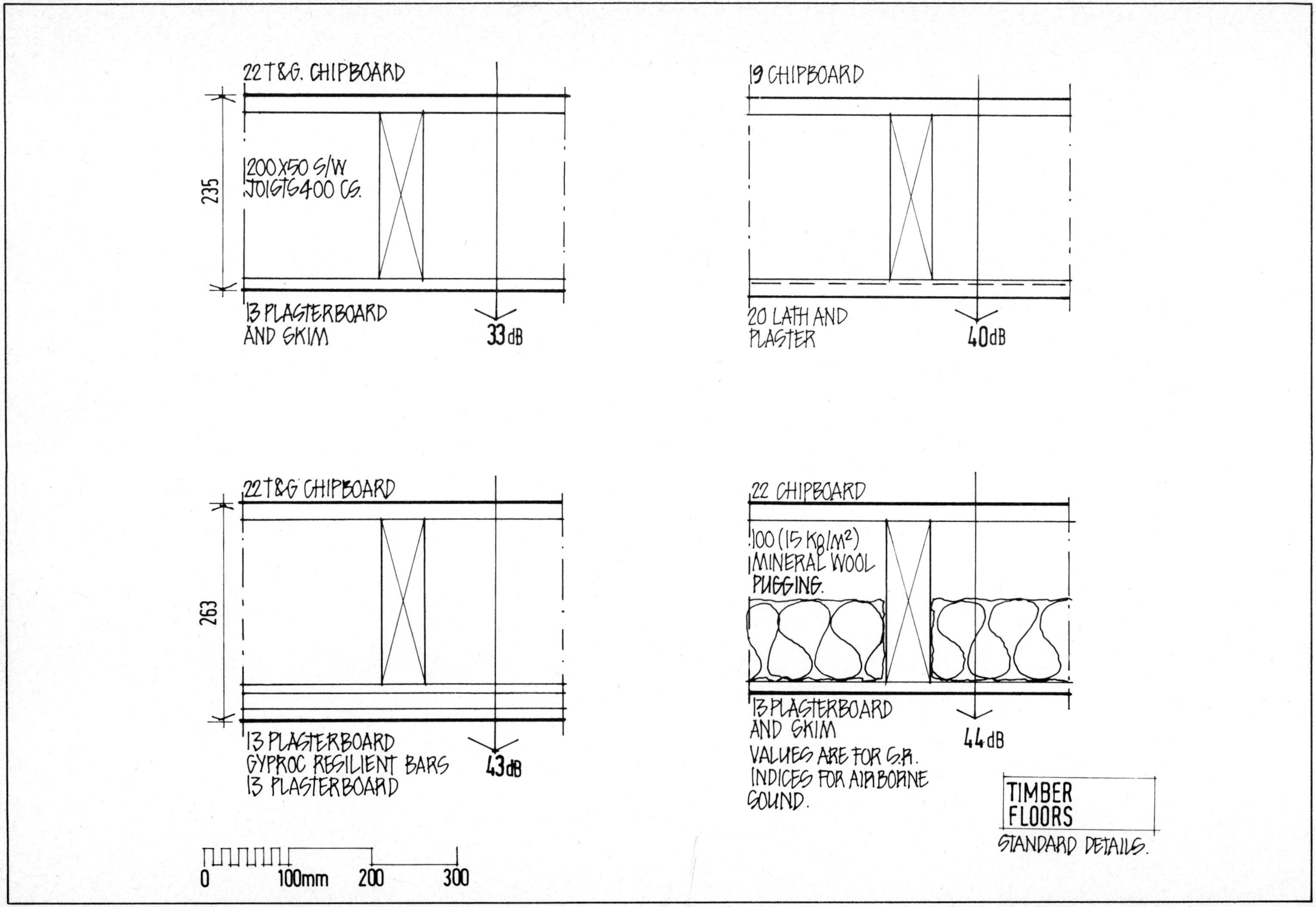
22 T&G. CHIPBOARD
200 X 50 S/W
JOISTS 400 CS.
235
13 PLASTERBOARD
AND SKIM
33 dB
19 CHIPBOARD
20 LATH AND
PLASTER
40 dB
22 T&G CHIPBOARD
263
13 PLASTERBOARD
GYPROC RESILIENT BARS
13 PLASTERBOARD
43 dB
22 CHIPBOARD
100 (15 Kg/M²)
MINERAL WOOL
PUGGING.
13 PLASTERBOARD
AND SKIM
44 dB
VALUES ARE FOR S.R.
INDICES FOR AIRBORNE
SOUND.
0
100mm
200
300
TIMBER
FLOORS
STANDARD DETAILS.

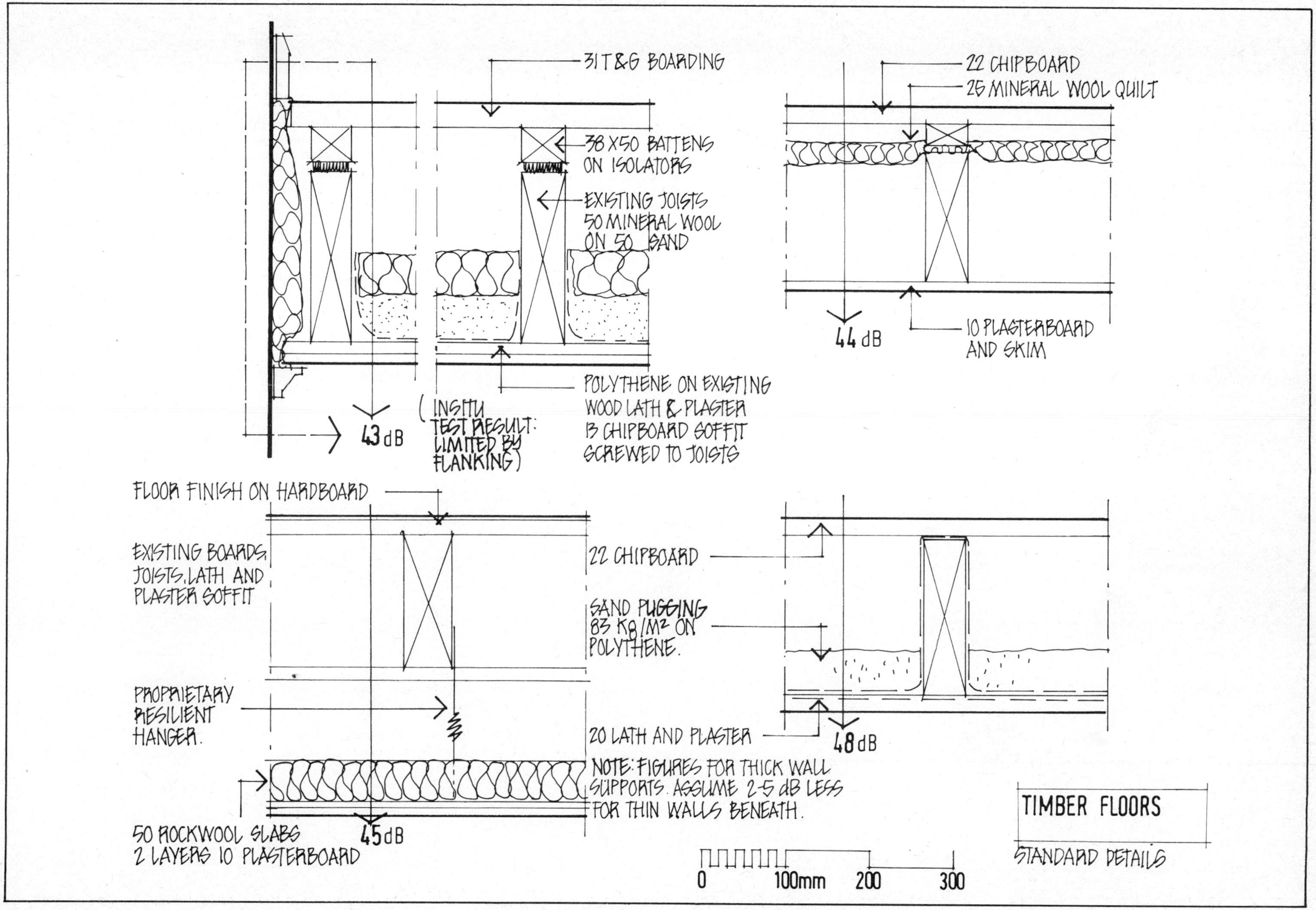
31 T&G BOARDING
38 X 50 BATTENS ON ISOLATORS
EXISTING JOISTS 50 MINERAL WOOL ON 50 SAND
POLYTHENE ON EXISTING WOOD LATH & PLASTER 13 CHIPBOARD SOFFIT SCREWED TO JOISTS
43 dB
(INSITU TEST RESULT: LIMITED BY FLANKING)
22 CHIPBOARD
25 MINERAL WOOL QUILT
10 PLASTERBOARD AND SKIM
44 dB
FLOOR FINISH ON HARDBOARD
EXISTING BOARDS, JOISTS, LATH AND PLASTER SOFFIT
PROPRIETARY RESILIENT HANGER.
50 ROCKWOOL SLABS 2 LAYERS 10 PLASTERBOARD
45 dB
22 CHIPBOARD
SAND PUGGING 83 Kg/M² ON POLYTHENE.
20 LATH AND PLASTER
48 dB
NOTE: FIGURES FOR THICK WALL SUPPORTS. ASSUME 2-5 dB LESS FOR THIN WALLS BENEATH.
0
100mm
200
300
TIMBER FLOORS
STANDARD DETAILS

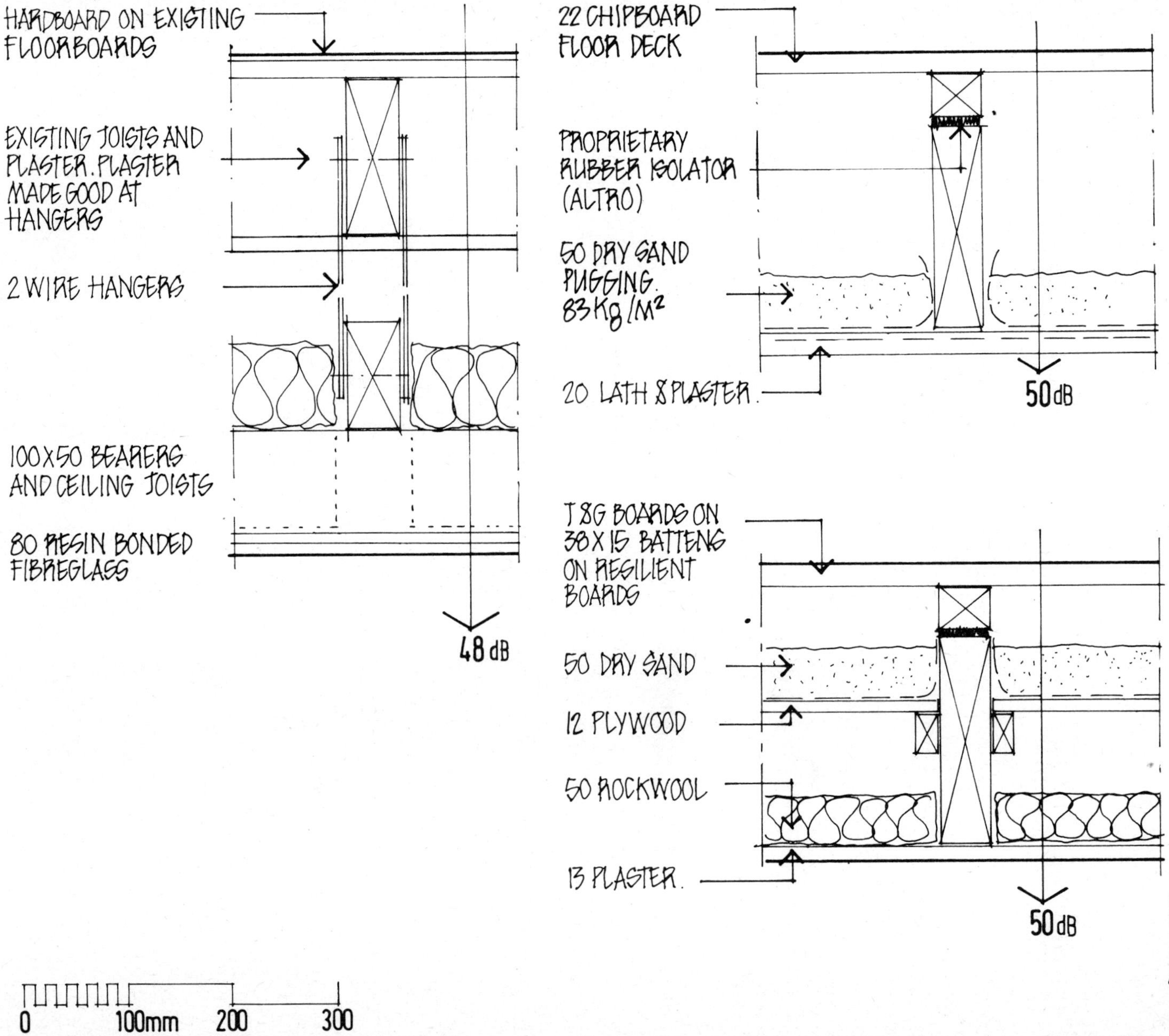

TIMBER FLOORS

STANDARD DETAILS

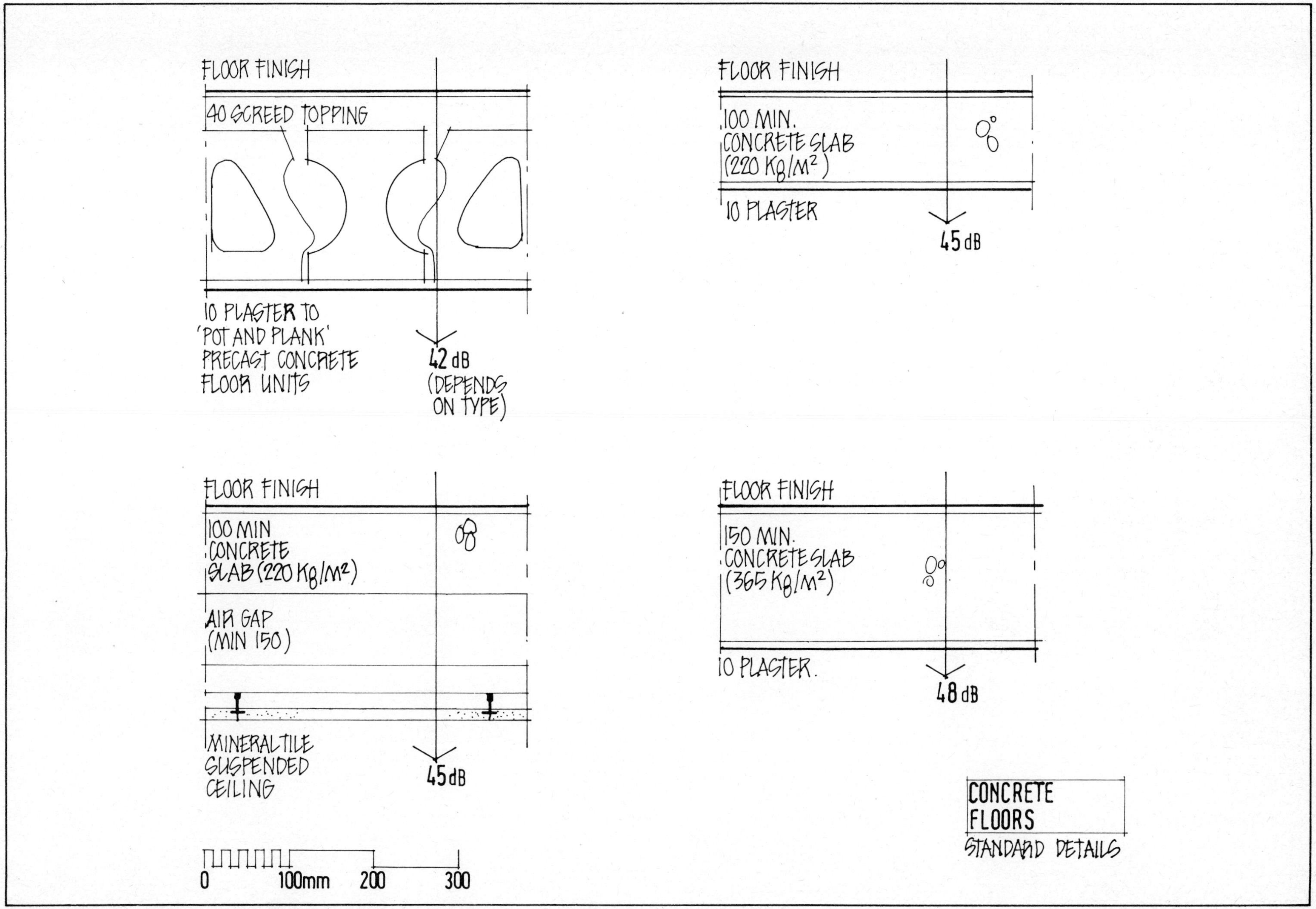
FLOOR FINISH
40 SCREED TOPPING
10 PLASTER TO 'POT AND PLANK' PRECAST CONCRETE FLOOR UNITS
42 dB (DEPENDS ON TYPE)
FLOOR FINISH
100 MIN. CONCRETE SLAB (220 Kg/M²)
10 PLASTER
45 dB
FLOOR FINISH
100 MIN CONCRETE SLAB (220 Kg/M²)
AIR GAP (MIN 150)
MINERAL TILE SUSPENDED CEILING
45 dB
FLOOR FINISH
150 MIN. CONCRETE SLAB (365 Kg/M²)
10 PLASTER.
48 dB
0
100mm
200
300
CONCRETE FLOORS
STANDARD DETAILS

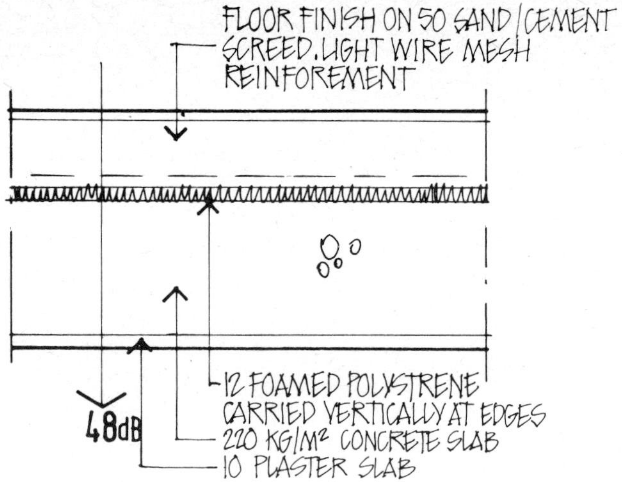

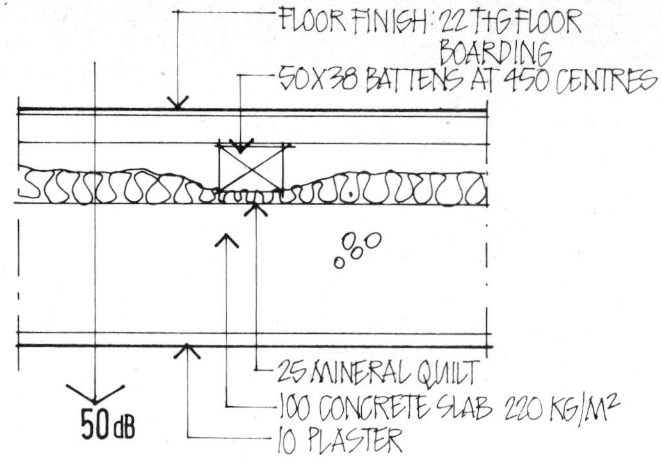

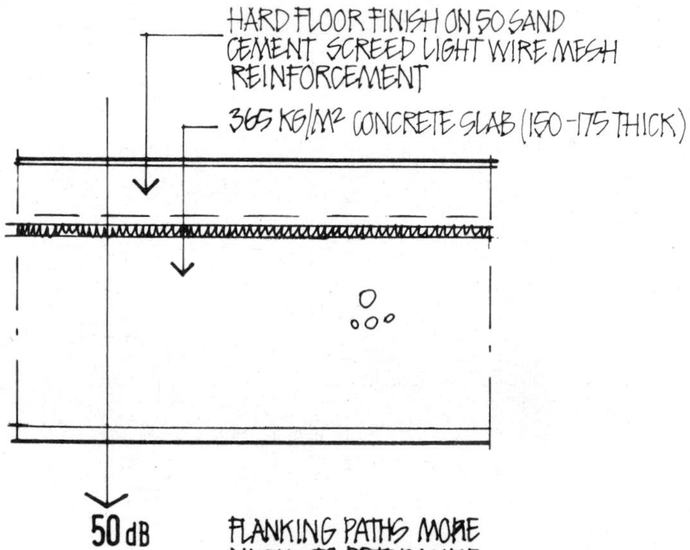

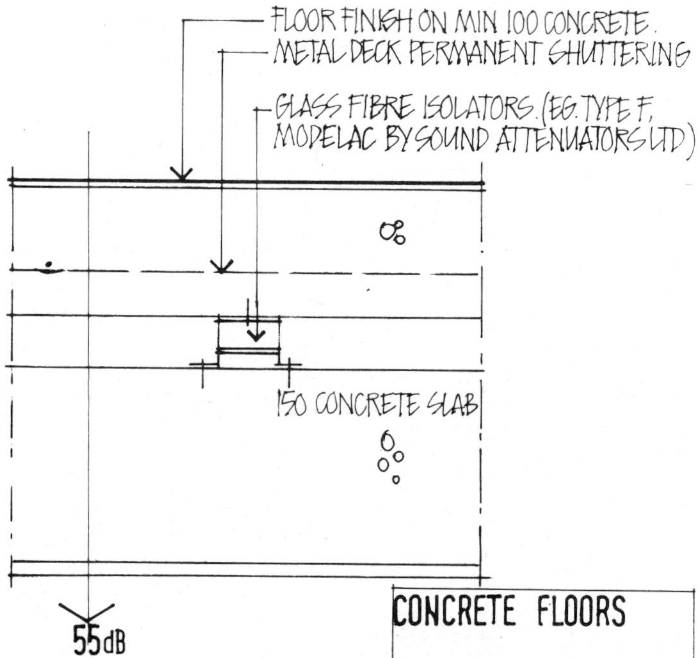

FLANKING PATHS MORE LIKELY TO DETERMINE PERFORMANCE LIMITS FOR FLOORS OVER 48 dB RATING.

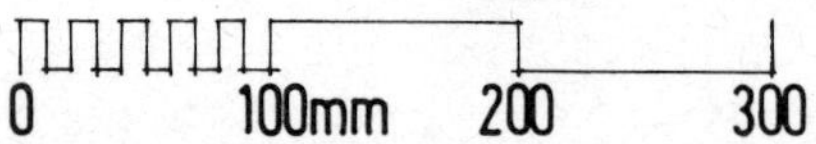

CONCRETE FLOORS

STANDARD DETAILS.

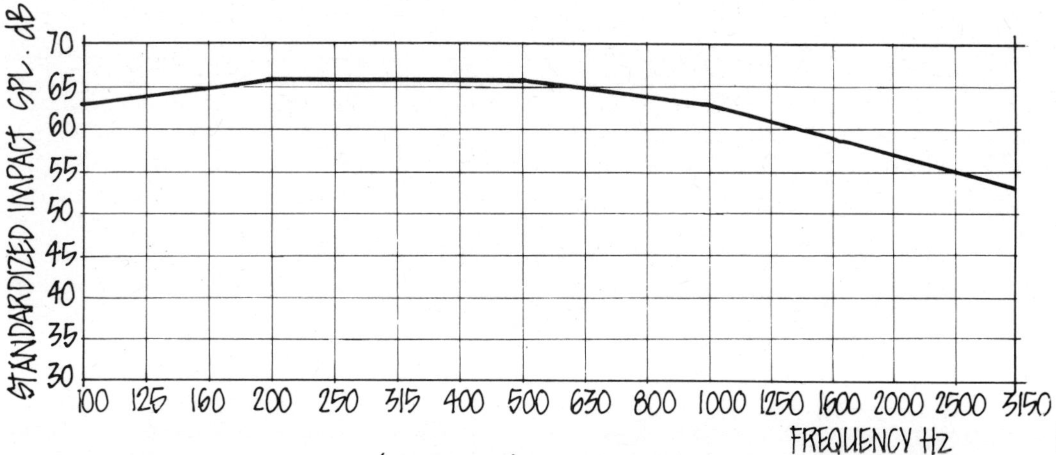

BUILDING REGULATIONS (1976 PART G) IMPACT SOUND TRANSMISSION
[NOTE: UNLIKE AIRBORNE SOUND A HIGHER VALUE IS A LESSER STANDARD]

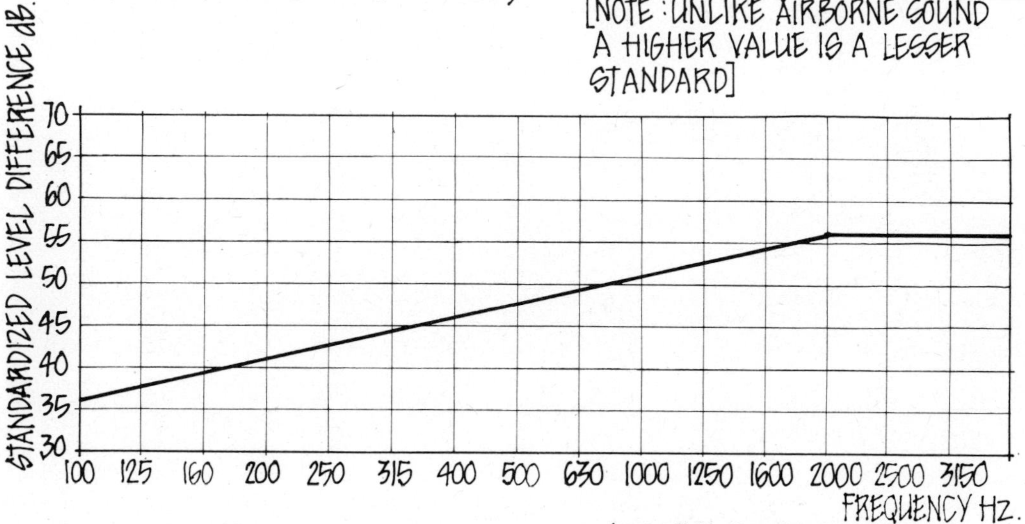

AIRBORNE SOUND INSULATION

FLOOR TYPE	AIRBORNE AAD		IMPACT AAD
	AVERAGE (dB)	RISK OF EXCEEDING 23dB (%)	
1	12	8	23
2	11	14	34
3	22	45	80
4	27	59	68

FLOOR TYPES

1. TIMBER RAFT FLOATED ON A SOLID CONCRETE BASE.
2. FLOATING CONCRETE SCREED ON A SOLID CONCRETE BASE.
3. LIGHTWEIGHT TIMBER RAFT AND PLATFORM FLOORS.
4. FLOATING CONCRETE SCREED ON HOLLOW CONCRETE BEAM OR BEAM AND HOLLOW BLOCK BASE

GIVEN AVERAGE WORKMANSHIP TYPE I FLOORS APPEAR TO GIVE THE MOST RELIABLE TEST RESULTS. OTHER TYPES MAY ALSO GIVE SATISFACTORY RESULTS BUT PERFORMANCE FIGURES ARE NOT AVAILABLE. TO COMPLY WITH GRADE I, THE SUM OF FAILURES AT ONE-THIRD OCTAVE MUST NOT BE GREATER THAN 23dB AVERAGE DIFFERENCE WITHIN 1·50 dB.

FLOORS
SOURCE:
BRE DIGEST 266

AIRBORNE SOUND INSULATION IN AN OFFICE BUILDING WITH A STEEL FRAME, LIGHT-WEIGHT SOLID FLOORS AND DOUBLE-LEAF WALLS. SUSPENDED CEILINGS

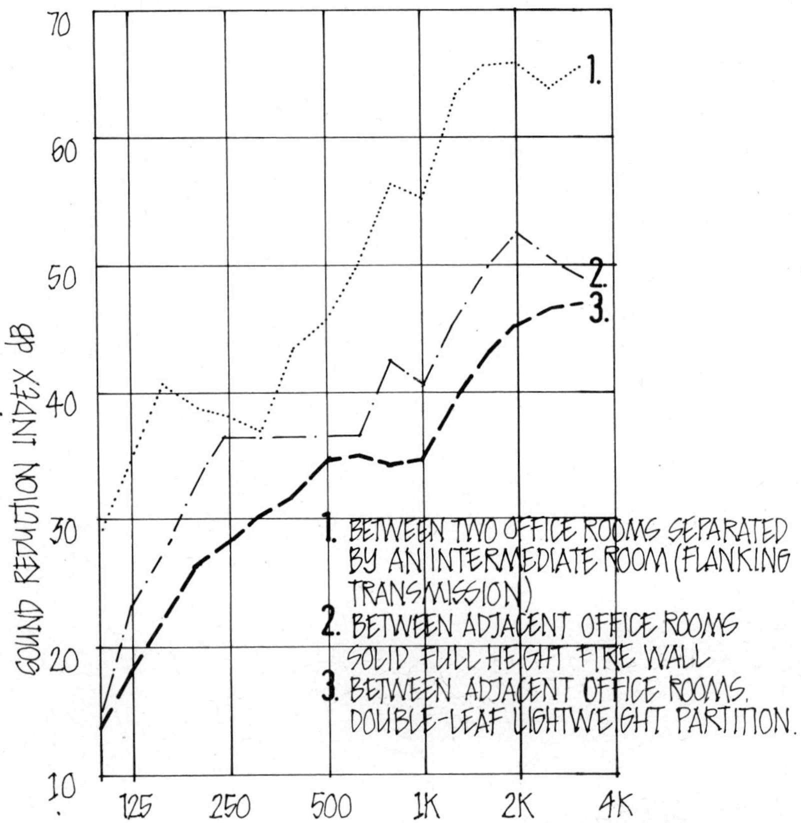

AIRBORNE SOUND INSULATION BETWEEN 2 ROOMS SEPARATED BY AN INTERMEDIATE FLOOR

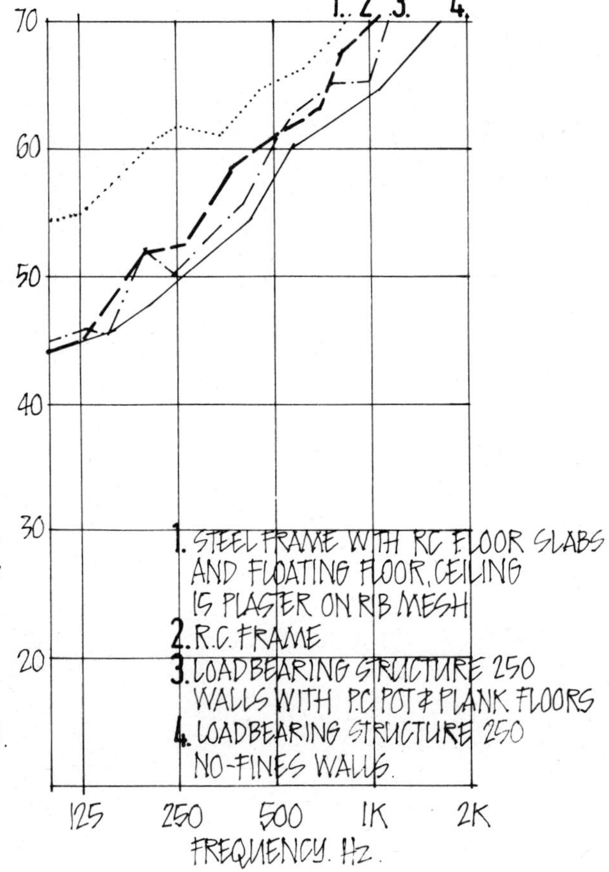

IMPROVEMENT IS ACHIEVED AT ALL FREQUENCIES IN SPITE OF FLANKING TRANSMISSION, BY INCREASING SOLIDITY OF SUBDIVISION AND SEPARATING NOISY AND NOISE-SENSITIVE ROOMS.

STRUCTURE
SOURCE: DR. LANG.
TECHNOLOGISCHE
GEWERBE MUSEUM. VIENNA

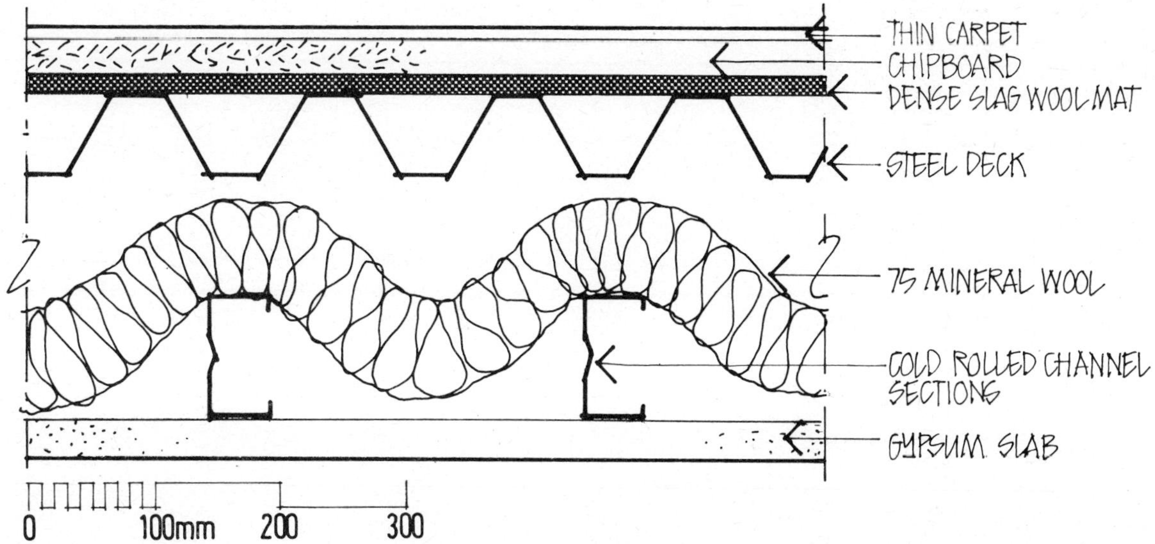

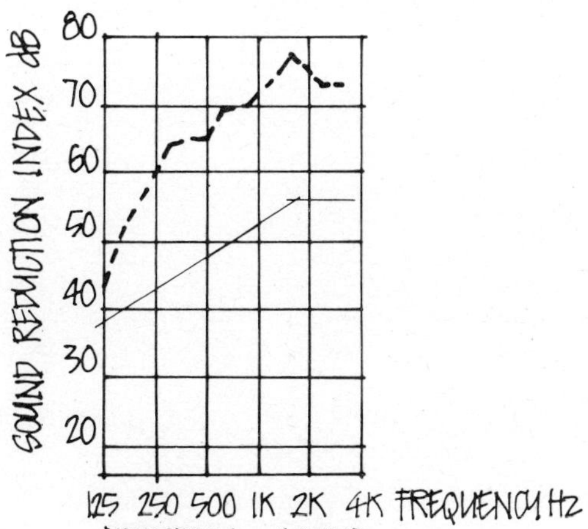

AIRBORNE SOUND
– BUILDING REGULATIONS DEEMED TO SATISFY PROVISION

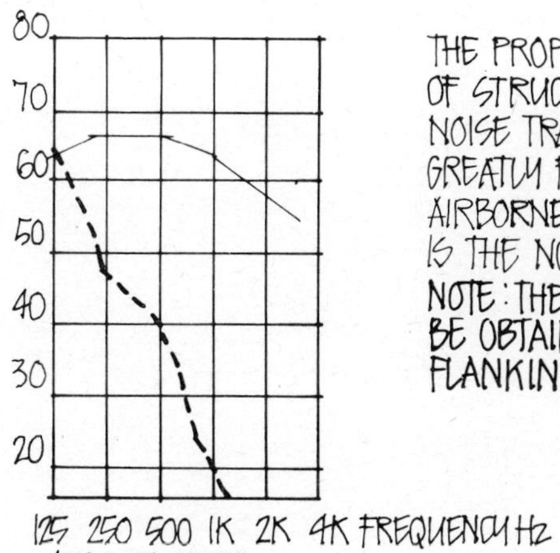

IMPACT SOUND
– BUILDING REGULATIONS DEEMED TO SATISFY PROVISION.

THE PROPERTY OF ELEMENTS OF STRUCTURE TO PREVENT NOISE TRANSMISSION DIFFERS GREATLY FOR WHETHER AIRBORNE OR IMPACT SOUND IS THE NOISE SOURCE.

NOTE: THESE WOULD ONLY BE BE OBTAINED WITH NEGLIGIBLE FLANKING.

FLOORS
SOURCE: DR. LANG
TECHNOLOGISCHES
GEWERBE MUSEUM: VIENNA.

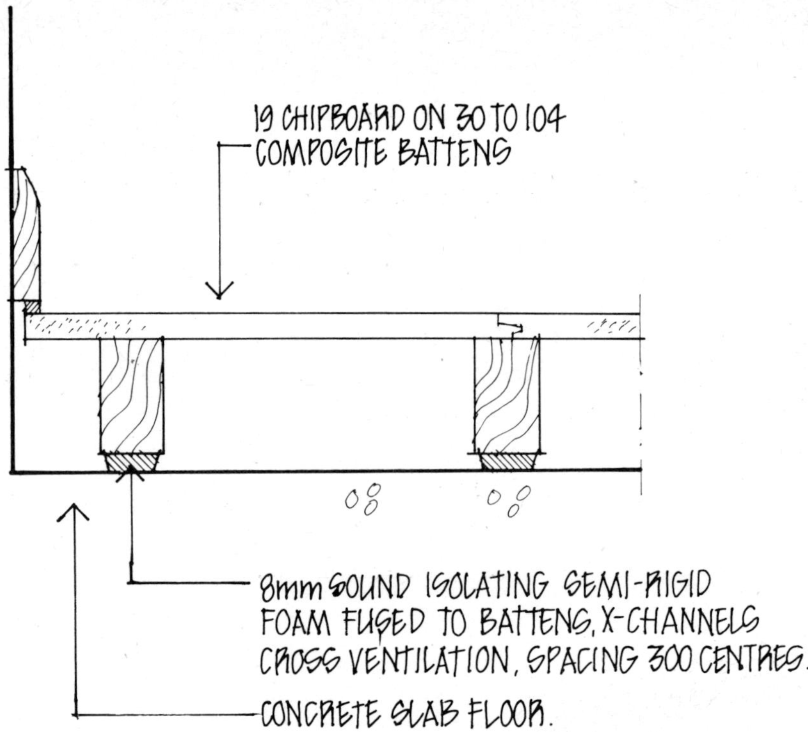

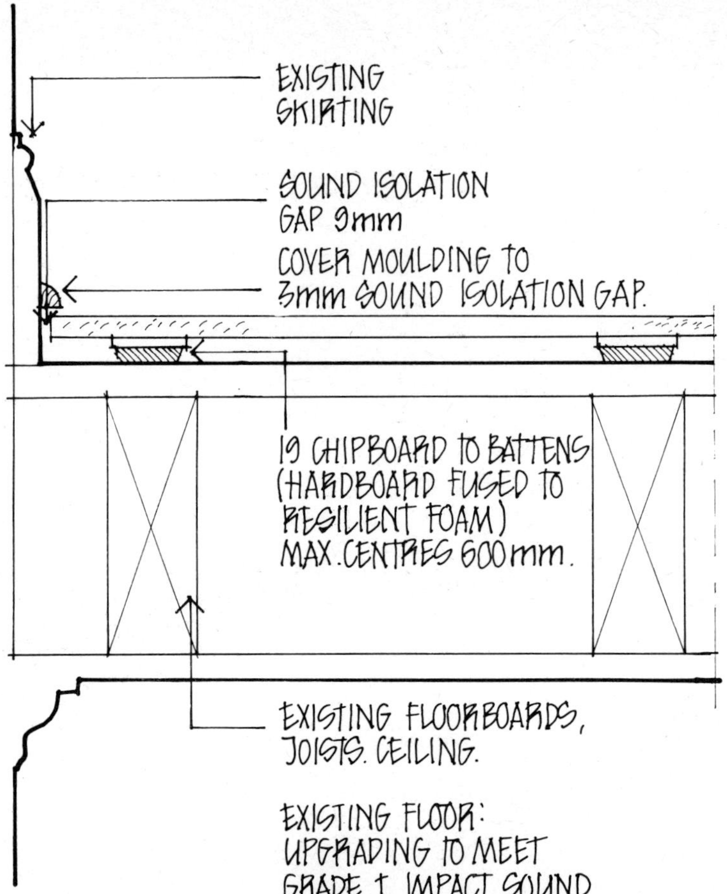

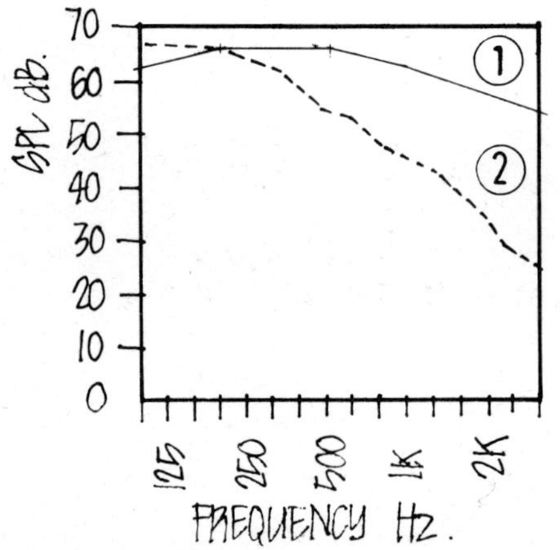

NEW CONSTRUCTION:
FLOATING FLOOR TO ACHIEVE GRADE I IMPACT SOUND INSULATION BY LOMBARD FLOORING SYSTEMS.

USE IN FLATS, MAISONETTES AND OPEN PLAN OFFICES.

(1) GRADE 1 BUILDING REGULATIONS

(2) NORMALISED IMPACT SOUND LEVEL ADVERSE DEVIATION 8·5 dB.

EXISTING FLOOR:
UPGRADING TO MEET GRADE I IMPACT SOUND INSULATION BY MODIFIED WESTBOURNE SYSTEM OF FLOORING.

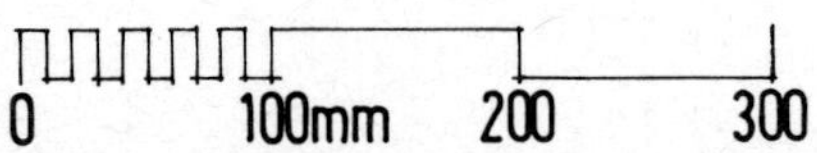

FLOORS
SOURCE: CONTIWOOD (DURABELLA) LTD.

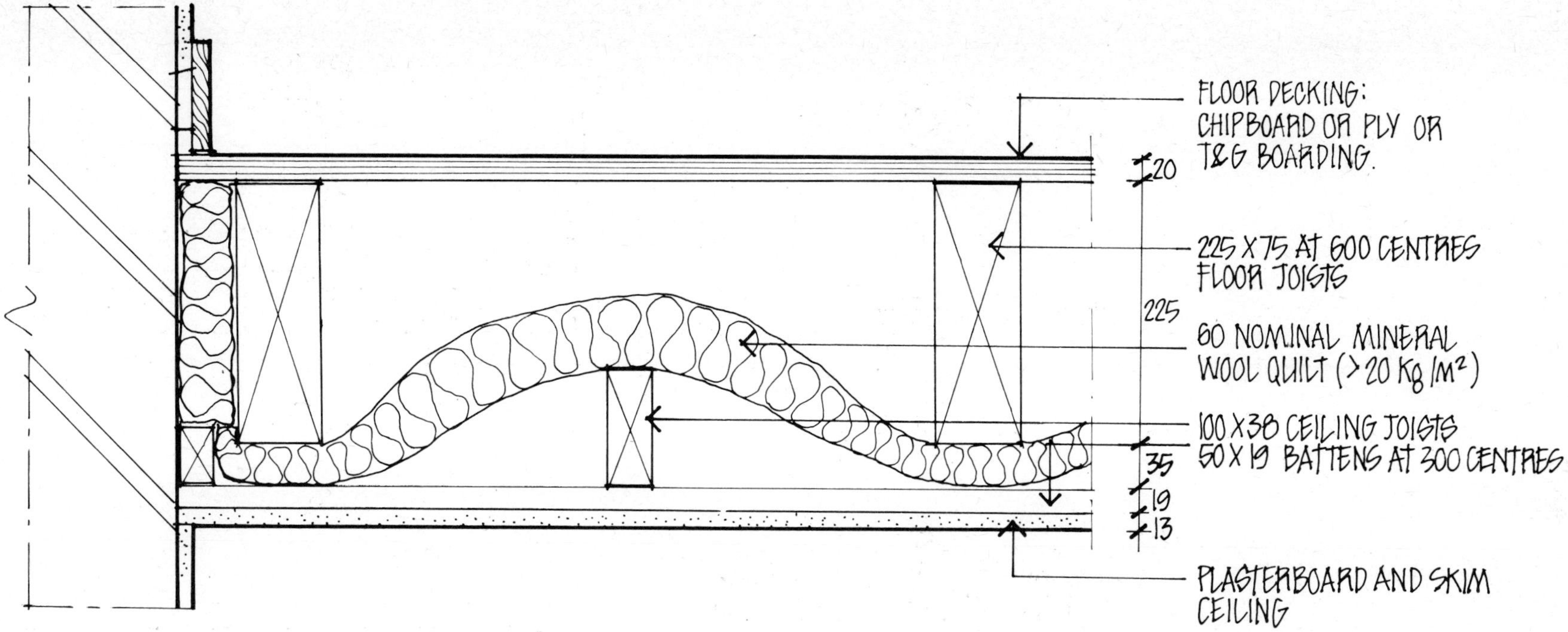

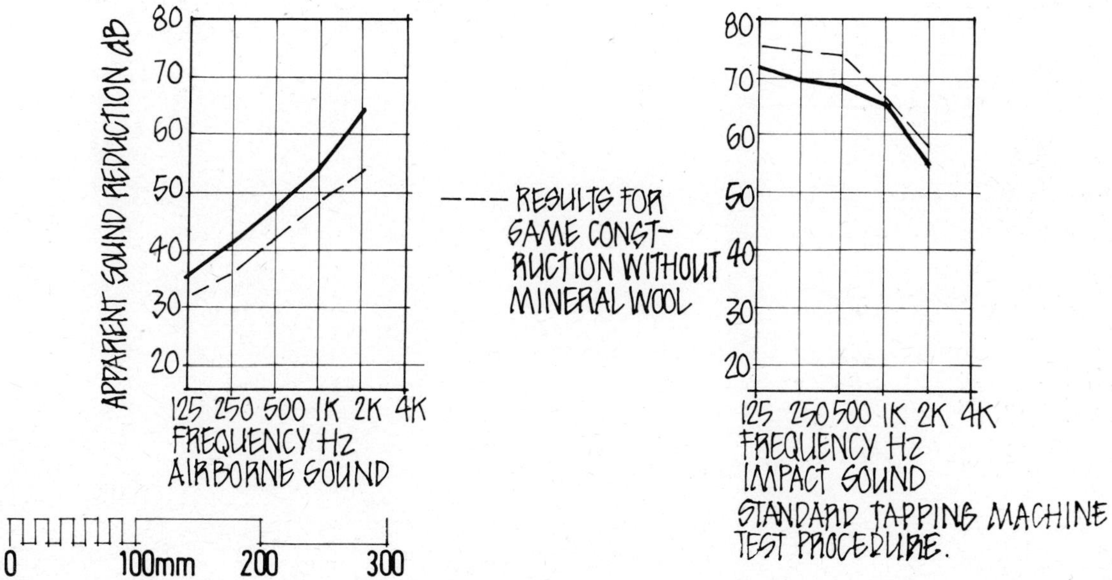

SUBSTANTIAL EDGE WALLS ASSUMED TO AVOID FLANKING
CEILING STRUCTURE SEPARATED FROM FLOOR STRUCTURE EXCEPT AT BEARING. WEIGHT: 48 Kg/M²

FLOORS

SOURCE: H.V. INSTITUTE AMSTERDAM

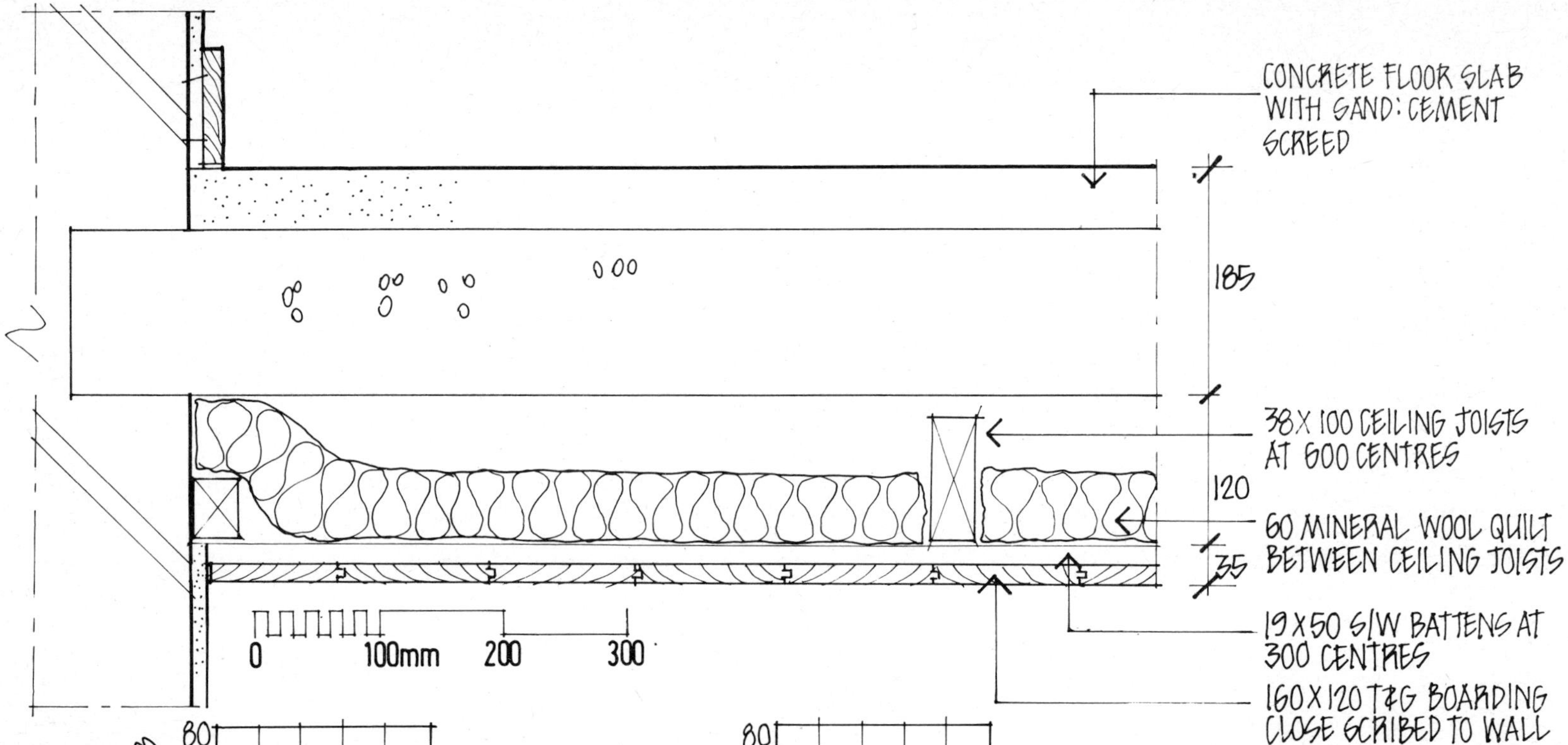

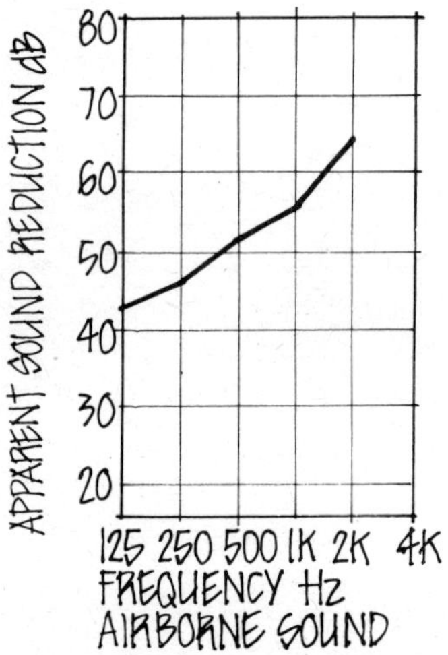

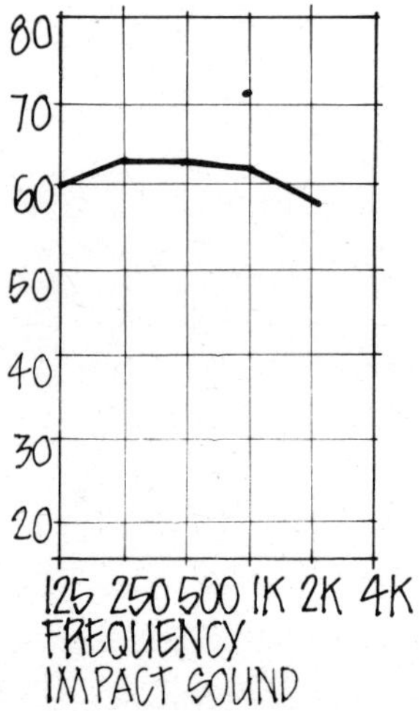

MASS COMPOSITE CONSTRUCTION AND DISCONTINUITY GIVE A GOOD OVERALL PERFORMANCE. BATTENS GIVE AIR SPACE BELOW AS WELL AS ABOVE QUILT. GOOD FOR ISOLATION OF UPPER FLOOR FROM NOISY ACTIVITIES BELOW (eg. COMMON ROOMS BELOW STUDENT BEDROOMS)
WEIGHT 420 Kg/M²

FLOORS
SOURCE: H.V. INSTITUTE AMSTERDAM

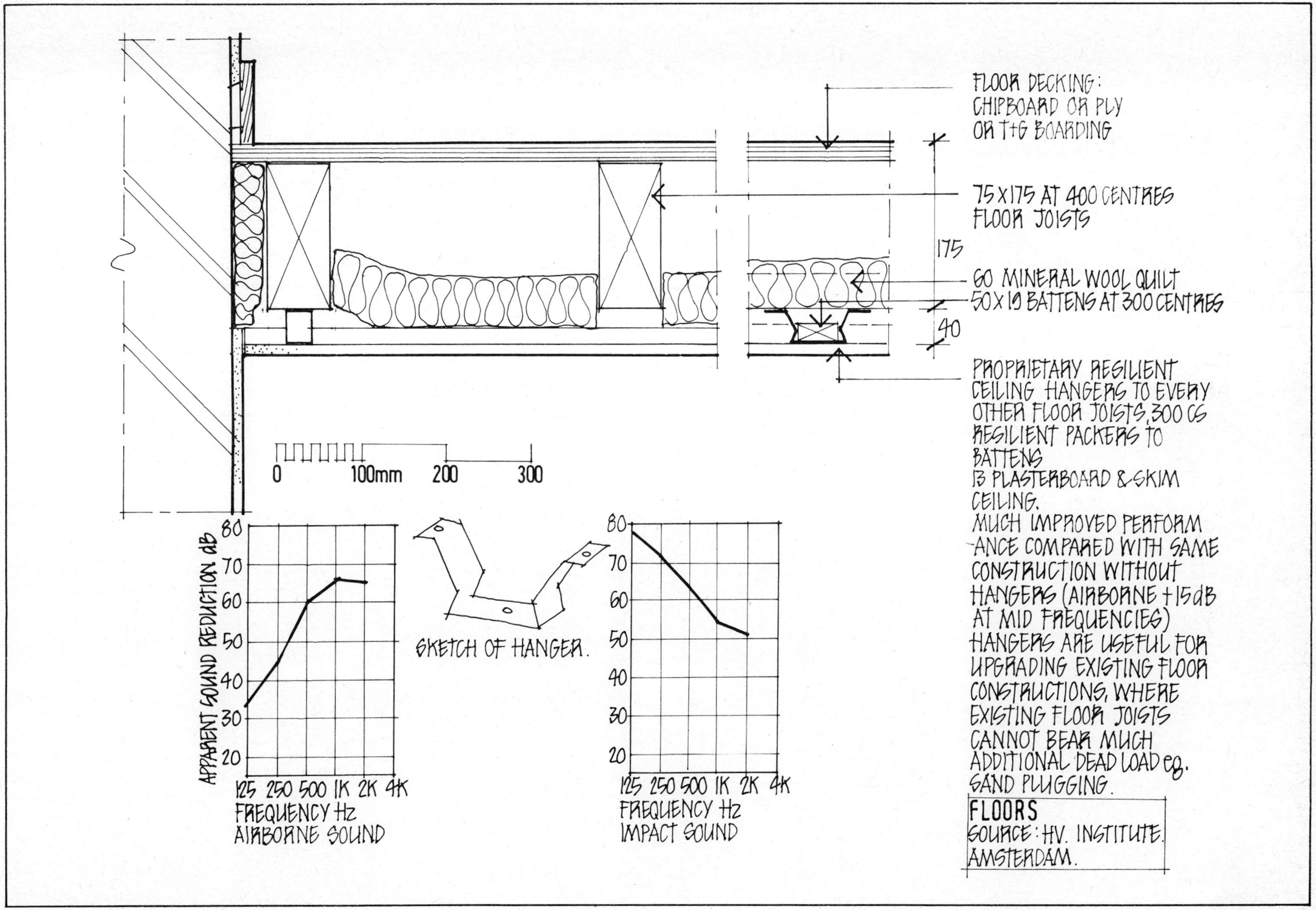
FLOOR DECKING:
CHIPBOARD OR PLY
OR T+G BOARDING
75 X 175 AT 400 CENTRES
FLOOR JOISTS
175
60 MINERAL WOOL QUILT
50 X 19 BATTENS AT 300 CENTRES
40
PROPRIETARY RESILIENT
CEILING HANGERS TO EVERY
OTHER FLOOR JOISTS, 300 CS
RESILIENT PACKERS TO
BATTENS
13 PLASTERBOARD & SKIM
CEILING.
MUCH IMPROVED PERFORM
-ANCE COMPARED WITH SAME
CONSTRUCTION WITHOUT
HANGERS (AIRBORNE +15 dB
AT MID FREQUENCIES)
HANGERS ARE USEFUL FOR
UPGRADING EXISTING FLOOR
CONSTRUCTIONS, WHERE
EXISTING FLOOR JOISTS
CANNOT BEAR MUCH
ADDITIONAL DEAD LOAD eg.
SAND PLUGGING.
FLOORS
SOURCE: HV. INSTITUTE.
AMSTERDAM.
0
100mm
200
300
APPARENT SOUND REDUCTION dB
80
70
60
50
40
30
20
125 250 500 1K 2K 4K
FREQUENCY Hz
AIRBORNE SOUND
SKETCH OF HANGER.
FREQUENCY Hz
IMPACT SOUND

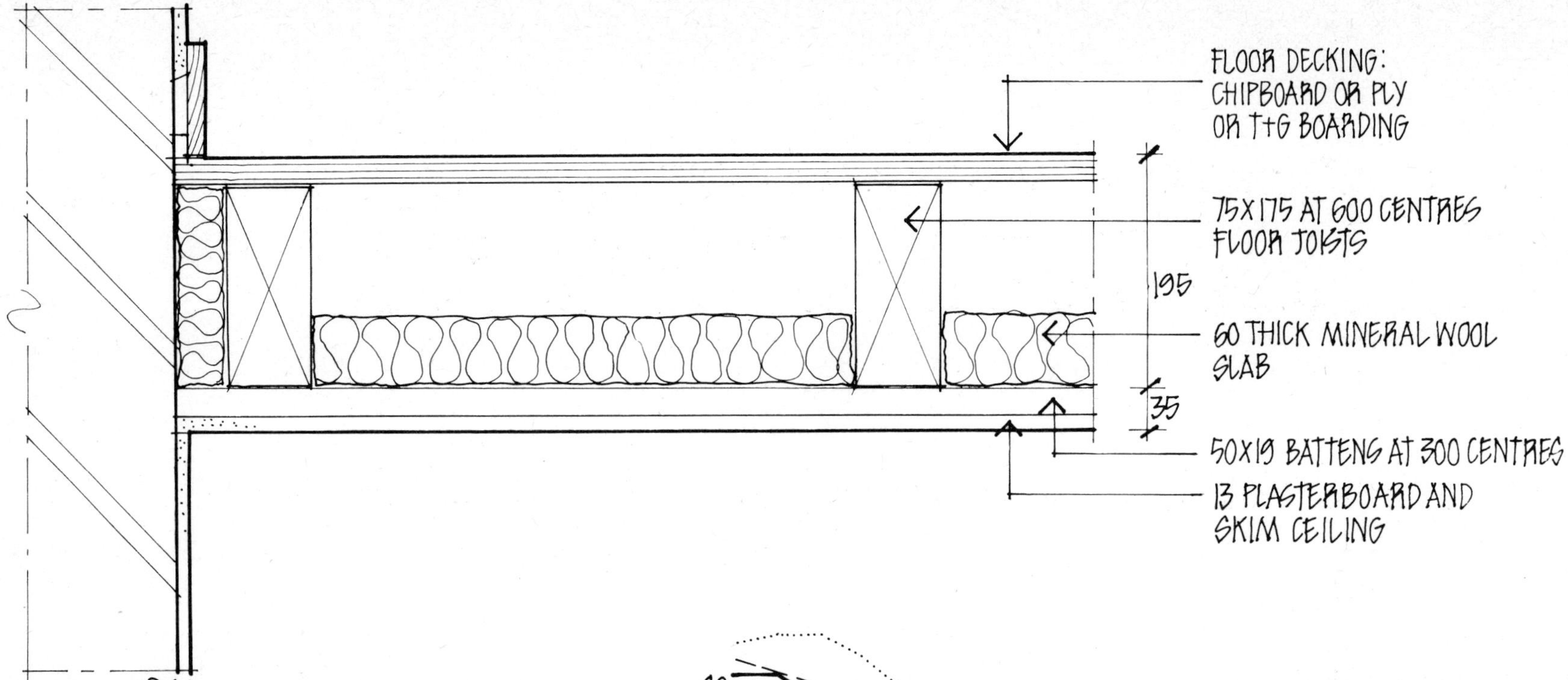

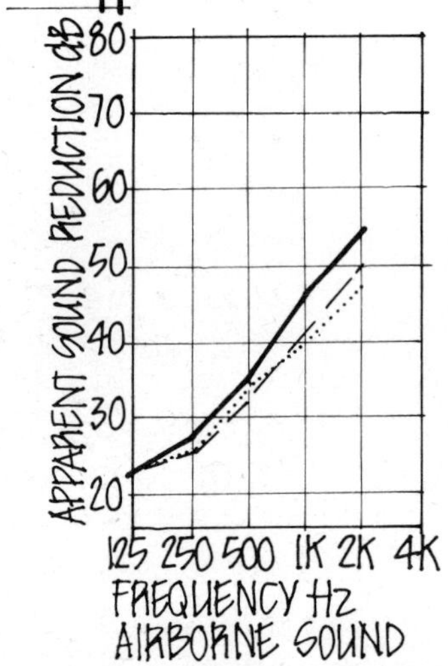

– – – RESULTS FOR SAME CONSTRUCTION WITHOUT MINERAL WOOL.

·········· RESULTS FOR CONSTRUCTION WITH 9.5 PLASTERBOARD CEILING DIRECT TO JOISTS. NO MINERAL WOOL.

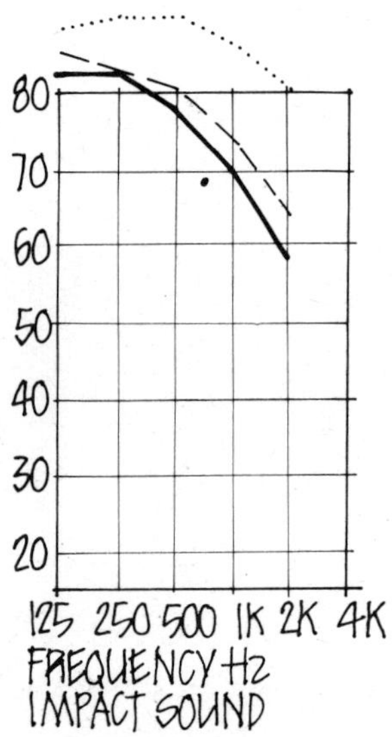

EVEN A SUBSTANTIAL TIMBER FLOOR CANNOT COMPARE IN AIRBORNE OR IMPACT SOUND PERFORMANCE WITH 'SPLIT' FLOOR AND CEILING CONSTRUCTION. ABSORPTIVE MATERIAL IN FLOOR VOID HELPS IN DAMPING THE TRACKING OF SOUND ACROSS. WEIGHT 39 Kg/m²

FLOORS
SOURCE: H.V. INSTITUTE AMSTERDAM.

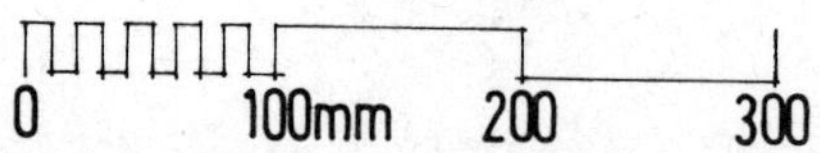

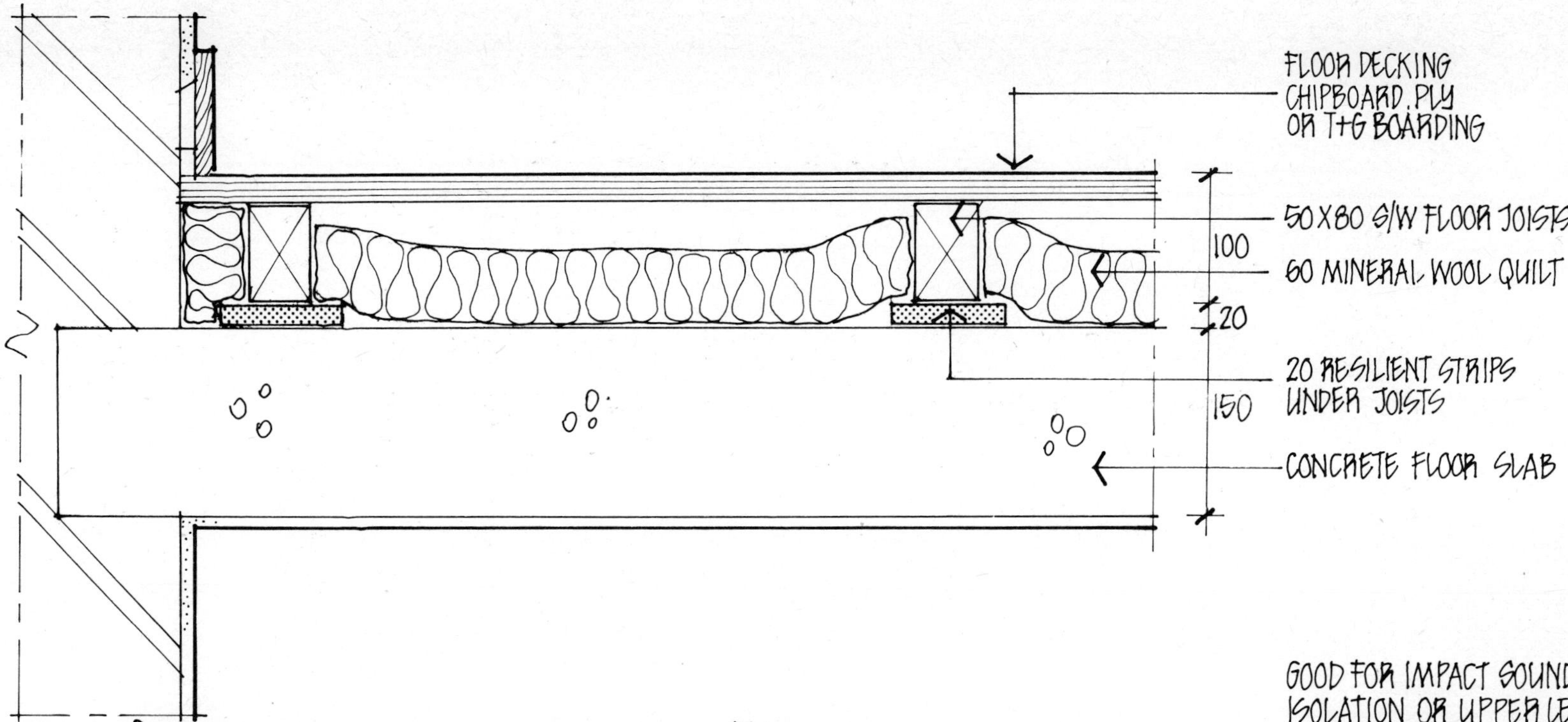

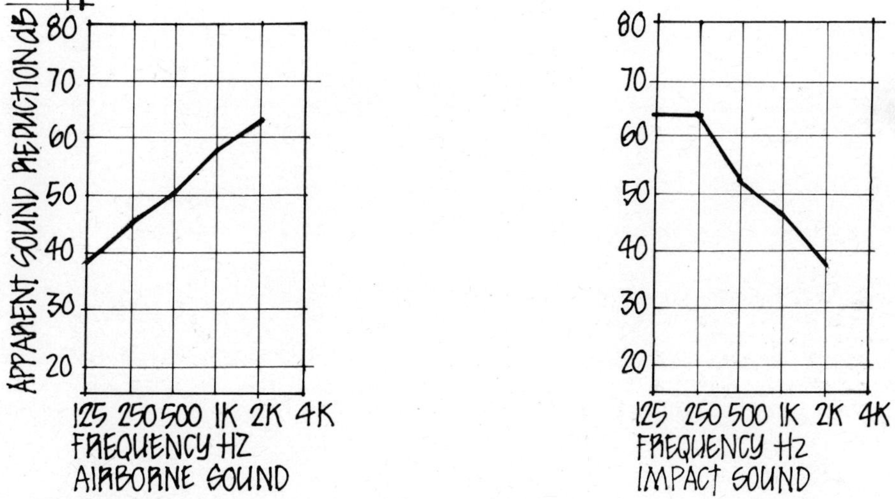

GOOD FOR IMPACT SOUND ISOLATION OR UPPER LEVEL NOISE SOURCE. BECAUSE OF FLANKING, IT IS NOT WORTH GOING TO THE TROUBLE OF TRIPLE FLOOR CONSTRUCTION (ie. FLOATING FLOOR/STRUCT-URAL SLAB/SEPARATE CEILING SUSPENSION) UNLESS ISOLATING WALLS TO SUCH ELEMENTS ARE ALSO USED. WEIGHT 350 Kg/m²

FLOORS
SOURCE: H.V. INSTITUTE AMSTERDAM

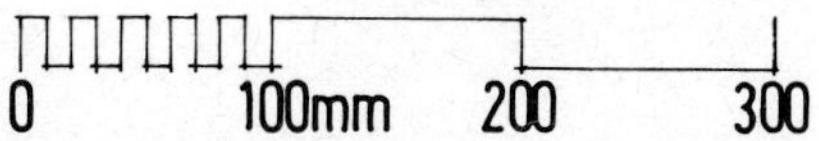

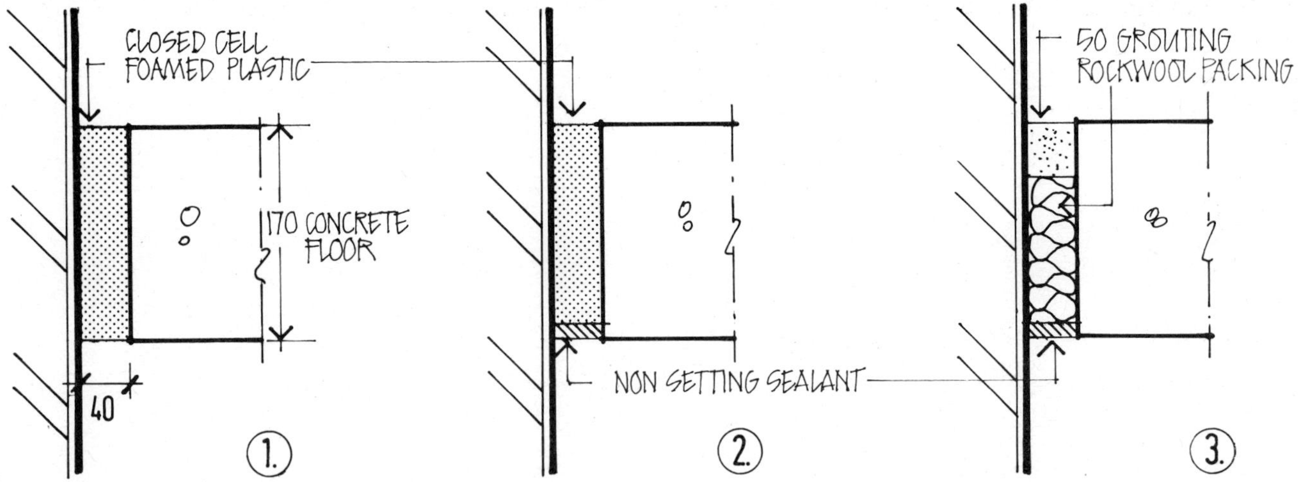

VERTICAL SECTIONS

SOUND LEVEL DIFFERENCE BETWEEN TWO ROOMS CAUSED BY A 2 METRE SLOT IN ONE SIDE OF A FLOOR (LABORATORY TEST). GOOD SEALING OF GAPS NECESSARY FOR SPEECH FREQUENCY PRIVACY, SEALING LESS CRITICAL AT LOW FREQUENCIES.

A SMALL STOREY-HEIGHT CRACK 0·2mm WIDE IN A WALL 10 M² IN AREA WOULD RESULT IN A 7dB LOSS IN A 50 dB-RATED WALL, BUT 1 dB ONLY REDUCTION OF INSULATION IN A 35 dB. PARTITION: LACK OF CRACKS OR GAPS IS PARTICULARLY IMPORTANT FOR SUBSTANTIAL SEPARATING ELEMENTS.

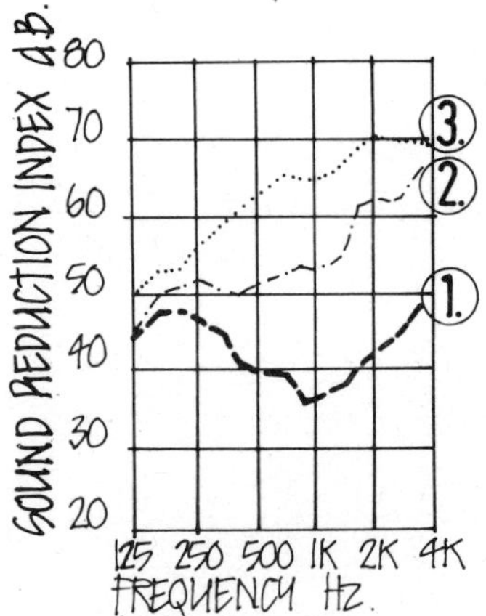

AIRBORNE SOUND

FLOORS
SOURCE: INSTITUT FÜR TECHNISCHE PHYSIK STUTTGART.

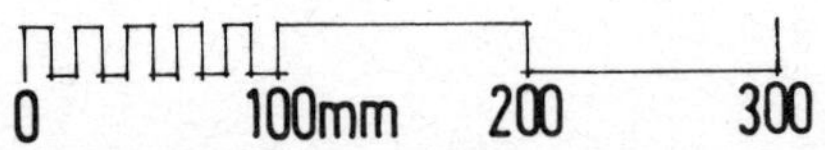

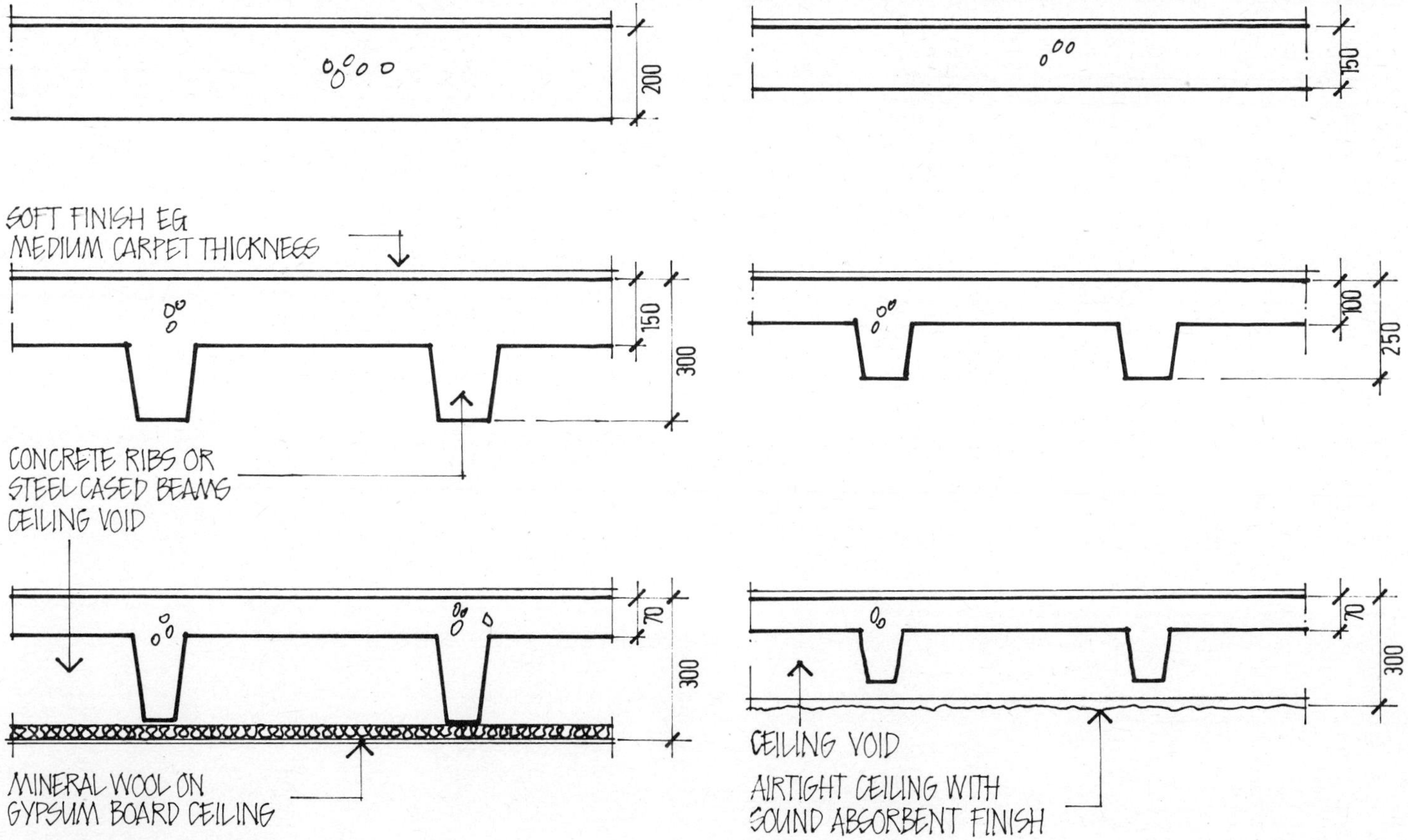

FLOOR STRUCTURES WITH
EQUAL IMPACT SOUND
INSULATION

LuTw = 63 dB
(LuTw IS WEIGHTED STANDARDIZED IMPACT
SOUND PRESSURE LEVEL. TO BS 5281 PART 7)
MEASURED IN ACCORDANCE WITH BS 2750

LuTw = 68 dB
SECTIONS NOT TO SCALE
FIGURES ASSUME
NEGLIBLE FLANKING.

FLOORS
SOURCE: DR. LANG
TECHNOLOGISCHES
GEWERBE MUSEUM, VIENNA

not to scale

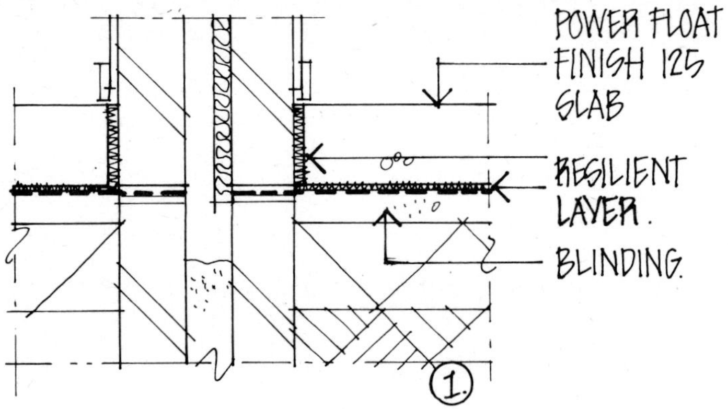

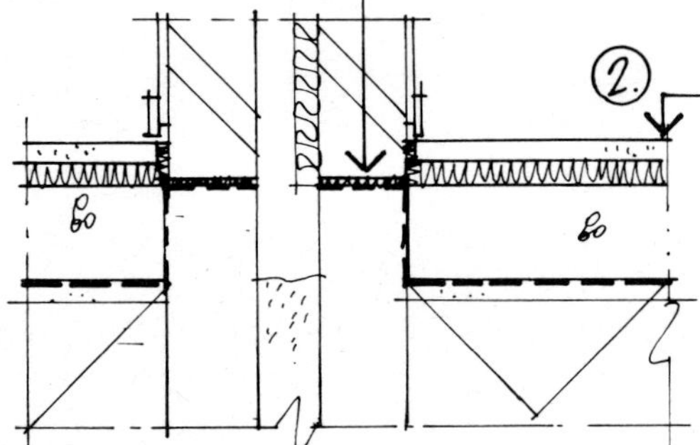

ELASTOMERIC LAYER OVER DPC

PROPRIETARY BUILT-UP FLOOR: 19 OR 22 CHIPBOARD ON EXPANDED POLYSTYRENE TO FLOOR SLAB

(1) IS POOR COMPARED WITH (3): RESILIENT QUILT IS SUBJECT TO EXTRA COMPRESSION LOAD AND WATER FROM CONCRETE MIX. ABSORBENT MATERIAL IN CAVITY CUTS DOWN TRACKING OF NOISE ACROSS CAVITY.

(1)–(4) SHOW VARYING MEANS OF ISOLATING FLOOR SLABS IN ADJACENT ROOMS TO PREVENT FLANKING. (eg. IN MUSIC ROOMS, CONFERENCE ROOMS)

(4) IS IMPRACTICAL TO INSTALL

LONG FIBRE GLASS WOOL AND MINERAL WOOL QUILTS HAVE BEEN SUGGESTED AS THE MOST PROVEN MATERIALS (BRE 103). FIBREBOARD AND HAIR FELT HAVE SHOWN UNSATISFACTORY COMPACTION UNDER CONTINUOUS LOAD. MATERIAL TO BE USED TO HAVE FOLLOWING FEATURES:

A.– DENSITY 15–25 Kg/M² MIN
B.– THICKNESS NOT LESS THAN 13mm
C.– BOARD PRECOMPRESSED TO HALF INITIAL THICKNESS WITH RAPID RECOVERY TO 9% INITIAL THICKNESS.

THIN MATERIALS ARE PROMISING IN PERFORMANCE: REDUC HAVE A MEMBRANE ONLY 1·5MM THICK. OTHER INSULATION PANELS eg. TRT PANELS CONSIST OF CHIPBOARD, HARDBOARD OR PLY EITHER SIDE OF A THIN VISCO ELASTIC LAYER WHICH HAS THE PROPERTY OF A HIGH DEGREE OF INTERNAL DAMPING.

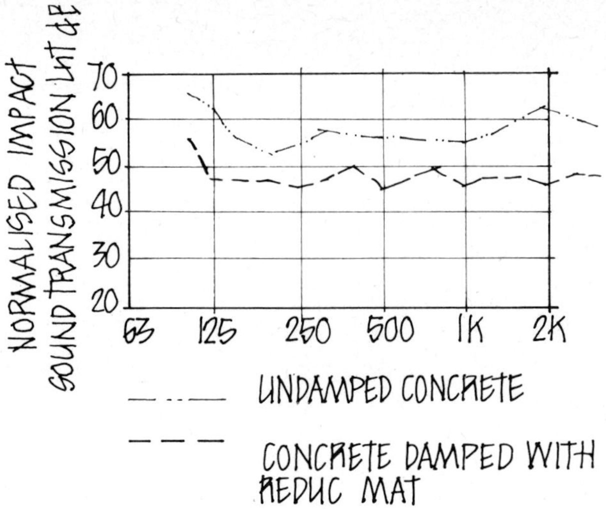

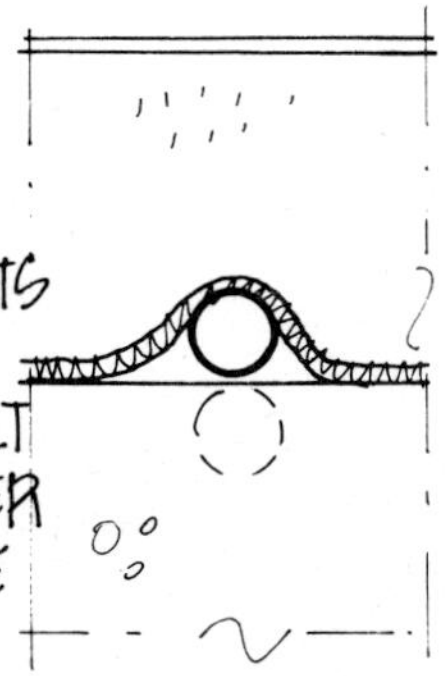

CONDUIT OR PIPE SET UNDER SCREED (NOT TO EXCEED 25 IN 50 SCREED)

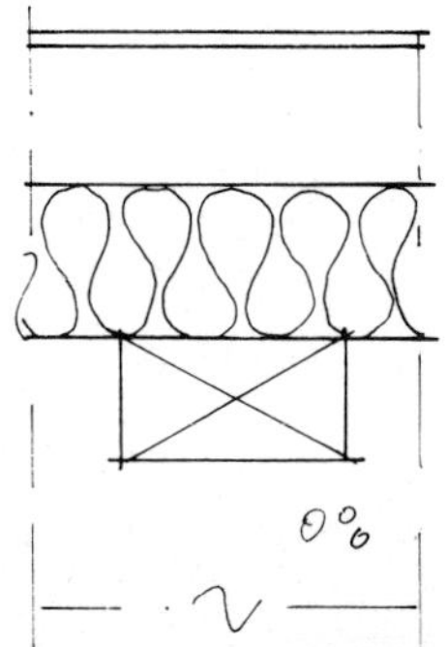

POCKETS UNDER EXPANDED POLY STYRENE NOT EXCEEDING 75 WIDTH

FLOORS

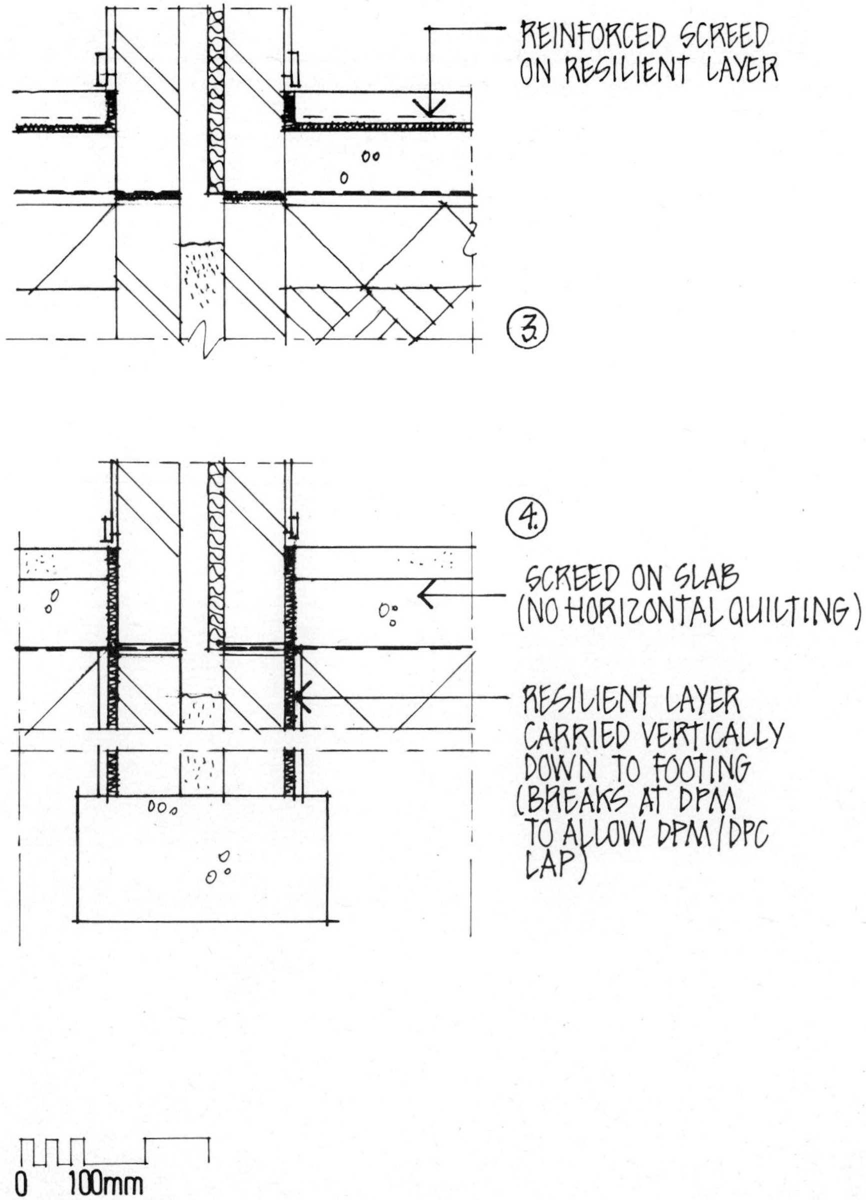
REINFORCED SCREED ON RESILIENT LAYER
3
4.
SCREED ON SLAB (NO HORIZONTAL QUILTING)
RESILIENT LAYER CARRIED VERTICALLY DOWN TO FOOTING (BREAKS AT DPM TO ALLOW DPM/DPC LAP)
0 100mm

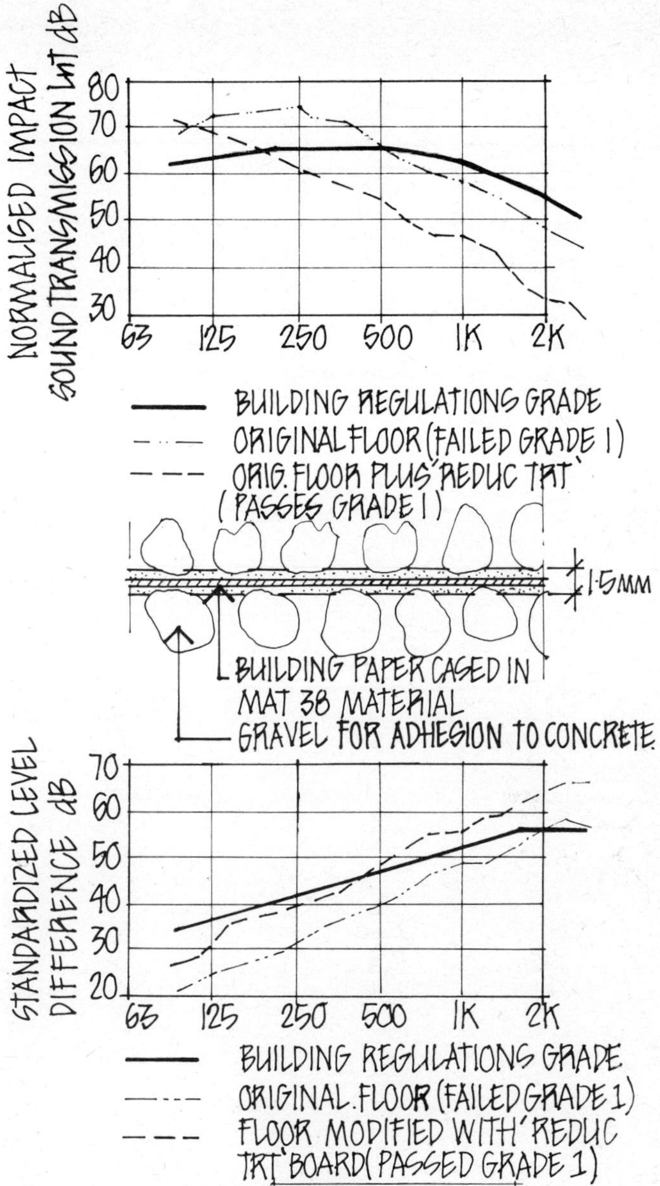
NORMALISED IMPACT SOUND TRANSMISSION LnT dB
80
70
60
50
40
30
63
125
250
500
1K
2K
BUILDING REGULATIONS GRADE
ORIGINAL FLOOR (FAILED GRADE 1)
ORIG. FLOOR PLUS 'REDUC TRT' (PASSES GRADE 1)
1·5MM
BUILDING PAPER CASED IN MAT 38 MATERIAL
GRAVEL FOR ADHESION TO CONCRETE
STANDARDIZED LEVEL DIFFERENCE dB
70
60
50
40
30
20
63
125
250
500
1K
2K
BUILDING REGULATIONS GRADE
ORIGINAL FLOOR (FAILED GRADE 1)
FLOOR MODIFIED WITH 'REDUC TRT' BOARD (PASSED GRADE 1)

FLOORS

DOORS

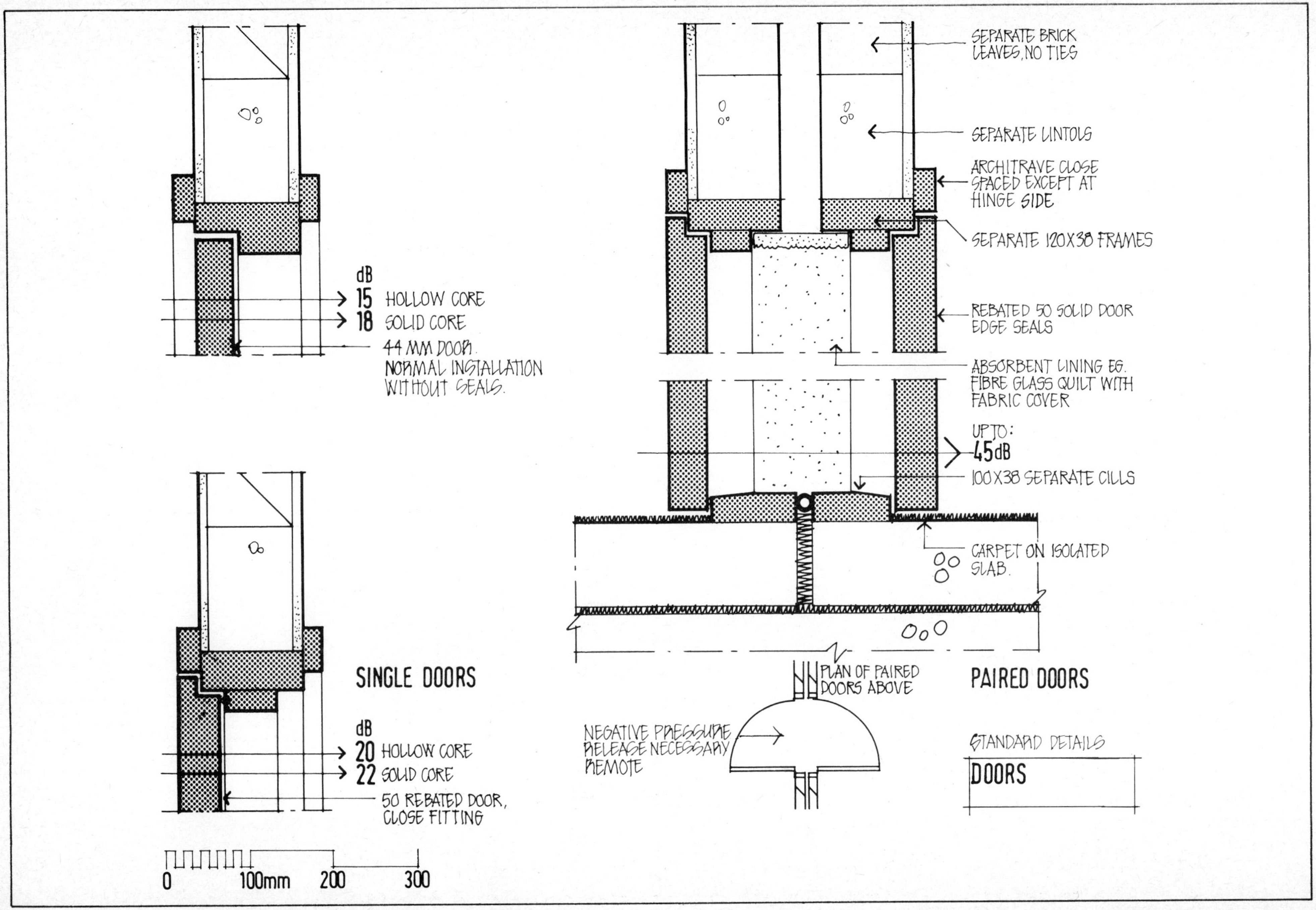
dB
15 HOLLOW CORE
18 SOLID CORE
44 MM DOOR. NORMAL INSTALLATION WITHOUT SEALS.
dB
20 HOLLOW CORE
22 SOLID CORE
50 REBATED DOOR, CLOSE FITTING
SINGLE DOORS
0 100mm 200 300
SEPARATE BRICK LEAVES, NO TIES
SEPARATE LINTOLS
ARCHITRAVE CLOSE SPACED EXCEPT AT HINGE SIDE
SEPARATE 120X38 FRAMES
REBATED 50 SOLID DOOR EDGE SEALS
ABSORBENT LINING EG. FIBRE GLASS QUILT WITH FABRIC COVER
UP TO: 45dB
100X38 SEPARATE CILLS
CARPET ON ISOLATED SLAB.
PLAN OF PAIRED DOORS ABOVE
NEGATIVE PRESSURE RELEASE NECESSARY REMOTE
PAIRED DOORS
STANDARD DETAILS
DOORS

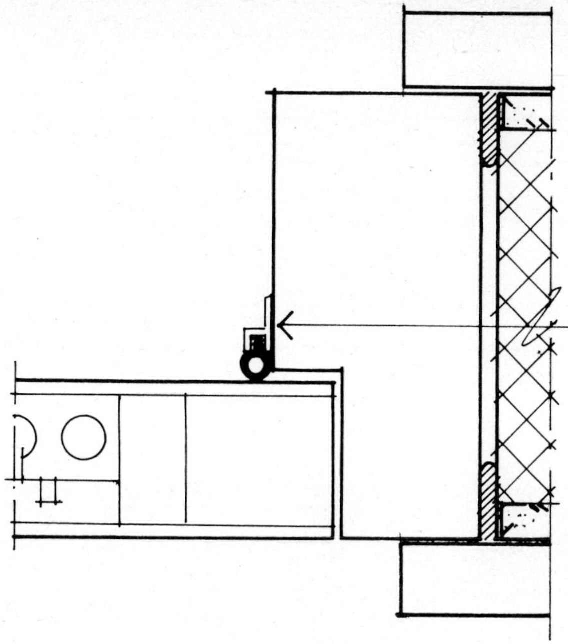

CLASS dB	33	40	42
THICKNESS	40	45	50
WEIGHT KG/M	23	26	32
DOOR FRAMING:			
TOP RAIL	40	40	60
BOTTOM RAIL	2x40	2x40	60
STILE	40+18	40+18	60+40
FACING: VENEERED HARDBOARD TO EITHER SIDE	2·10	3·20	3·20

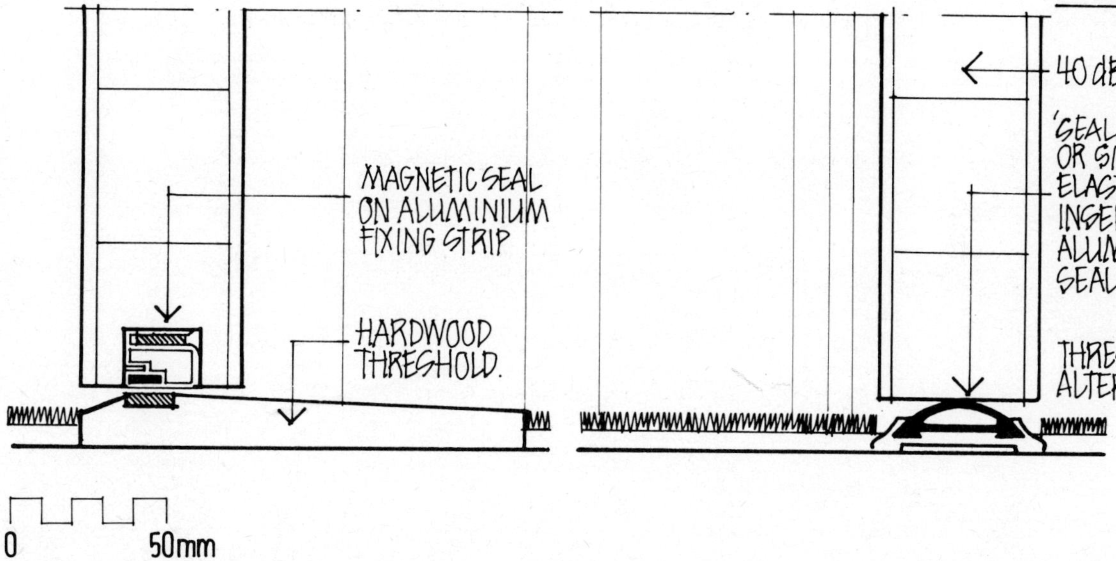

DOORS
SOURCE:
SOUND ACOUSTICS LTD

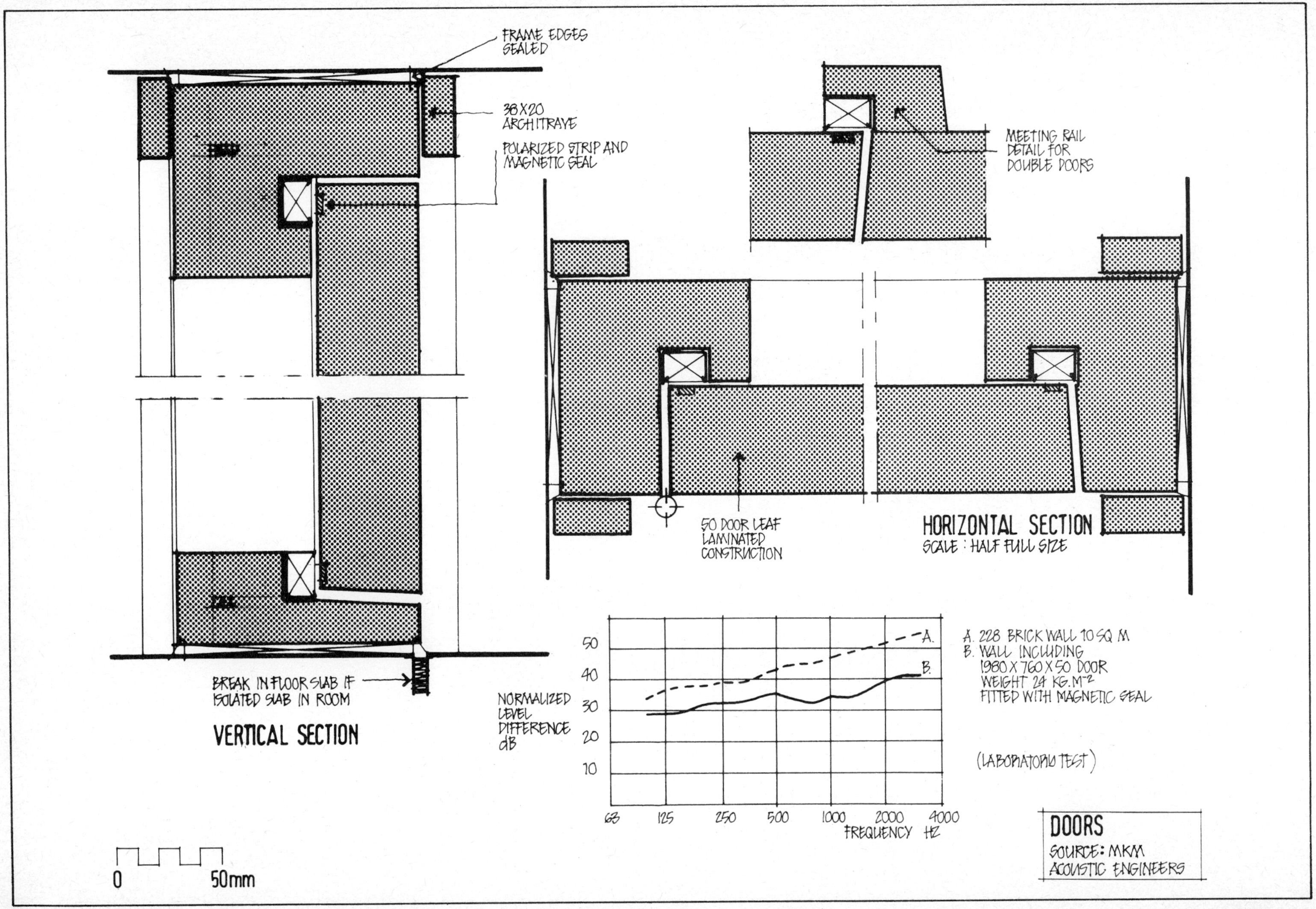
FRAME EDGES SEALED
38 X 20 ARCHITRAVE
POLARIZED STRIP AND MAGNETIC SEAL
BREAK IN FLOOR SLAB IF ISOLATED SLAB IN ROOM
VERTICAL SECTION
MEETING RAIL DETAIL FOR DOUBLE DOORS
50 DOOR LEAF LAMINATED CONSTRUCTION
HORIZONTAL SECTION
SCALE : HALF FULL SIZE
NORMALIZED LEVEL DIFFERENCE dB
50
40
30
20
10
63
125
250
500
1000
2000
4000
FREQUENCY HZ
A.
B.
A. 228 BRICK WALL 10 SQ M
B. WALL INCLUDING 1980 X 760 X 50 DOOR WEIGHT 24 KG.M^{-2} FITTED WITH MAGNETIC SEAL
(LABORATORY TEST)
0
50mm
DOORS
SOURCE: MKM ACOUSTIC ENGINEERS

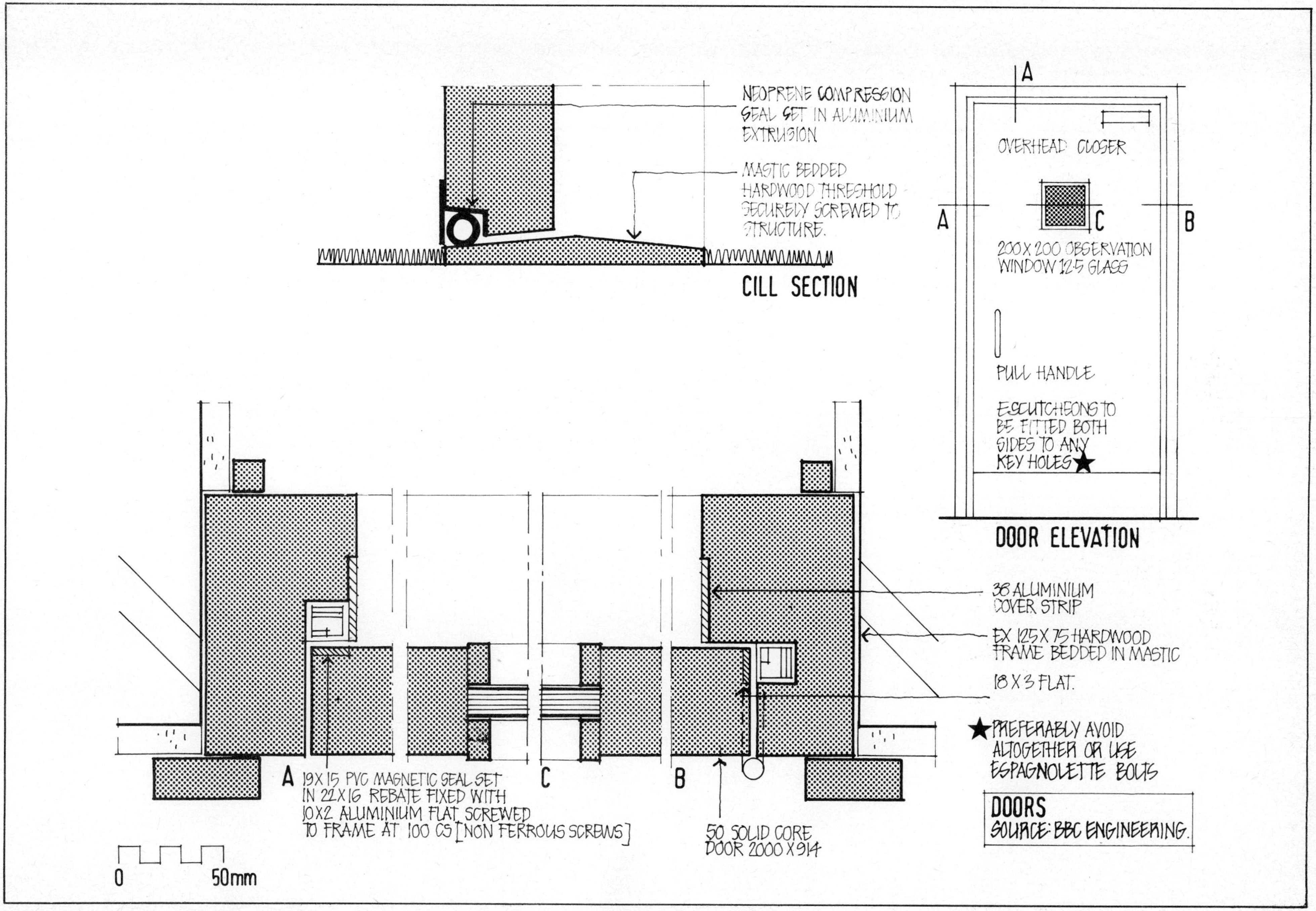
NEOPRENE COMPRESSION SEAL SET IN ALUMINIUM EXTRUSION
MASTIC BEDDED HARDWOOD THRESHOLD SECURELY SCREWED TO STRUCTURE.
CILL SECTION
A
A
B
C
OVERHEAD CLOSER
200 X 200 OBSERVATION WINDOW 12.5 GLASS
PULL HANDLE
ESCUTCHEONS TO BE FITTED BOTH SIDES TO ANY KEY HOLES★
DOOR ELEVATION
38 ALUMINIUM COVER STRIP
EX 125 X 75 HARDWOOD FRAME BEDDED IN MASTIC
18 X 3 FLAT.
★PREFERABLY AVOID ALTOGETHER OR USE ESPAGNOLETTE BOLTS
A
C
B
19 X 15 PVC MAGNETIC SEAL SET IN 22 X 16 REBATE FIXED WITH 10 X 2 ALUMINIUM FLAT SCREWED TO FRAME AT 100 CS [NON FERROUS SCREWS]
50 SOLID CORE DOOR 2000 X 914
0
50mm
DOORS
SOURCE: BBC ENGINEERING.

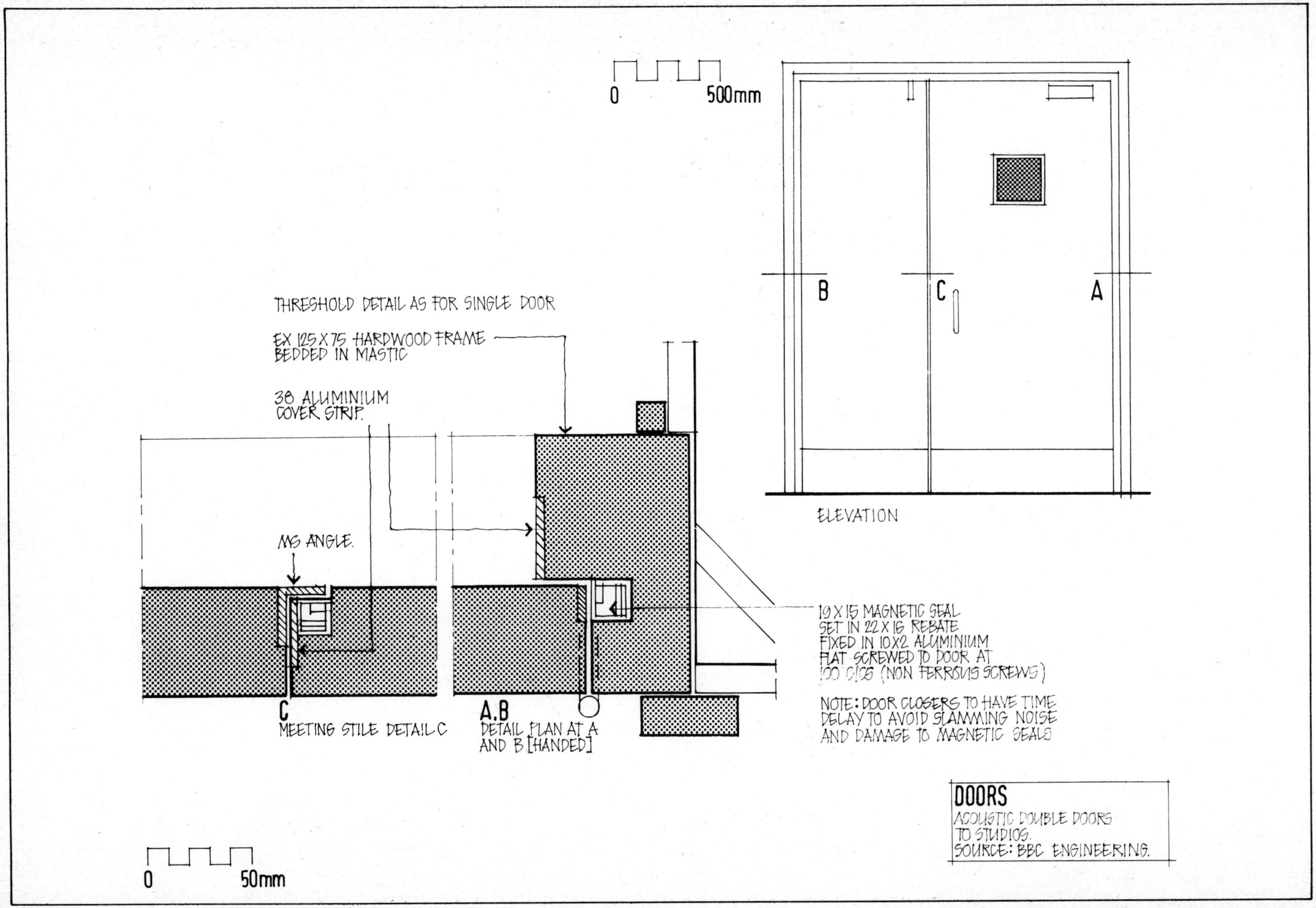
0 500mm
ELEVATION
B
C
A
THRESHOLD DETAIL AS FOR SINGLE DOOR
EX 125 X 75 HARDWOOD FRAME BEDDED IN MASTIC
38 ALUMINIUM COVER STRIP.
MS ANGLE.
C
MEETING STILE DETAIL C
A.B
DETAIL PLAN AT A AND B [HANDED]
19 X 15 MAGNETIC SEAL SET IN 22 X 16 REBATE FIXED IN 10X2 ALUMINIUM FLAT SCREWED TO DOOR AT 100 C/CS (NON FERROUS SCREWS)
NOTE: DOOR CLOSERS TO HAVE TIME DELAY TO AVOID SLAMMING NOISE AND DAMAGE TO MAGNETIC SEALS
0 50mm
DOORS
ACOUSTIC DOUBLE DOORS TO STUDIOS.
SOURCE: BBC ENGINEERING.

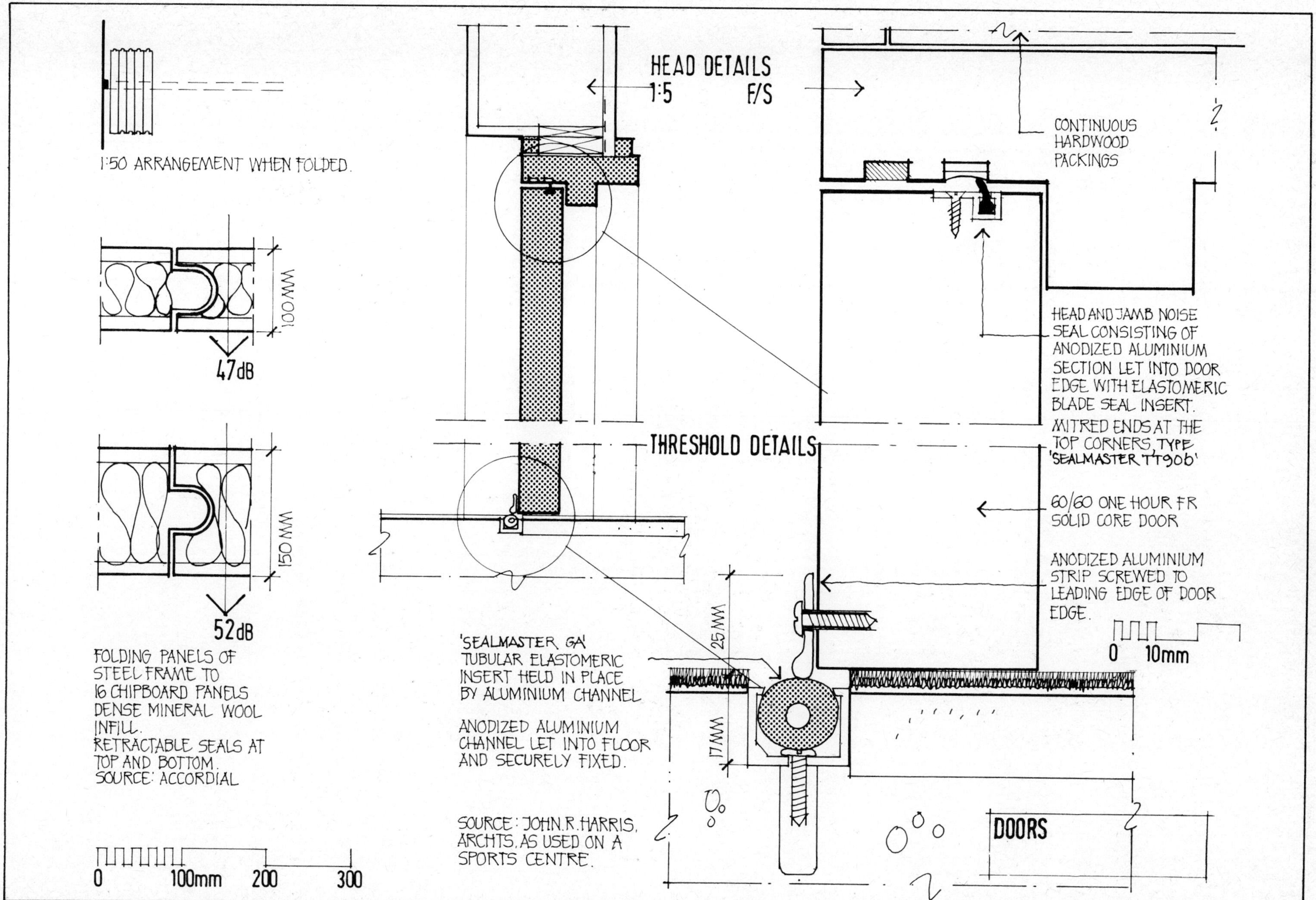
1:50 ARRANGEMENT WHEN FOLDED.
100MM
47dB
150 MM
52dB
FOLDING PANELS OF STEEL FRAME TO 16 CHIPBOARD PANELS DENSE MINERAL WOOL INFILL.
RETRACTABLE SEALS AT TOP AND BOTTOM.
SOURCE: ACCORDIAL
0 100mm 200 300
HEAD DETAILS
1:5 F/S
CONTINUOUS HARDWOOD PACKINGS
HEAD AND JAMB NOISE SEAL CONSISTING OF ANODIZED ALUMINIUM SECTION LET INTO DOOR EDGE WITH ELASTOMERIC BLADE SEAL INSERT.
MITRED ENDS AT THE TOP CORNERS, TYPE 'SEALMASTER T'T906'
THRESHOLD DETAILS
60/60 ONE HOUR FR SOLID CORE DOOR
ANODIZED ALUMINIUM STRIP SCREWED TO LEADING EDGE OF DOOR EDGE.
0 10mm
'SEALMASTER GA' TUBULAR ELASTOMERIC INSERT HELD IN PLACE BY ALUMINIUM CHANNEL
ANODIZED ALUMINIUM CHANNEL LET INTO FLOOR AND SECURELY FIXED.
25 MM
17MM
SOURCE: JOHN.R.HARRIS, ARCHTS. AS USED ON A SPORTS CENTRE.
DOORS

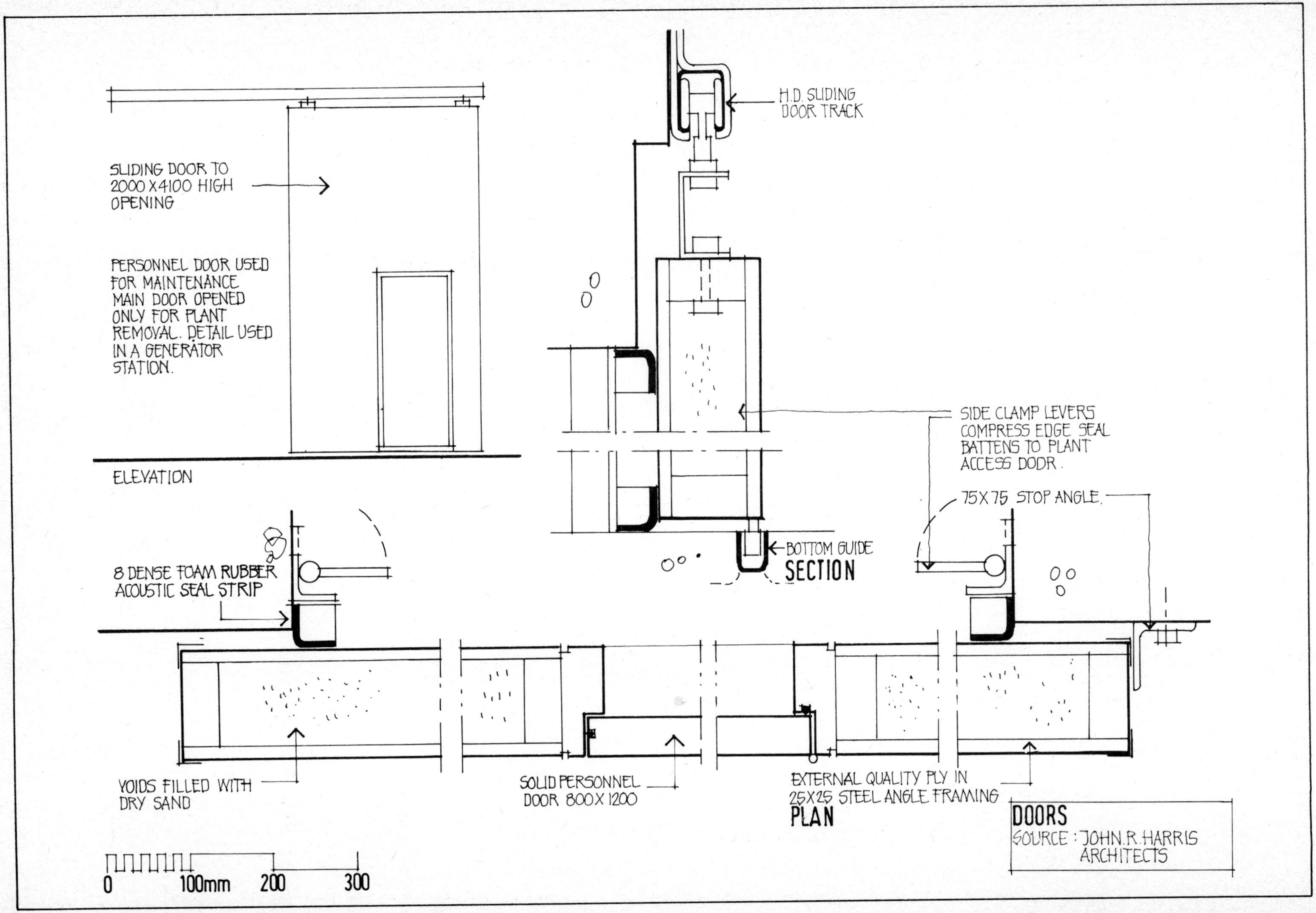
H.D. SLIDING DOOR TRACK
SLIDING DOOR TO 2000 X 4100 HIGH OPENING
PERSONNEL DOOR USED FOR MAINTENANCE MAIN DOOR OPENED ONLY FOR PLANT REMOVAL. DETAIL USED IN A GENERATOR STATION.
SIDE CLAMP LEVERS COMPRESS EDGE SEAL BATTENS TO PLANT ACCESS DOOR.
ELEVATION
75 X 75 STOP ANGLE.
BOTTOM GUIDE
SECTION
8 DENSE FOAM RUBBER ACOUSTIC SEAL STRIP
VOIDS FILLED WITH DRY SAND
SOLID PERSONNEL DOOR 800 X 1200
EXTERNAL QUALITY PLY IN 25X25 STEEL ANGLE FRAMING
PLAN
DOORS
SOURCE : JOHN R. HARRIS ARCHITECTS
0
100mm
200
300

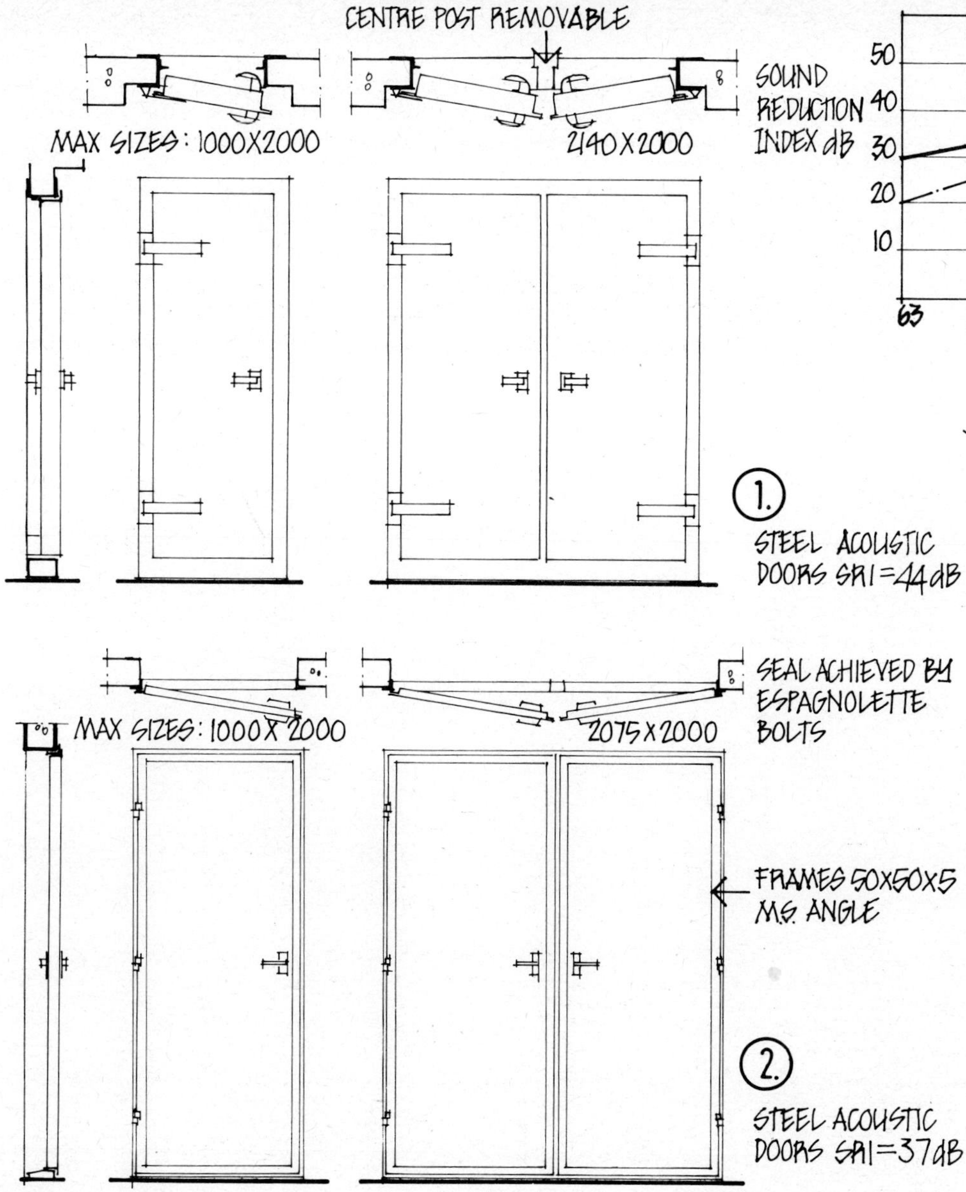

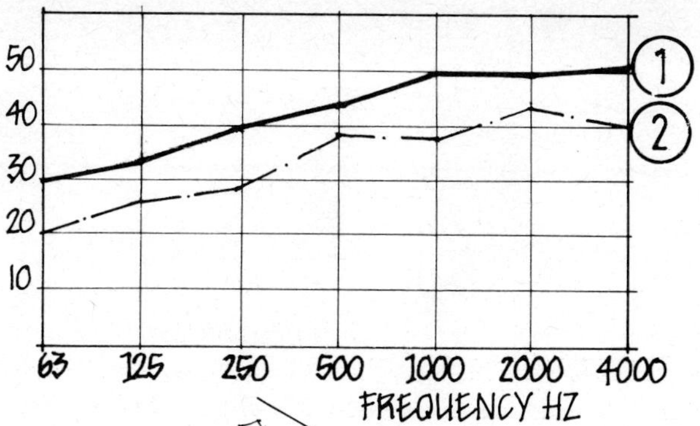

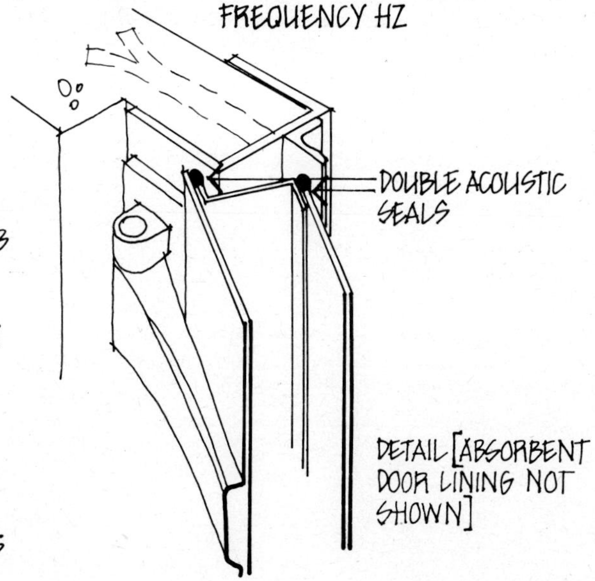

not to scale

DOORS
SOURCE:
SOUND ATTENUATORS LTD

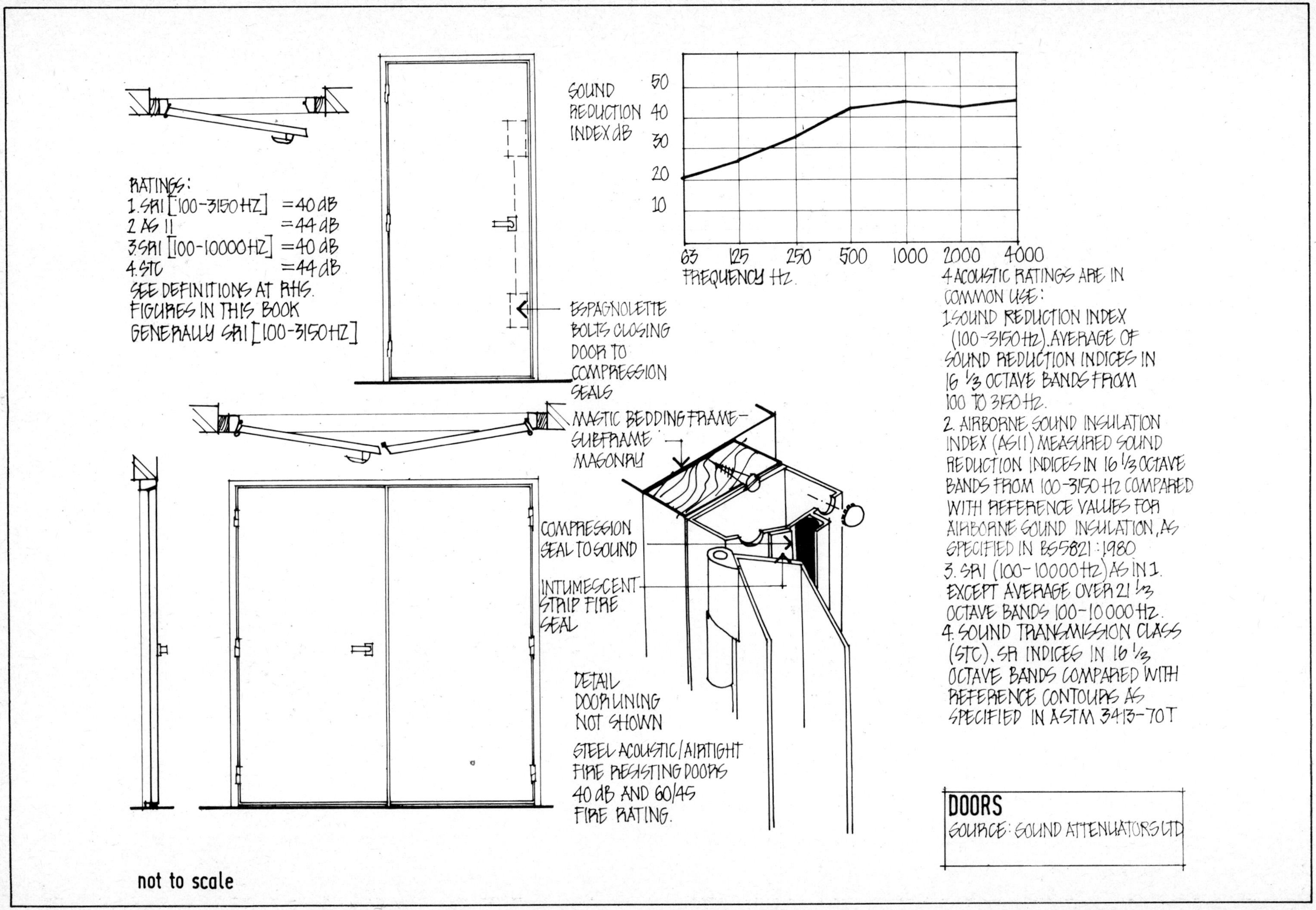
RATINGS:
1. SRI [100–3150 HZ] = 40 dB
2. AS II = 44 dB
3. SRI [100–10000 HZ] = 40 dB
4. STC = 44 dB
SEE DEFINITIONS AT RHS.
FIGURES IN THIS BOOK GENERALLY SRI [100–3150 HZ]
SOUND REDUCTION INDEX dB
50
40
30
20
10
63
125
250
500
1000
2000
4000
FREQUENCY Hz.
ESPAGNOLETTE BOLTS CLOSING DOOR TO COMPRESSION SEALS
MASTIC BEDDING FRAME–SUBFRAME MASONRY
COMPRESSION SEAL TO SOUND
INTUMESCENT STRIP FIRE SEAL
DETAIL DOOR LINING NOT SHOWN
STEEL ACOUSTIC/AIRTIGHT FIRE RESISTING DOORS 40 dB AND 60/45 FIRE RATING.
4 ACOUSTIC RATINGS ARE IN COMMON USE:
1. SOUND REDUCTION INDEX (100–3150 Hz). AVERAGE OF SOUND REDUCTION INDICES IN 16 1/3 OCTAVE BANDS FROM 100 TO 3150 Hz.
2. AIRBORNE SOUND INSULATION INDEX (ASII) MEASURED SOUND REDUCTION INDICES IN 16 1/3 OCTAVE BANDS FROM 100–3150 Hz COMPARED WITH REFERENCE VALUES FOR AIRBORNE SOUND INSULATION, AS SPECIFIED IN BS5821: 1980
3. SRI (100–10000 Hz) AS IN 1. EXCEPT AVERAGE OVER 21 1/3 OCTAVE BANDS 100–10 000 Hz.
4. SOUND TRANSMISSION CLASS (STC). SR INDICES IN 16 1/3 OCTAVE BANDS COMPARED WITH REFERENCE CONTOURS AS SPECIFIED IN ASTM 3413–70T
DOORS
SOURCE: SOUND ATTENUATORS LTD
not to scale

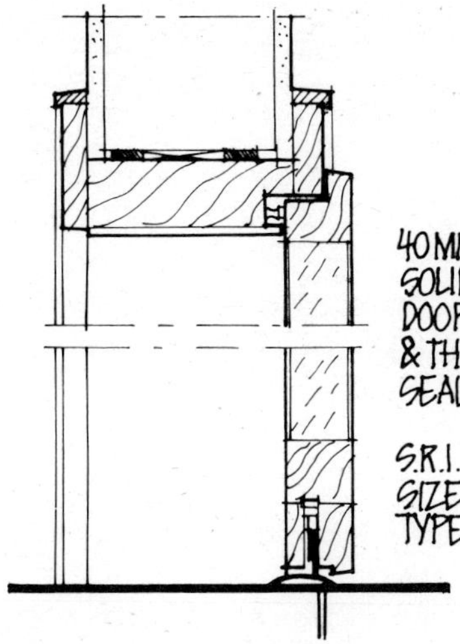

40 MM. THICK SOLID CORE DOOR TO EDGE & THRESHOLD SEALS

S.R.I. 33 dB
SIZE 1000 X 2000
TYPE TH 4

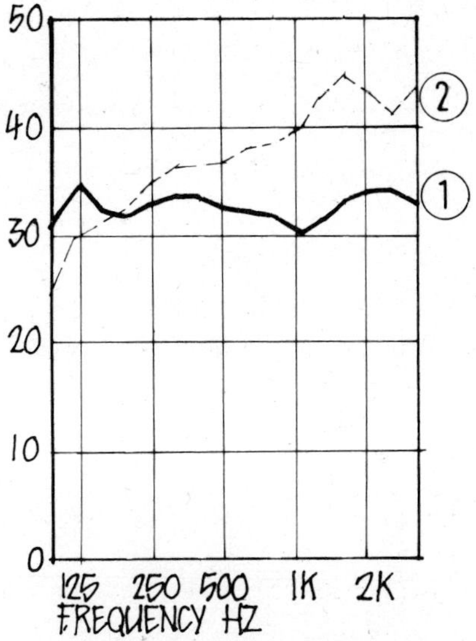

SOURCE: JORDAN. DENMARK (GERMAN MANUFACTURERS)

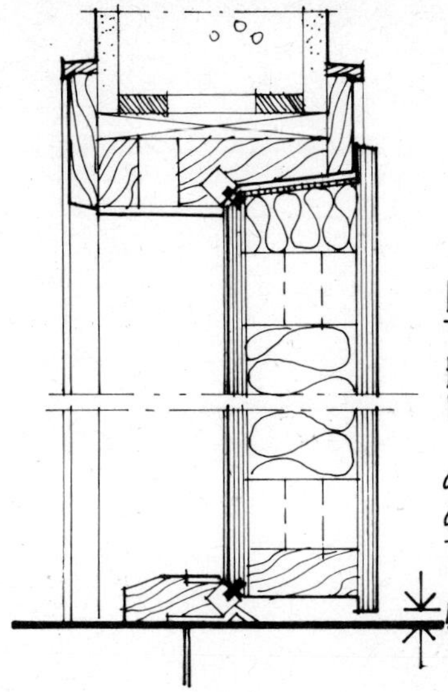

100 MM. THICK DOOR: TIMBER FRAME. PLY FACED MINERAL WOOL LINING

S.R.I. 38 dB
SIZE 1000 X 2000
TYPE TH 10

1

2

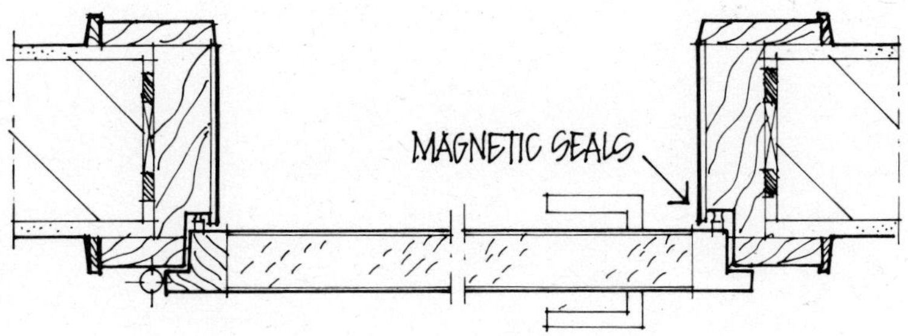

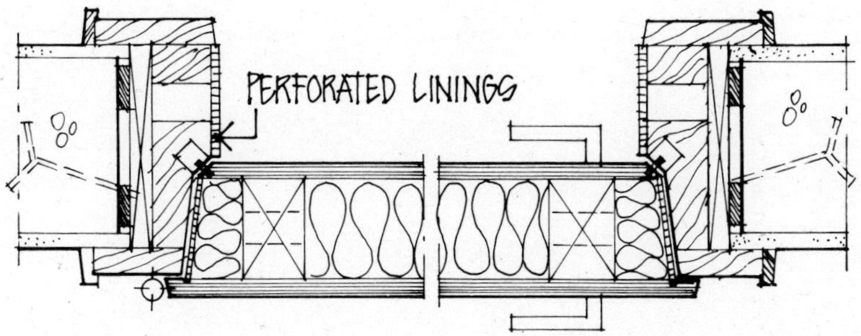

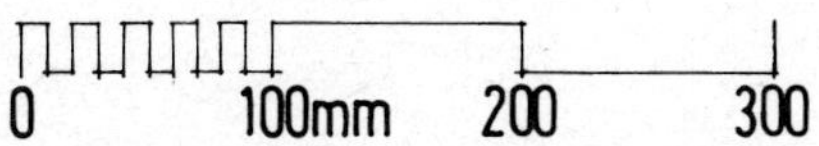

DOORS

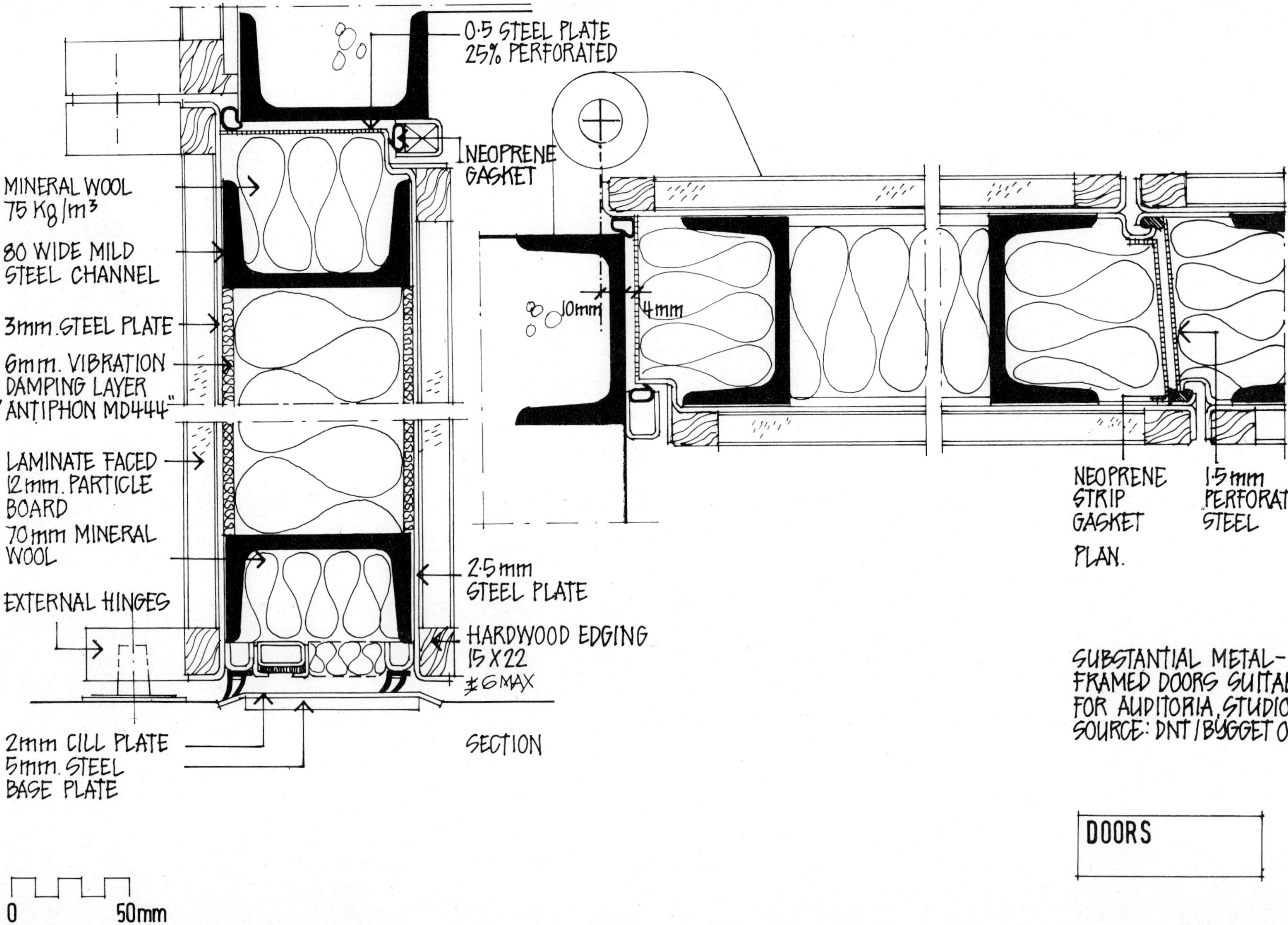

SUBSTANTIAL METAL-FRAMED DOORS SUITABLE FOR AUDITORIA, STUDIOS
SOURCE: DNT/BYGGET OSLO.

DOORS

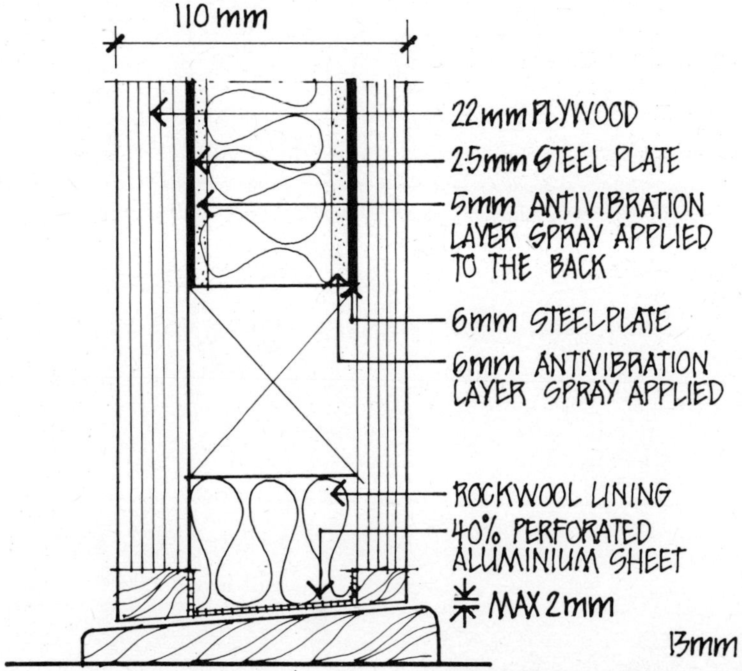

SUBSTANTIAL TIMBER FRAMED DOOR, TEST RATED AT 45 dB

SOURCE : JORDAN DENMARK

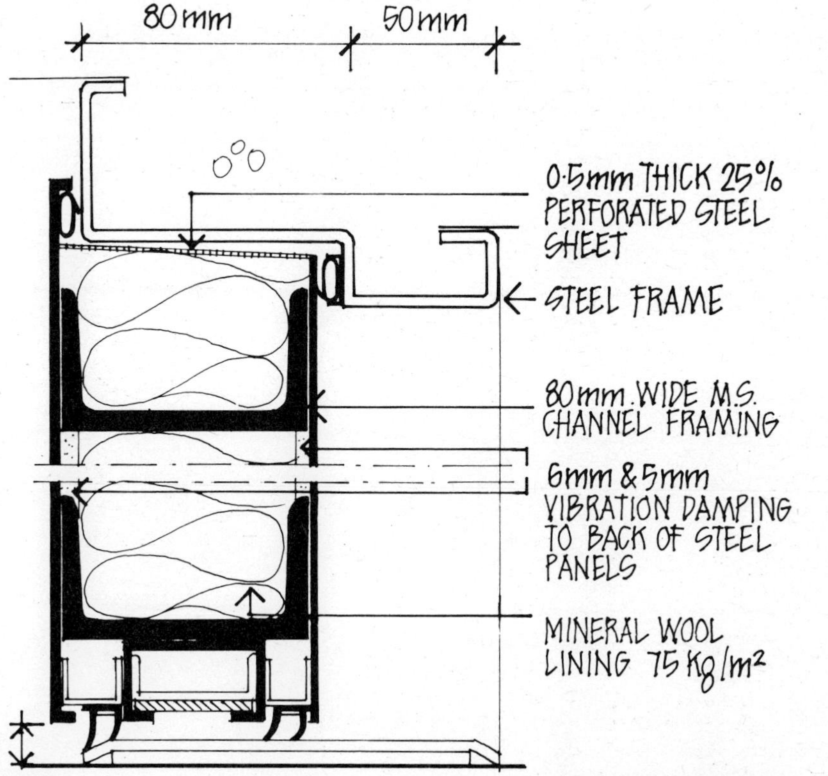

STEEL FACED & FRAMED DOOR WITH DOUBLE SEAL REBATES
SOURCE: DNT- BYGGET OSLO

DOORS

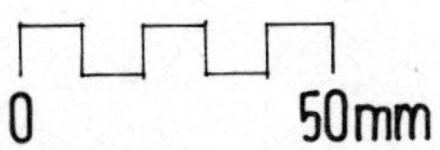

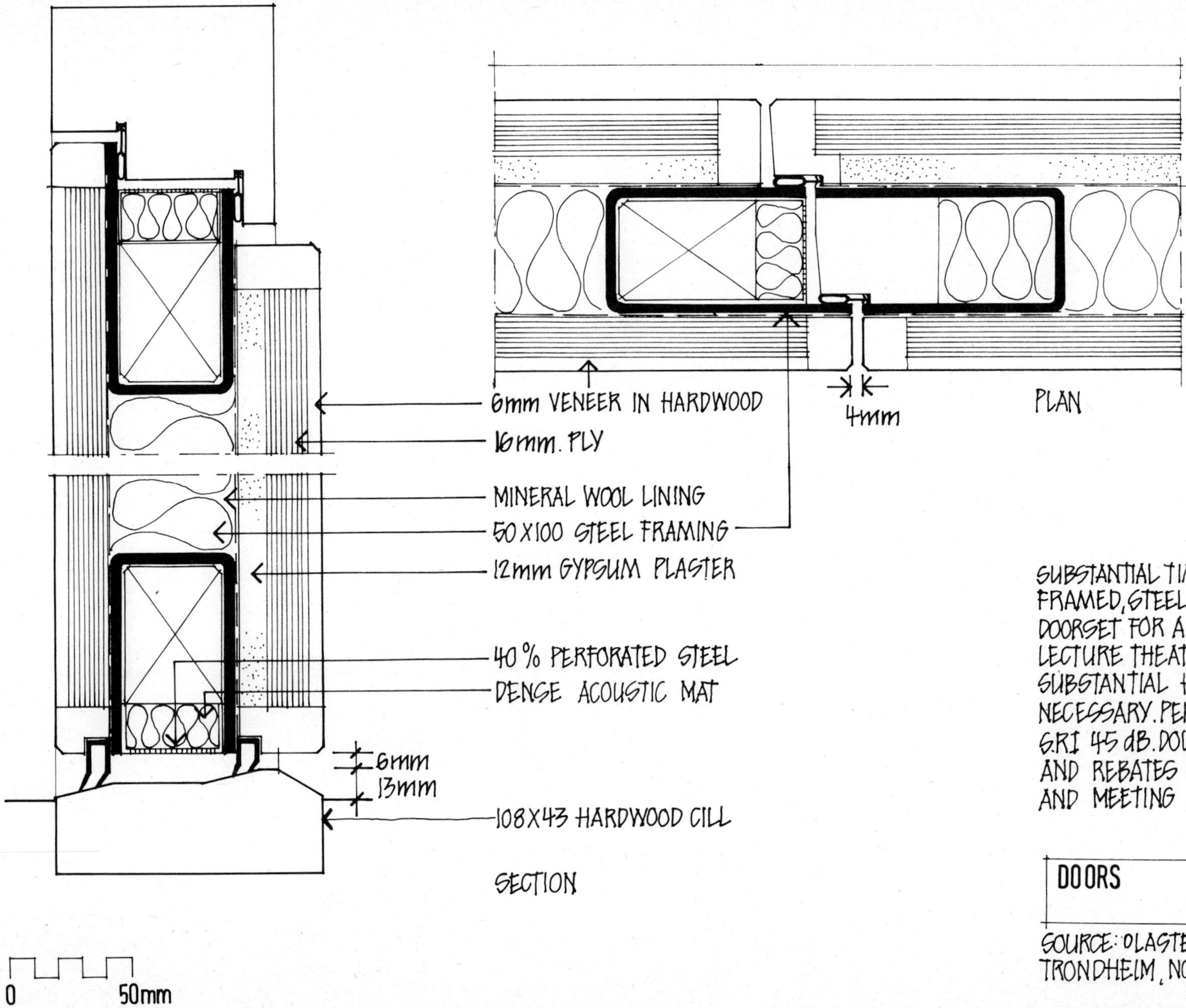

SUBSTANTIAL TIMBER FRAMED, STEEL CORED DOORSET FOR AUDITORIA, LECTURE THEATRES. SUBSTANTIAL HINGES NECESSARY. PERFORMANCE S.R.I. 45 dB. DOUBLE SEALS AND REBATES AT JAMBS AND MEETING STILES

DOORS

SOURCE: OLASTEEN ARCHTS, TRONDHEIM, NORWAY.

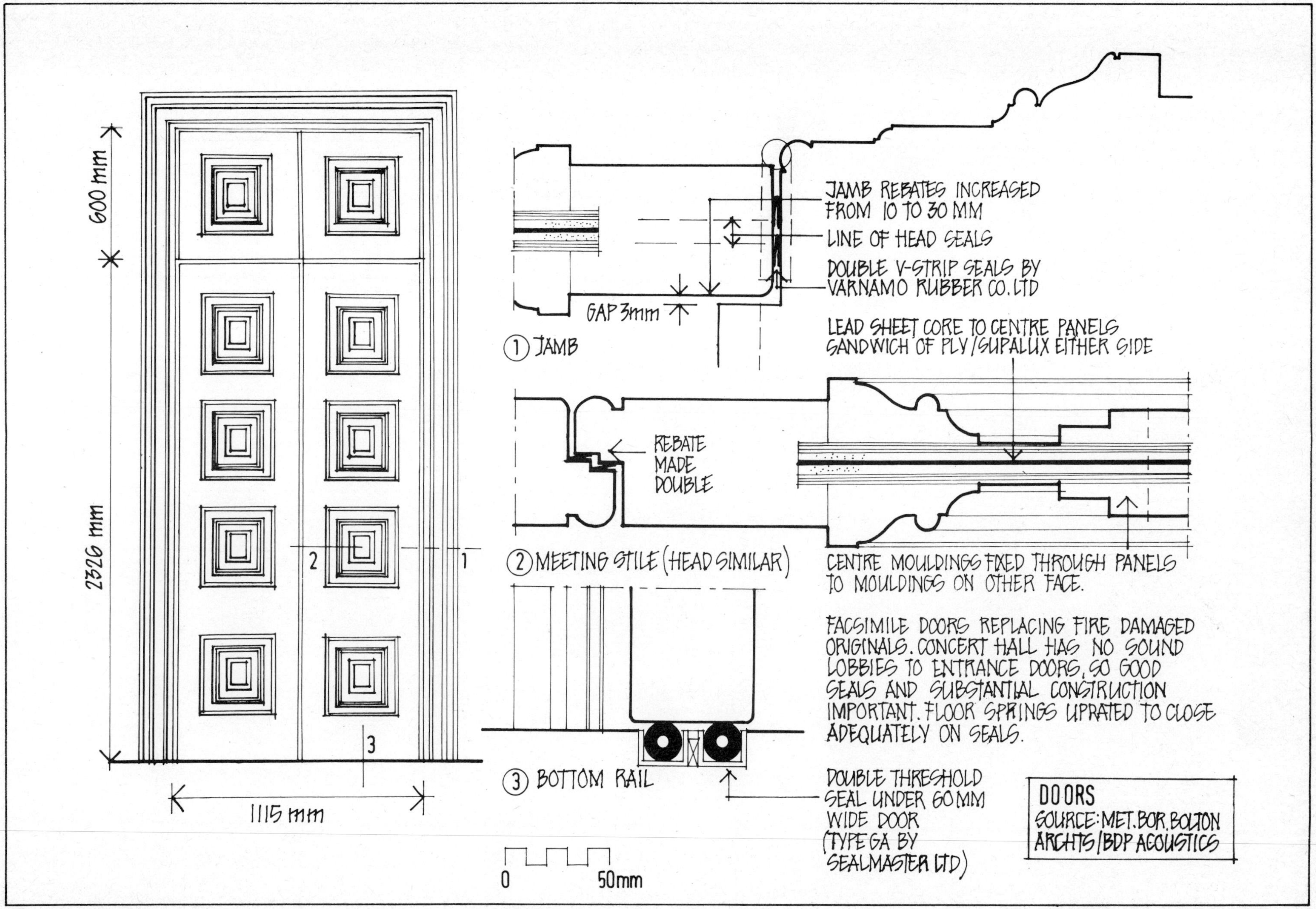
600 mm
2326 mm
1115 mm
2
1
3
JAMB REBATES INCREASED
FROM 10 TO 30 MM
LINE OF HEAD SEALS
DOUBLE V-STRIP SEALS BY
VARNAMO RUBBER CO. LTD
GAP 3mm
1 JAMB
LEAD SHEET CORE TO CENTRE PANELS
SANDWICH OF PLY/SUPALUX EITHER SIDE
REBATE
MADE
DOUBLE
2 MEETING STILE (HEAD SIMILAR)
CENTRE MOULDINGS FIXED THROUGH PANELS
TO MOULDINGS ON OTHER FACE.
FACSIMILE DOORS REPLACING FIRE DAMAGED
ORIGINALS. CONCERT HALL HAS NO SOUND
LOBBIES TO ENTRANCE DOORS, SO GOOD
SEALS AND SUBSTANTIAL CONSTRUCTION
IMPORTANT. FLOOR SPRINGS UPRATED TO CLOSE
ADEQUATELY ON SEALS.
3 BOTTOM RAIL
DOUBLE THRESHOLD
SEAL UNDER 60 MM
WIDE DOOR
(TYPE GA BY
SEALMASTER LTD)
DOORS
SOURCE: MET. BOR. BOLTON
ARCHTS/BDP ACOUSTICS
0
50mm

SEALMASTER PEFA TYPE BLADE SEAL REBATED INTO A SLIDING DOOR, SLIDING AGAINST A PVC STRIP ON THE FRAME

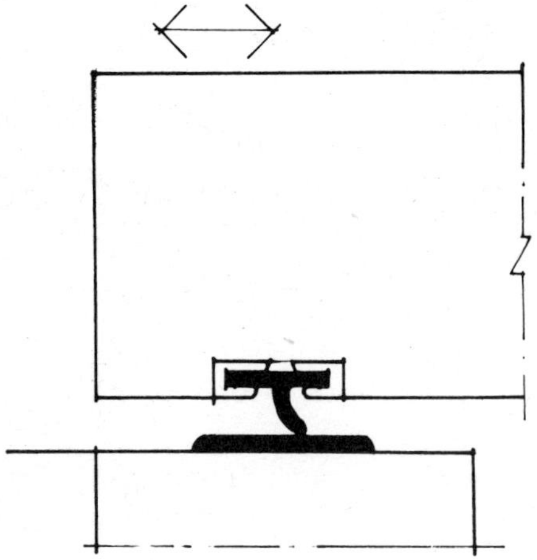

SLIDING AND TWO-WAY SWING DOORS CANNOT HAVE REBATED FRAMES AND SO GOOD SEALS ARE IMPORTANT FOR ANYTHING MORE THAN NOMINAL ACOUSTIC PERFORMANCE

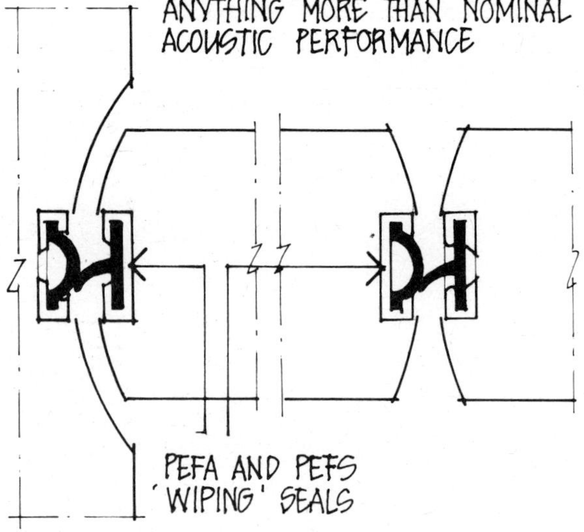

PEFA AND PEFS 'WIPING' SEALS

SOURCE: SEALMASTER LTD.

ADVANTAGEOUS OVER SOME 'O' COMPRESSION SEALS IN THAT ONLY LIGHT CLOSING PRESSURE NECESSARY: SUITABLE FOR LIGHT DOORS AND SLIDING SASH WINDOWS.

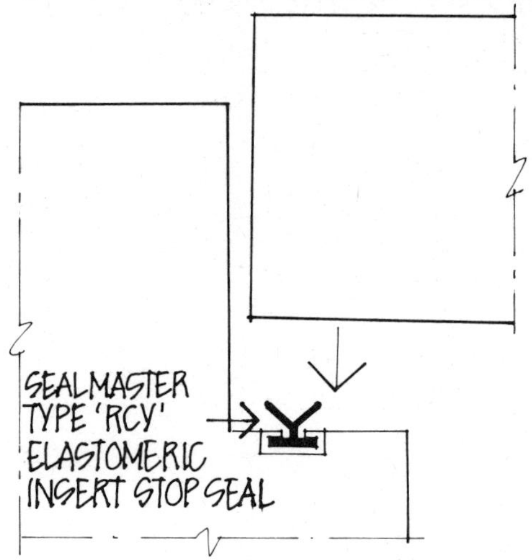

SEALMASTER TYPE 'RCY' ELASTOMERIC INSERT STOP SEAL

JAMB PLAN DETAILS

ONE-WAY SWING DOOR.

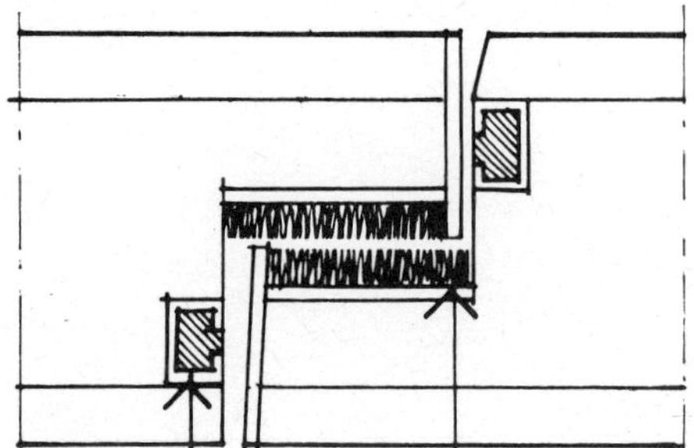

INTUMESCENT STRIPS
CARPET INSERTS
CARPET ABSORBS SOUND AT REBATE AND BUFFERS CLOSING IMPACT.

TWO-WAY SWING DOOR
SOURCE: ARUP ACOUSTICS

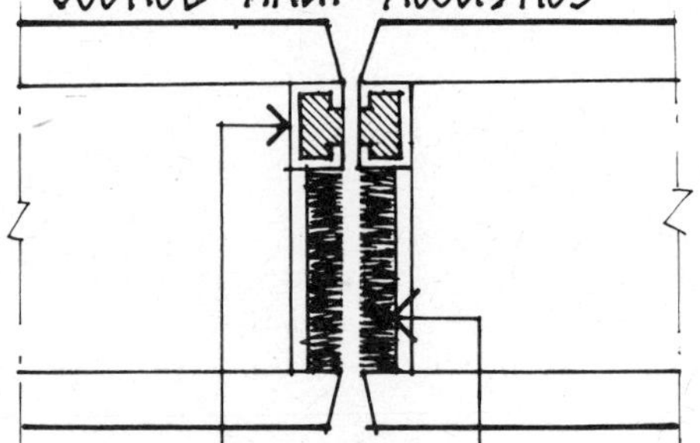

INTUMESCENT STRIPS
CARPET INSERTS AND PACKING

DOORS

0 10mm

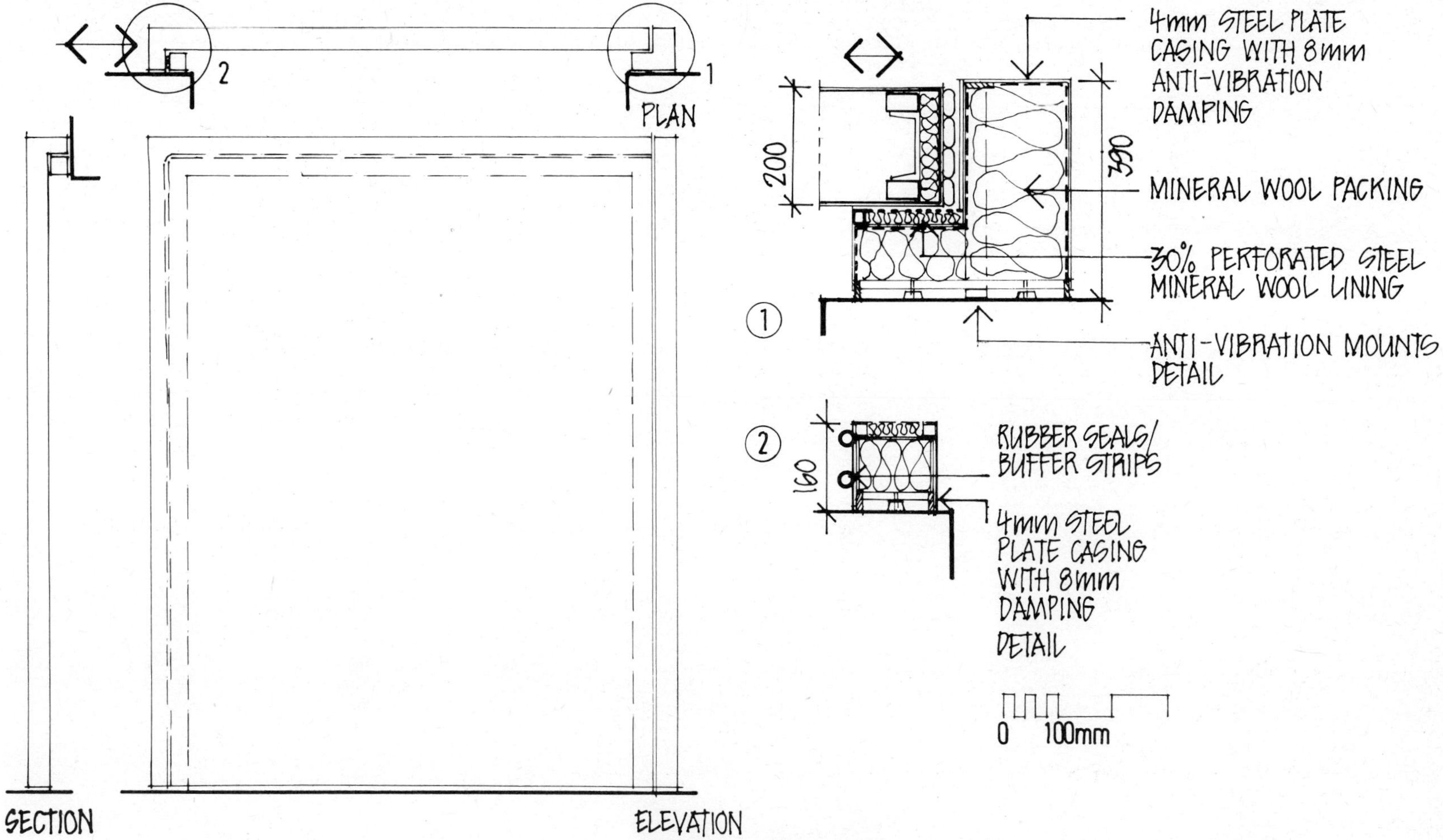

SPECIAL SLIDING DOOR.
SUITABLE FOR STUDIOS, AUDITORIA EQUIPMENT ACCESS.

DOORS
SOURCE: DNT-BYGGNET OSLO

not to scale

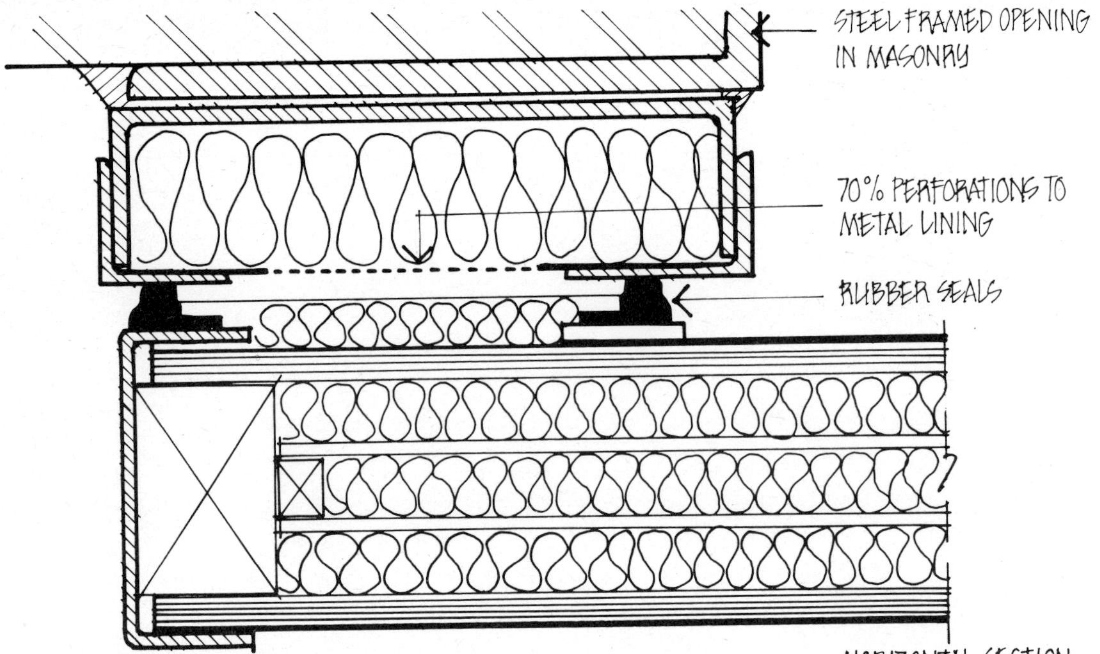

HORIZONTAL SECTION
DOOR BLADE AND FRAME

DOOR OPENING : 1340 X 2090 HIGH
OPERATION : ELECTRICAL- AS DOOR CLOSES, DROPS AND MOVES 45° TOWARDS FRAME TO SEAL AT SILL AND FRAME
SEALS : RUBBER GASKETS
DOOR : 90MM THICK. 18MM ROCKWOOL CORE, 10MM PLYWOOD AND 2MM STEEL SHEET EITHER SIDE

GOOD PERFORMANCE THROUGH THE FREQUENCY RANGE, ALTHOUGH LOW FREQUENCY PERFORMANCE LESS THAN FOR THE 152 STUDIO DOOR ILLUSTRATED.

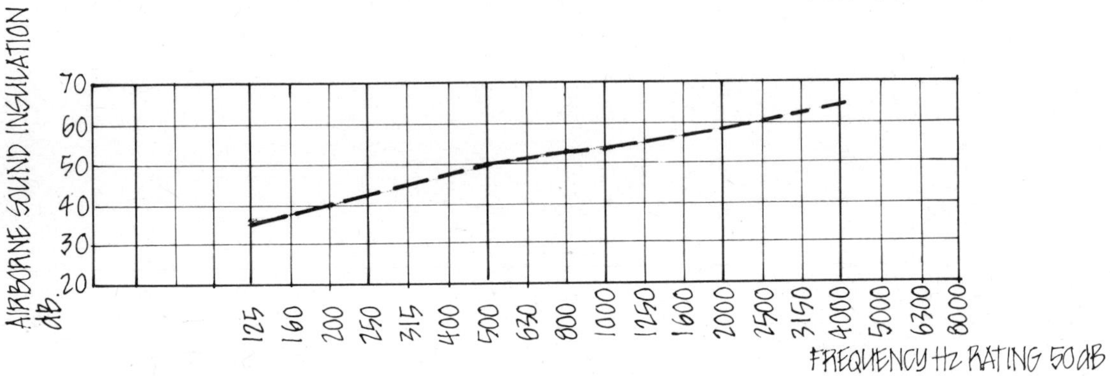

0 50mm

STUDIO DOORS
SOURE: RESEARCH INSTITUTE FOR ENVIRONMENTAL HYGIENE, DELFT / MARKNS

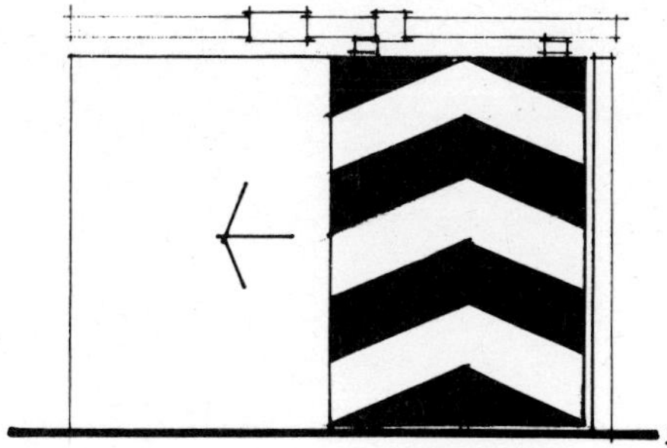

ELEVATION NTS.

DOOR OPENING : 300 X 4940 HIGH
OPERATION : ELECTRICAL SLIDING
PANEL WEIGHT : 1820 Kg (4000 lbs)
SEALS : DOUBLE FOAM-FILLED 37 DIAMETER GASKETS PLUS LEADING EDGE SEALS

DOOR : 152MM THICK WITH GLASS FIBRE CORE, ROCKWOOL 0·80 MM LEAD SHEETS CLAD WITH 1·60MM (16 GAUGE) GALVANIZED STEEL

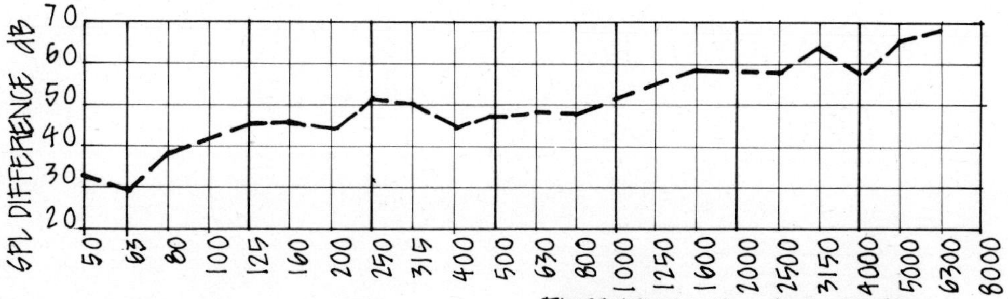

GOOD SEALS, THICK LAMINAR CONSTRUCTION AND WEIGHT GIVE GOOD SOUND REDUCTION PROPERTIES AT ALL FREQUENCIES

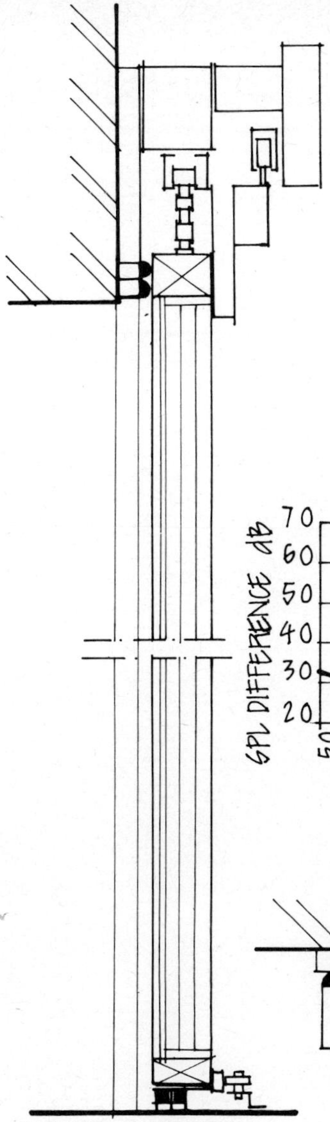

VERTICAL SECTION

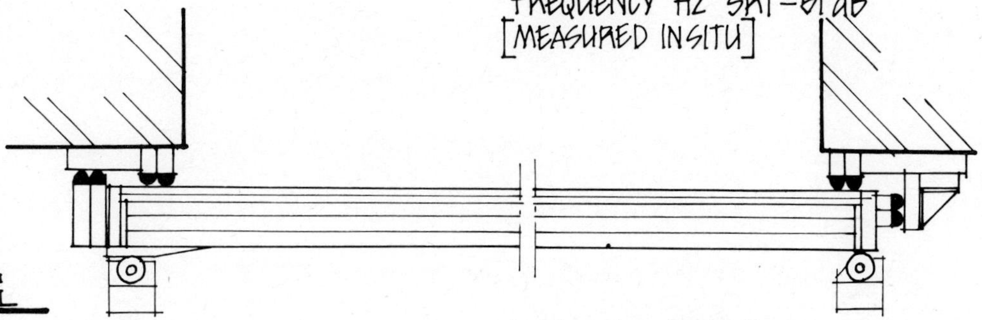

HORIZONTAL SECTION

STUDIO DOORS

SOURCE : BBC/CLARK DOOR LTD

not to scale

WINDOWS

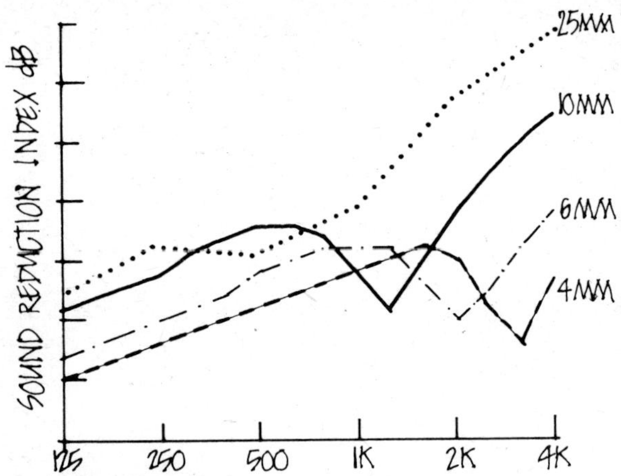

SINGLE GLAZING

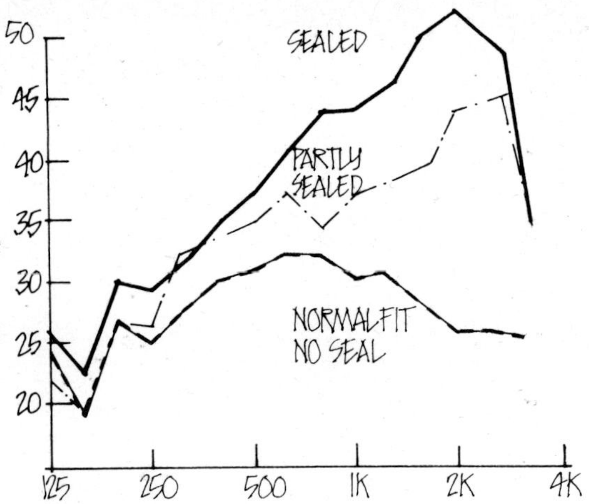

EFFECTS OF SEALING DOUBLE WINDOWS

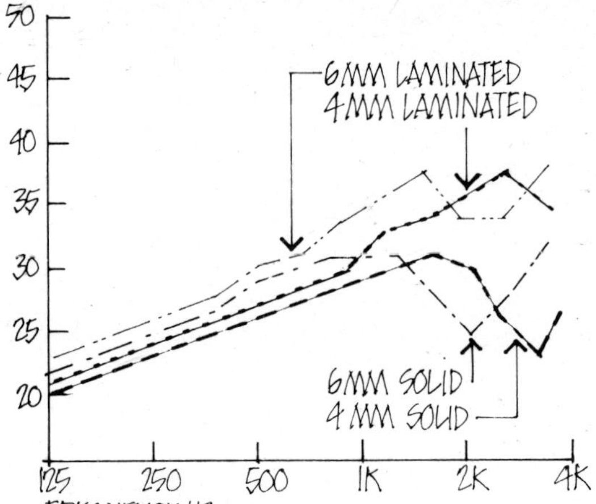

EFFECT OF LAMINATION

SOUND INSULATION OF WINDOWS

		SOUND REDUCTION AV.100 – 3150 HZ
1.	OPEN WINDOW	UP TO 10 dB
2.	OPENABLE WINDOW CLOSED BUT NOT WEATHER-STRIPPED.	UP TO 20 dB
3.	SINGLE FIXED OR SEALED OPENABLE WINDOW 6MM GLASS	UP TO 25 dB
4.	FIXED WINDOW 12 GLASS	UP TO 30 dB
5.	FIXED WINDOW 24 GLASS	UP TO 35 dB
6.	DOUBLE GLAZED WINDOW. 150-200 SPACE OPENABLE BUT WEATHER STRIPPED	UP TO 40 dB
7.	DOUBLE WINDOWS SPACED 300-400 MM ONE FRAME FIXED. ABSORPTION IN REVEALS	UP TO 45 dB.

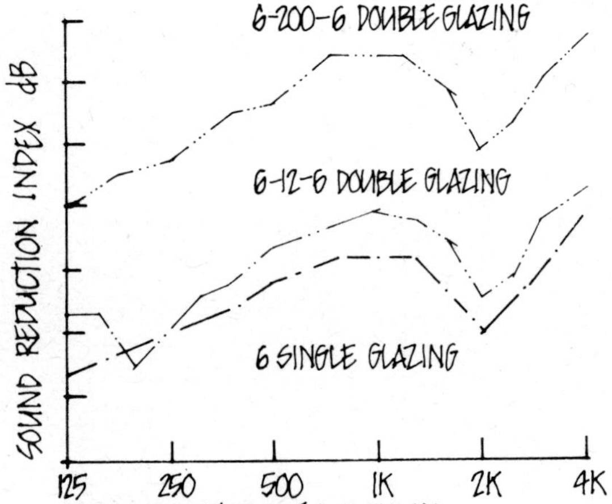

EFFECT OF AIRSPACE WIDTH ON DOUBLE GLAZING

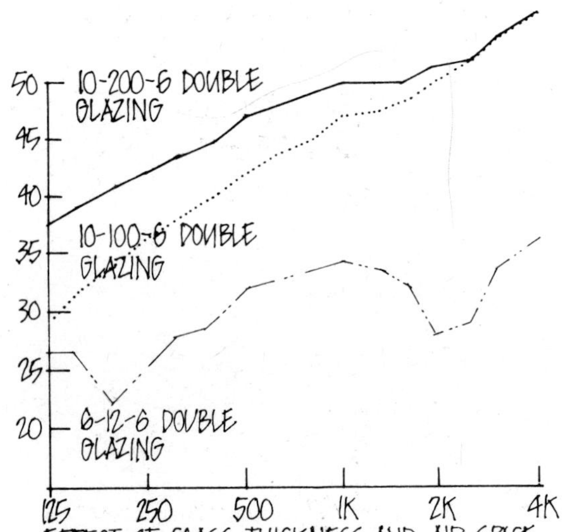

EFFECT OF GLASS THICKNESS AND AIR SPACE ON DOUBLE GLAZING

WINDOWS

SOURCE:
PILKINGTON FLAT GLASS LTD/
BRS DIGEST 128

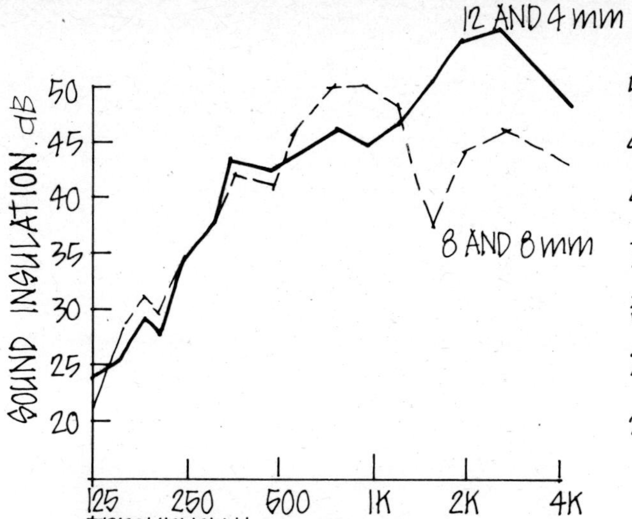

FREQUENCY Hz
EFFECT OF VARYING PANELS THICKNESS (BOTH TO DOUBLE GLAZING WITH PANE SEPARATION 160 mm)

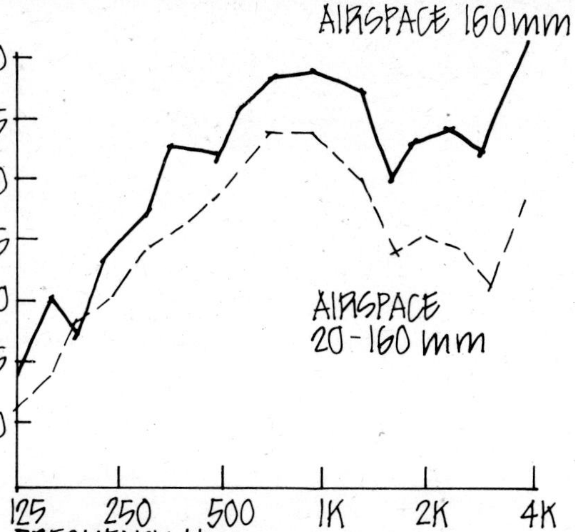

FREQUENCY Hz
EFFECT OF NON-PARALLEL PANES IN DOUBLE GLAZING.

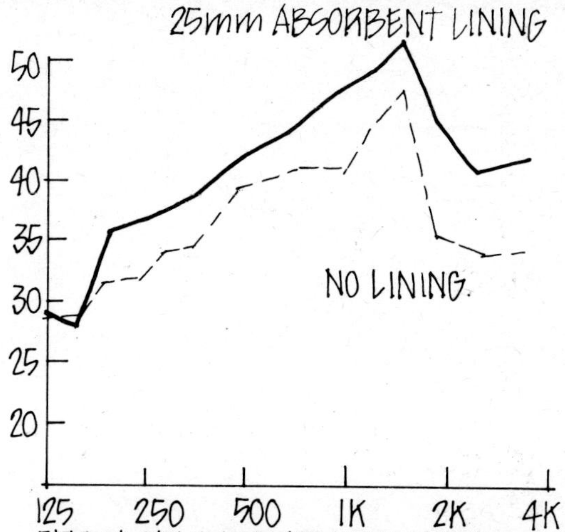

EFFECT OF LINING AIRSPACE EDGES OF DOUBLE WINDOW. 26mm PANES SEPARATION 100 mm
SOURCE: PILKINGTONS ADVISORY SERVICE.

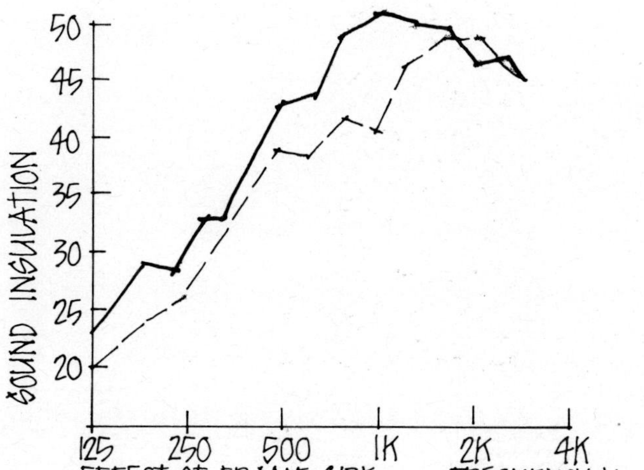

EFFECT OF FRAME SIZE   FREQUENCY Hz.

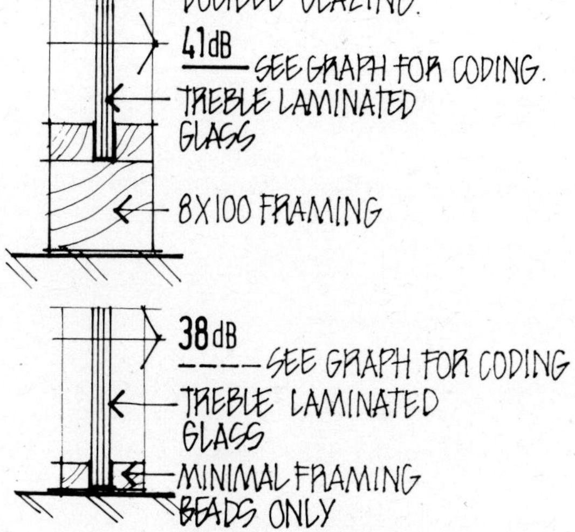

SOURCE: DR. RUCKWARD BAM. BERLIN.

WINDOWS

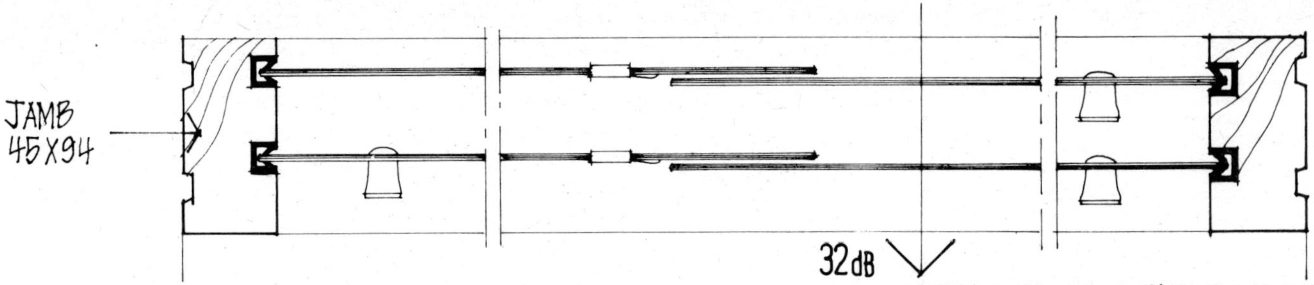

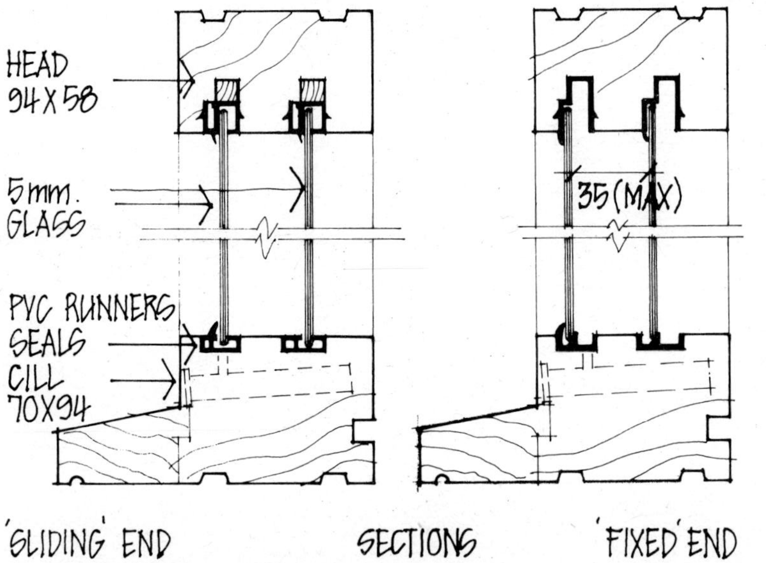

QUEBEC 35 MK III. TYPE WINDOW BY SASHLESS WINDOW CO. LTD.

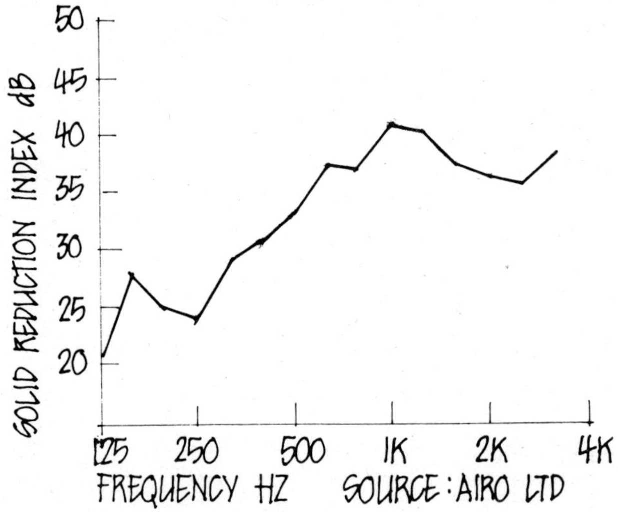

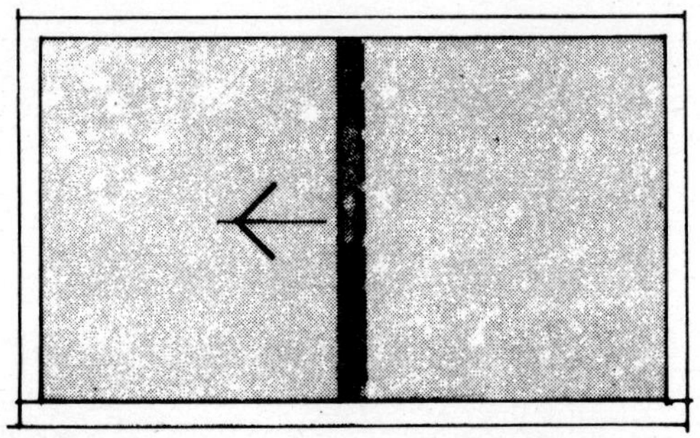

ELEVATION

not to scale

TEST CONDITION: 900X1500 WINDOW IN 250 BRICKWORK.
WHERE NATURAL LIGHT IS NOT ESSENTIAL IN A SINGLE GLAZED WINDOW PANEL, PILKINGTONS 'ARMOURCLAD' COMPOSITE GLASS / INSULATION PANELS MAY BE CONSIDERED:

6mm. GLASS	/25 INSULATION	32 dB. RATING
6	65	37
10	25	34
10	65	38

WINDOWS

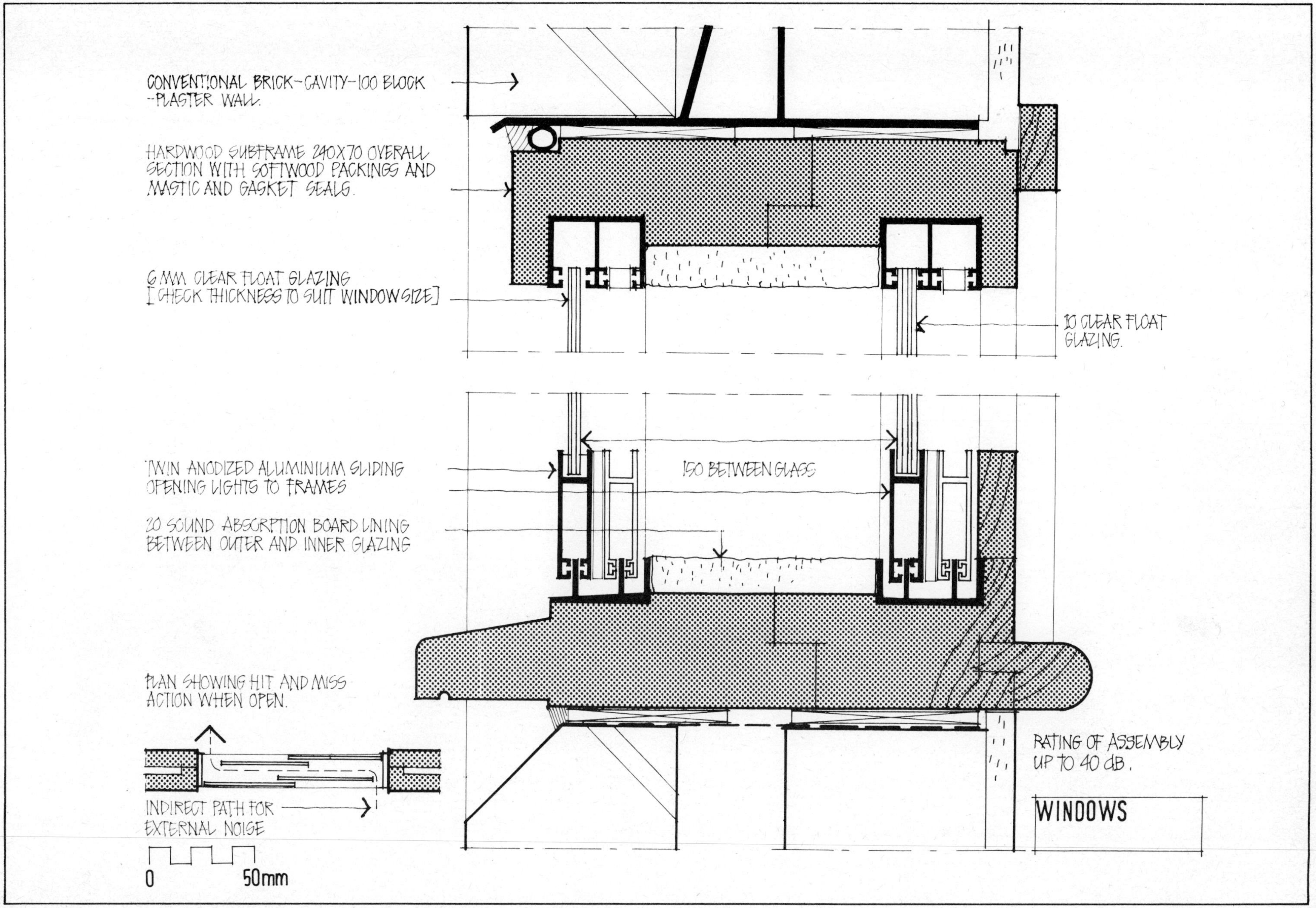
CONVENTIONAL BRICK-CAVITY-100 BLOCK -PLASTER WALL.
HARDWOOD SUBFRAME 240X70 OVERALL SECTION WITH SOFTWOOD PACKINGS AND MASTIC AND GASKET SEALS.
6 MM CLEAR FLOAT GLAZING [CHECK THICKNESS TO SUIT WINDOW SIZE]
10 CLEAR FLOAT GLAZING.
TWIN ANODIZED ALUMINIUM SLIDING OPENING LIGHTS TO FRAMES
150 BETWEEN GLASS
20 SOUND ABSORPTION BOARD LINING BETWEEN OUTER AND INNER GLAZING
PLAN SHOWING HIT AND MISS ACTION WHEN OPEN.
INDIRECT PATH FOR EXTERNAL NOISE
0
50mm
RATING OF ASSEMBLY UP TO 40 dB.
WINDOWS

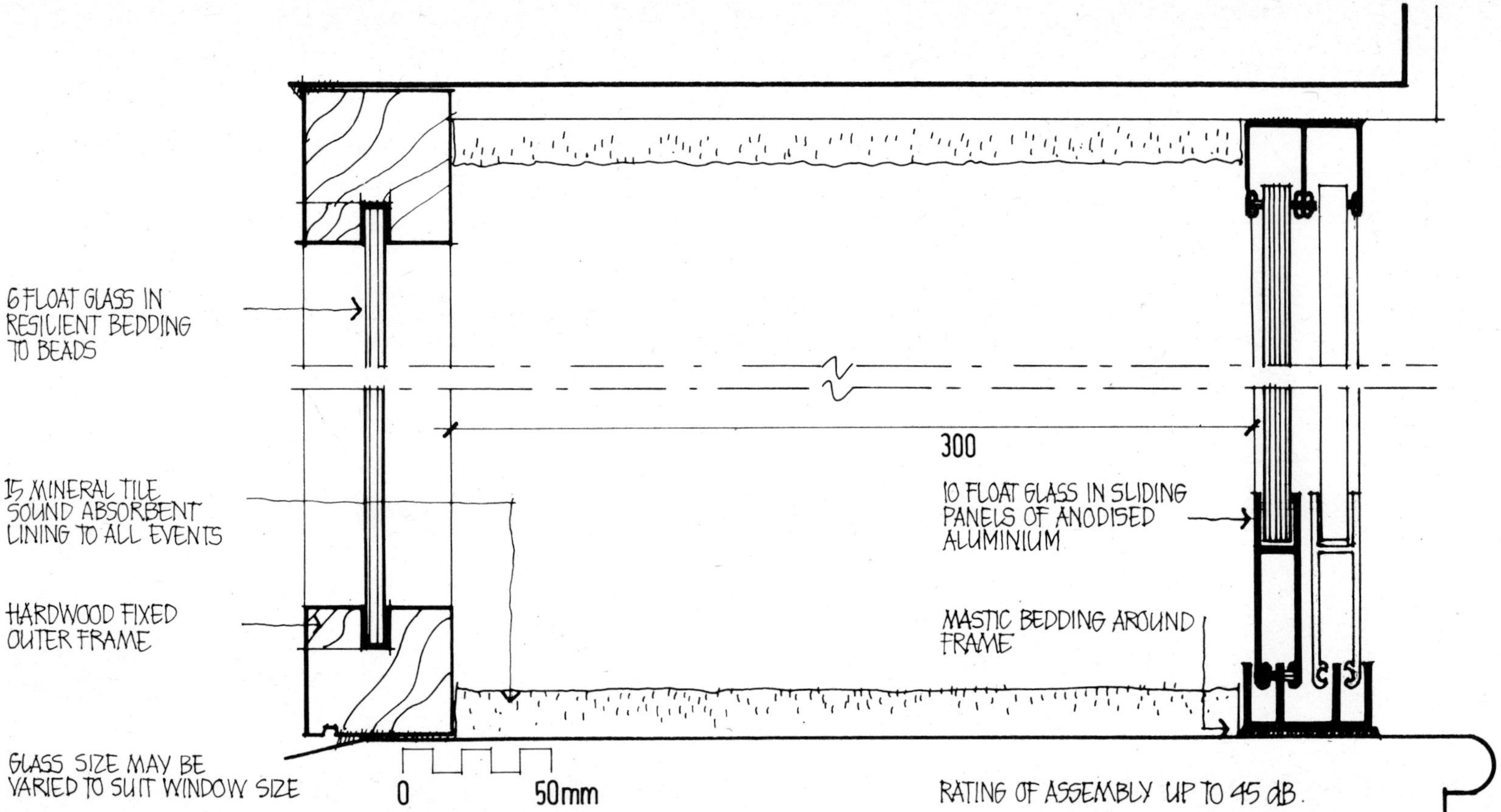

RATING OF ASSEMBLY UP TO 45 dB.

NOTE: INNER FRAME COULD ALSO BE USED TO UPGRADE EXISTING WINDOW OPENINGS. AN ORIEL MAY HAVE TO BE FORMED BECAUSE THE 300 AIR GAP MEANS THE WINDOW ASSEMBLY IS WIDER THAN THE CAVITY WALL AROUND.

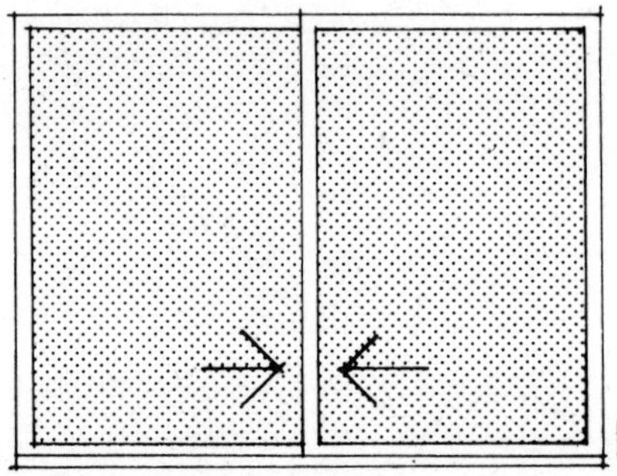

ELEVATION

WINDOWS

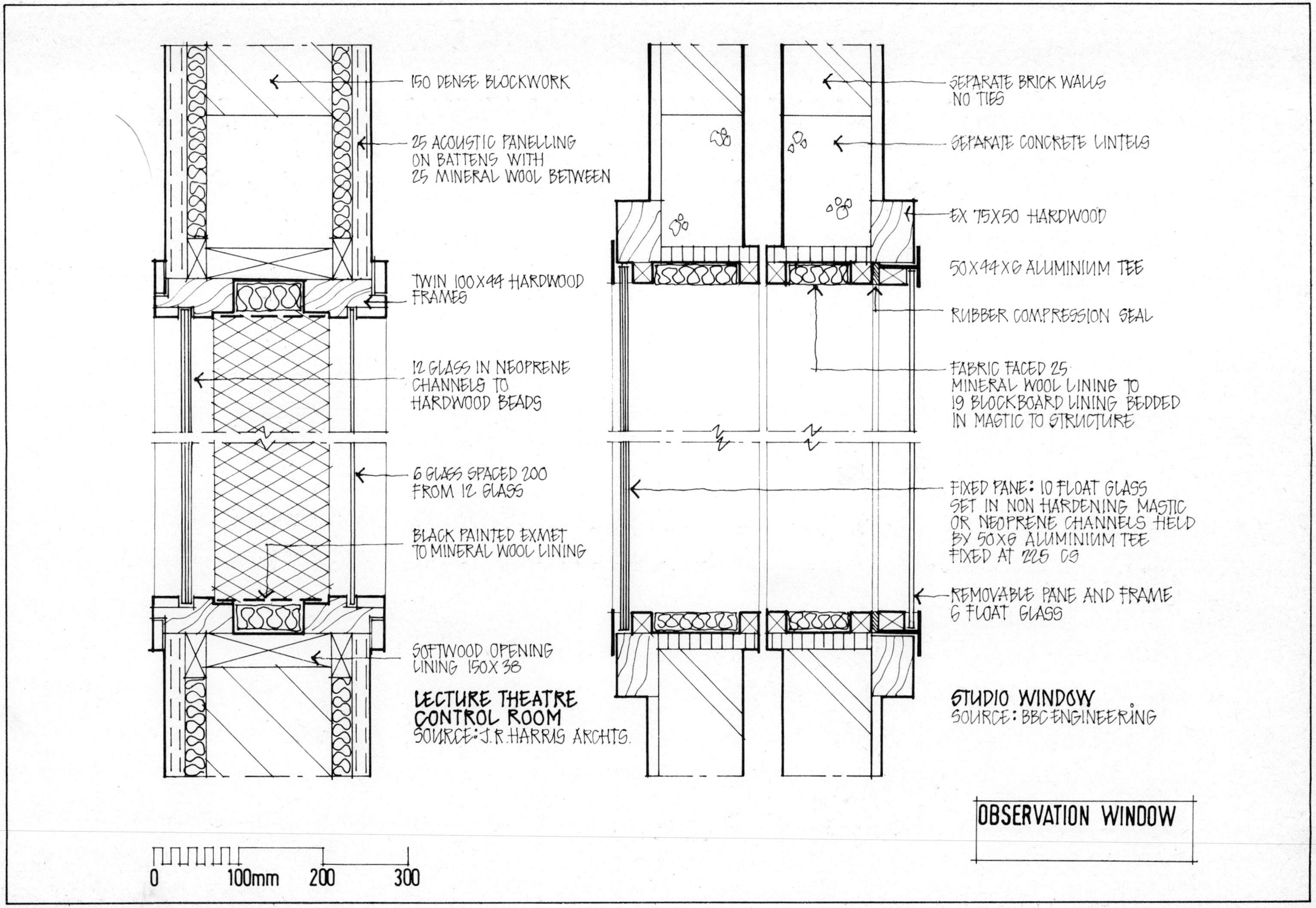
150 DENSE BLOCKWORK
25 ACOUSTIC PANELLING ON BATTENS WITH 25 MINERAL WOOL BETWEEN
TWIN 100X44 HARDWOOD FRAMES
12 GLASS IN NEOPRENE CHANNELS TO HARDWOOD BEADS
6 GLASS SPACED 200 FROM 12 GLASS
BLACK PAINTED EXMET TO MINERAL WOOL LINING
SOFTWOOD OPENING LINING 150X38
LECTURE THEATRE CONTROL ROOM
SOURCE: J.R.HARRIS ARCHTS.
SEPARATE BRICK WALLS NO TIES
SEPARATE CONCRETE LINTELS
EX 75X50 HARDWOOD
50X44X6 ALUMINIUM TEE
RUBBER COMPRESSION SEAL
FABRIC FACED 25 MINERAL WOOL LINING TO 19 BLOCKBOARD LINING BEDDED IN MASTIC TO STRUCTURE
FIXED PANE: 10 FLOAT GLASS SET IN NON HARDENING MASTIC OR NEOPRENE CHANNELS HELD BY 50X6 ALUMINIUM TEE FIXED AT 225 CS
REMOVABLE PANE AND FRAME 6 FLOAT GLASS
STUDIO WINDOW
SOURCE: BBC ENGINEERING
OBSERVATION WINDOW
0
100mm
200
300

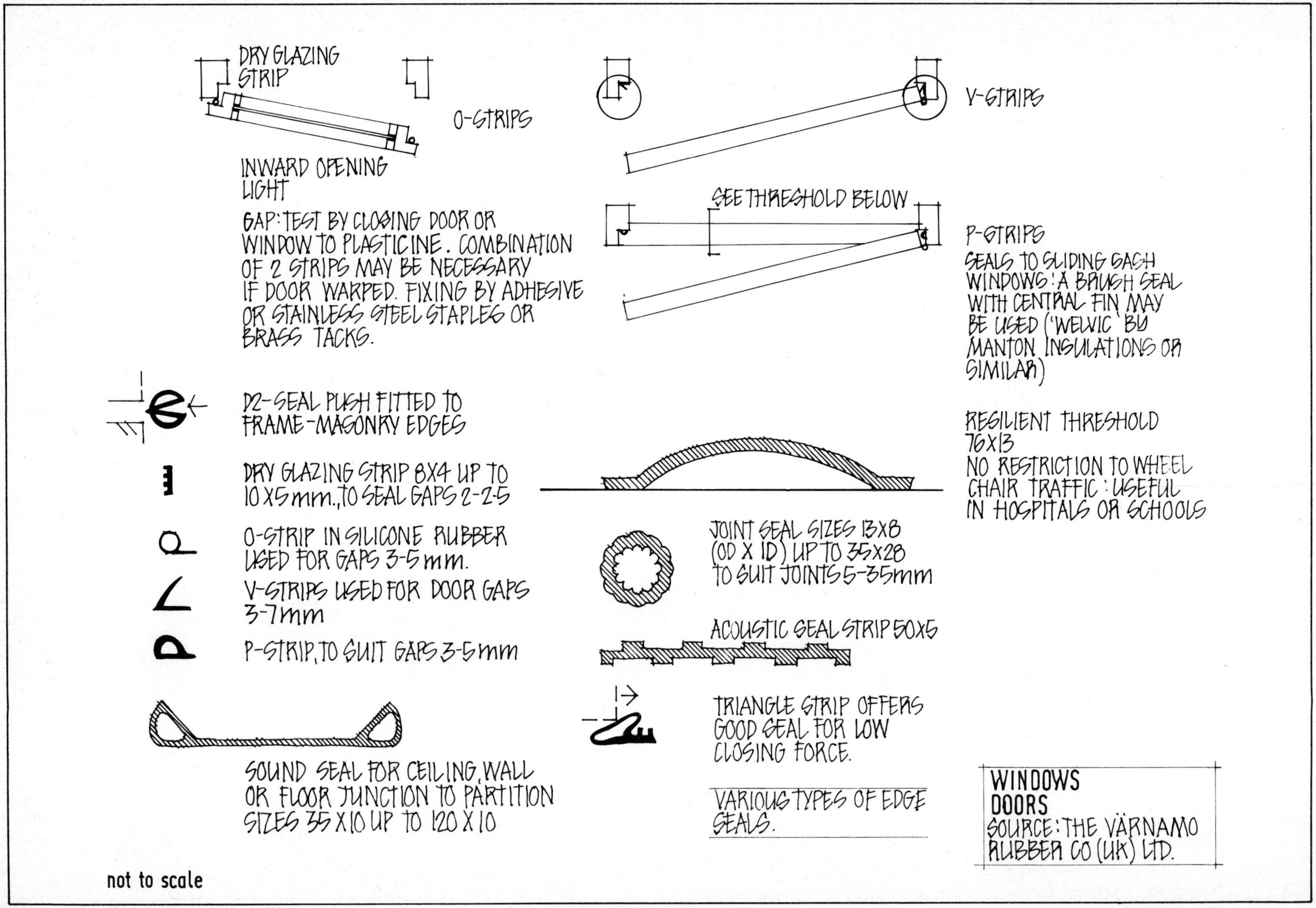
DRY GLAZING STRIP
O-STRIPS
INWARD OPENING LIGHT
GAP: TEST BY CLOSING DOOR OR WINDOW TO PLASTICINE. COMBINATION OF 2 STRIPS MAY BE NECESSARY IF DOOR WARPED. FIXING BY ADHESIVE OR STAINLESS STEEL STAPLES OR BRASS TACKS.
V-STRIPS
SEE THRESHOLD BELOW
P-STRIPS
SEALS TO SLIDING SASH WINDOWS: A BRUSH SEAL WITH CENTRAL FIN MAY BE USED ('WELVIC' BY MANTON INSULATIONS OR SIMILAR)
D2-SEAL PUSH FITTED TO FRAME-MASONRY EDGES
DRY GLAZING STRIP 8X4 UP TO 10X5 mm., TO SEAL GAPS 2-2.5
O-STRIP IN SILICONE RUBBER USED FOR GAPS 3-5 mm.
V-STRIPS USED FOR DOOR GAPS 3-7mm
P-STRIP, TO SUIT GAPS 3-5mm
RESILIENT THRESHOLD 76X13
NO RESTRICTION TO WHEEL CHAIR TRAFFIC: USEFUL IN HOSPITALS OR SCHOOLS
JOINT SEAL SIZES 13X8 (OD X ID) UP TO 35X28 TO SUIT JOINTS 5-35mm
ACOUSTIC SEAL STRIP 50X5
TRIANGLE STRIP OFFERS GOOD SEAL FOR LOW CLOSING FORCE.
SOUND SEAL FOR CEILING, WALL OR FLOOR JUNCTION TO PARTITION SIZES 35 X 10 UP TO 120 X 10
VARIOUS TYPES OF EDGE SEALS.
WINDOWS
DOORS
SOURCE: THE VÄRNAMO RUBBER CO (UK) LTD.
not to scale

LININGS

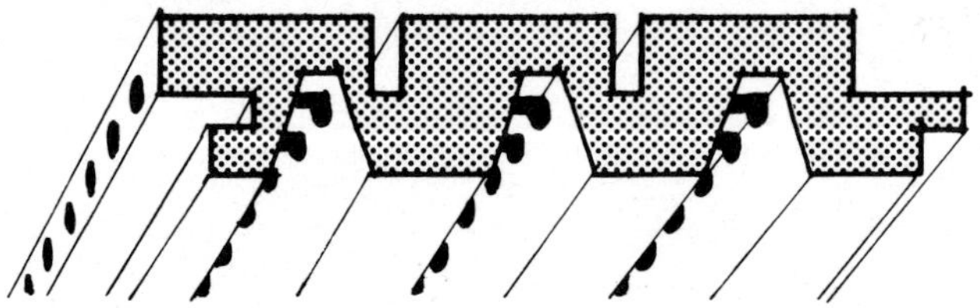

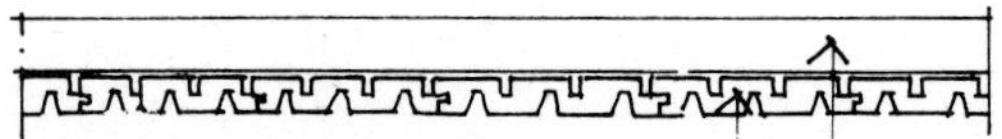

APPLIED ACOUSTICS VENABLES
PROFILED PROPRIETARY
HARDWOOD PANELS
PATTERN PERFORATED
AS SKETCH

50 X 50 BATTENS TO CEILING (OR WALL)
0·1 0·35 0·80 0·40 0·25 0·35

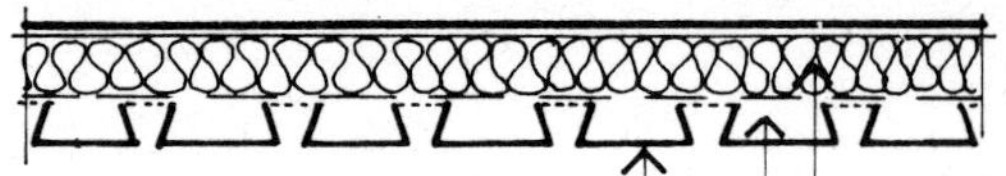

PROPRIETARY VINYL
COVERED STEEL SLATS.
(A.A-V. STEELTONE)
GLASS FIBRE MEMBRANE
25 MINERAL WOOL QUILT
– 0·34 0·88 0·70 0·50 0·57

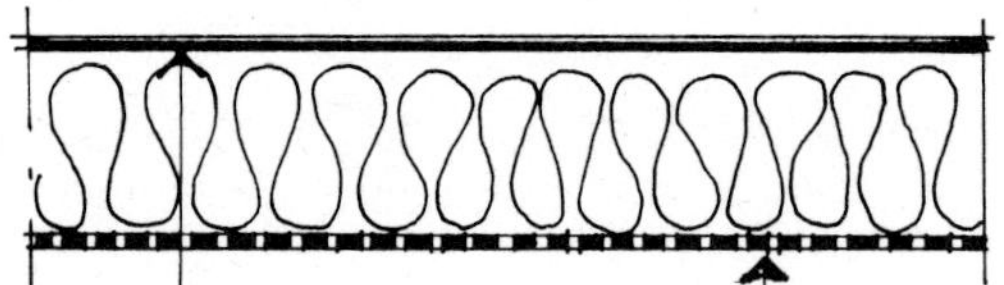

ONE SIDE NOT
PERFORATED
PROPRIETARY (HAUSERMAN)
PARTITION WITH OPTIONAL ONE
FACE PERFORATED
(NOTE : PERFORATIONS REDUCE
SOUND INSULATION VALUE
OF PARTITION TO 35 dB OR LESS)
0·46 (100–3150 Hz) 0·91 (400–1250 Hz) 0·84 (1600–5000 Hz)

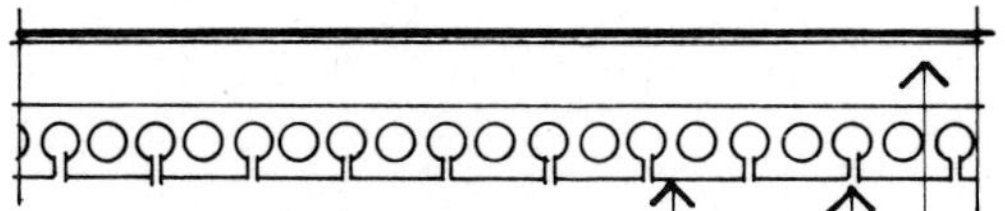

26 EXTRUDED CHIPBOARD
PANELS WITH DECORATIVE VENEER
3 CONTINUOUS SLOTS
38X25 SOFTWOOD BATTENS WITH
GLASSFIBRE INSULATION BETWEEN
0·15 0·35 0·40 0·60 0·85 0·55

STANDARD DETAILS.

WALL FINISHES / PARTITIONS
SOUND ABSORPTION.

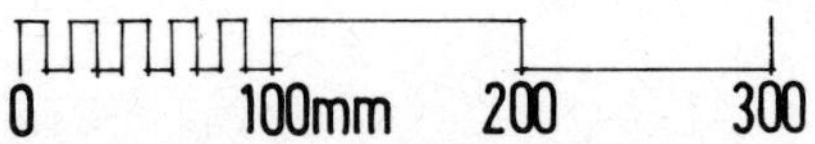

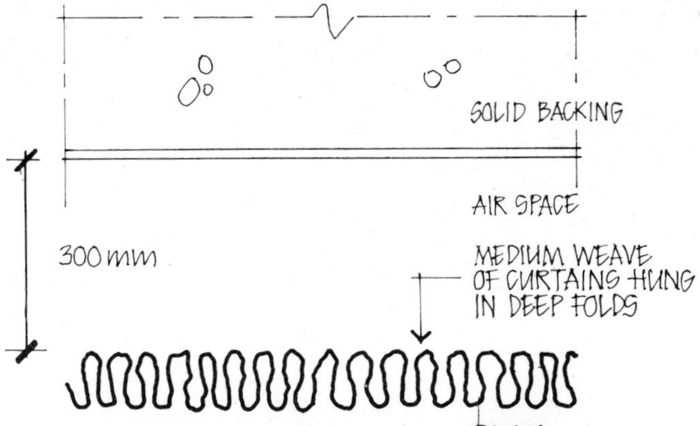

PLAN

(1)
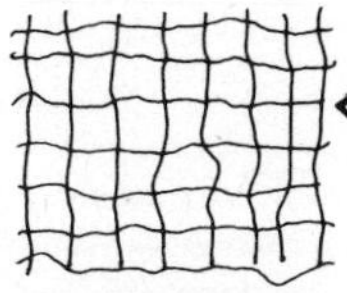

TOO LOOSE A WEAVE DENSITY AND MATERIAL IS ACOUSTICALLY TRANSPARENT

(2)
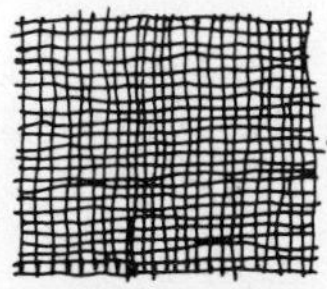

MEDIUM WEAVE GIVES MUCH BETTER ABSORPTION THAN 1. OR 2. TEST BY TRYING TO BLOW THROUGH MATERIAL. IT SHOULD BE POSSIBLE TO BLOW THROUGH THE FABRIC BUT WITH SOME RESISTANCE: FABRIC HAS OPTIMUM FLOW RESISTANCE TO SOUND

(3)
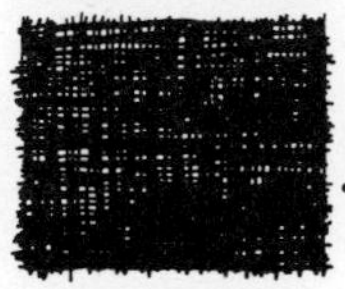

TOO TIGHT A WEAVE AND THE MATERIAL IS ONLY SLIGHTLY SOUND ABSORPTIVE.

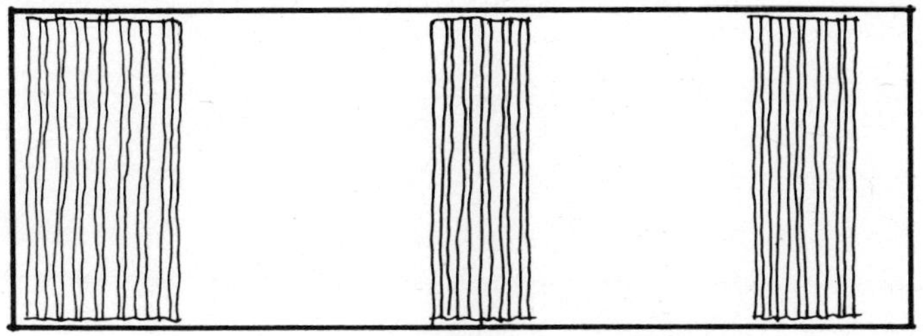

ROOM ELEVATION

AREAS OF CURTAINS DISTRIBUTED AROUND ARE MORE EFFECTIVE IN TERMS OF SOUND ABSORPTION AND GIVE MORE EVEN SOUND QUALITY THAN IF CONCENTRATED IN ONE AREA

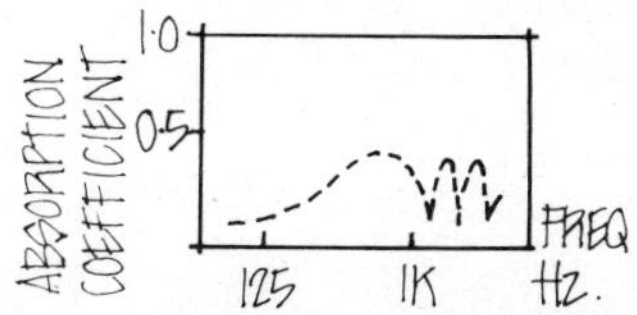

ABSORPTION CHARACTERISTIC OF THIN CURTAIN NOT HUNG IN FOLDS SPACED OFF WALL

BEST USE OF CURTAIN IS AS IN PLAN, WITH MEDIUM WEAVE. FREE STANDING DIVIDER CURTAINS ARE LESS EFFECTIVE. CURTAINS ARE A SIMPLE METHOD OF PROVIDING AN ADAPTABLE AMOUNT OF ABSORPTION EG. IN MUSIC PRACTICE ROOMS. CURTAIN LINING ALSO IMPROVES SOUND ABSORPTION.

CURTAINS

not to scale

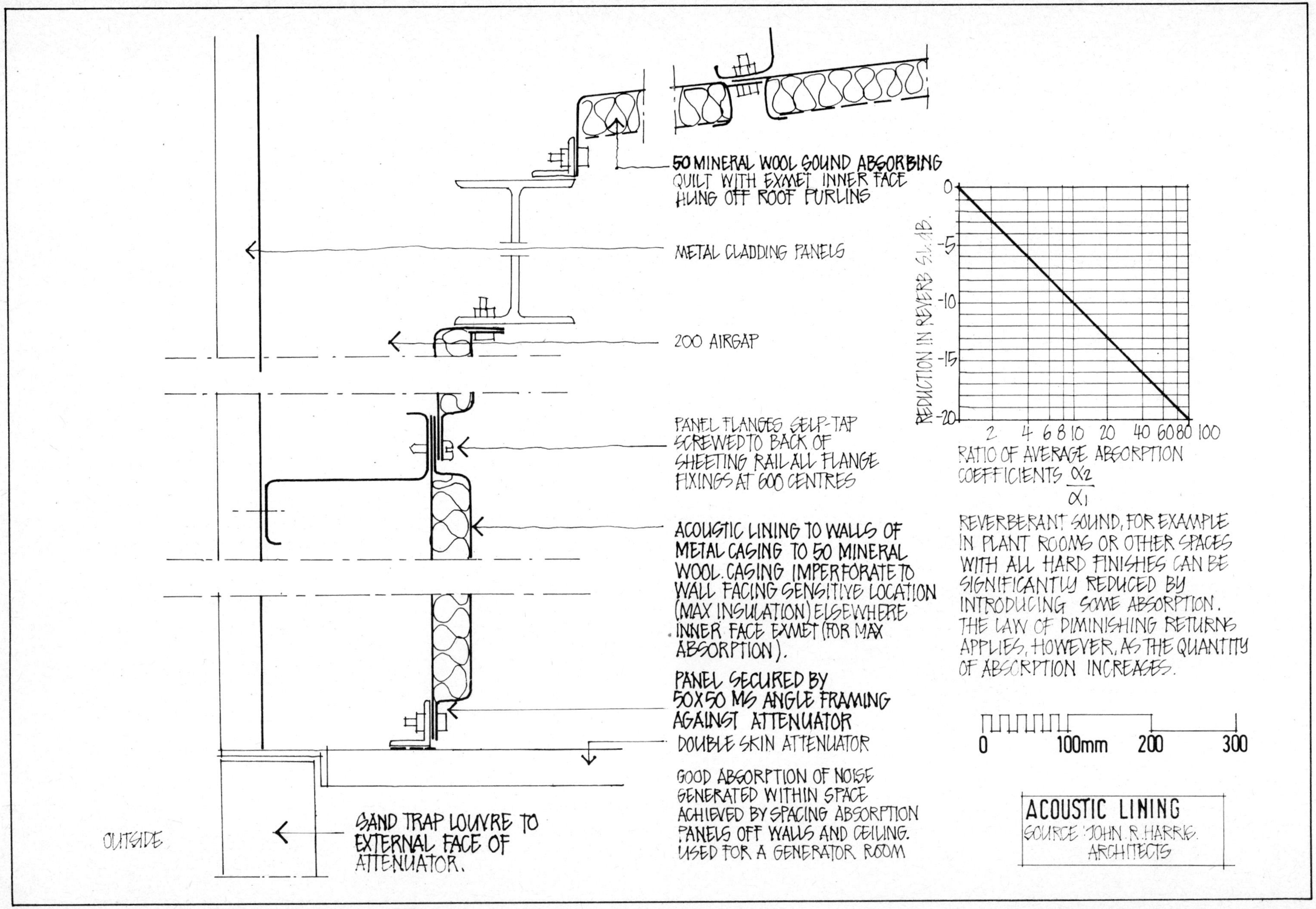
50 MINERAL WOOL SOUND ABSORBING QUILT WITH EXMET INNER FACE HUNG OFF ROOF PURLINS
METAL CLADDING PANELS
200 AIRGAP
PANEL FLANGES SELF-TAP SCREWED TO BACK OF SHEETING RAIL ALL FLANGE FIXINGS AT 600 CENTRES
ACOUSTIC LINING TO WALLS OF METAL CASING TO 50 MINERAL WOOL. CASING IMPERFORATE TO WALL FACING SENSITIVE LOCATION (MAX INSULATION) ELSEWHERE INNER FACE EXMET (FOR MAX ABSORPTION).
PANEL SECURED BY 50X50 MS ANGLE FRAMING AGAINST ATTENUATOR
DOUBLE SKIN ATTENUATOR
GOOD ABSORPTION OF NOISE GENERATED WITHIN SPACE ACHIEVED BY SPACING ABSORPTION PANELS OFF WALLS AND CEILING. USED FOR A GENERATOR ROOM
SAND TRAP LOUVRE TO EXTERNAL FACE OF ATTENUATOR.
OUTSIDE
REDUCTION IN REVERB SL dB.
0
-5
-10
-15
-20
2
4
6
8
10
20
40
60
80
100
RATIO OF AVERAGE ABSORPTION COEFFICIENTS α2/α1
REVERBERANT SOUND, FOR EXAMPLE IN PLANT ROOMS OR OTHER SPACES WITH ALL HARD FINISHES CAN BE SIGNIFICANTLY REDUCED BY INTRODUCING SOME ABSORPTION. THE LAW OF DIMINISHING RETURNS APPLIES, HOWEVER, AS THE QUANTITY OF ABSORPTION INCREASES.
0
100mm
200
300
ACOUSTIC LINING
SOURCE: JOHN R. HARRIS. ARCHITECTS

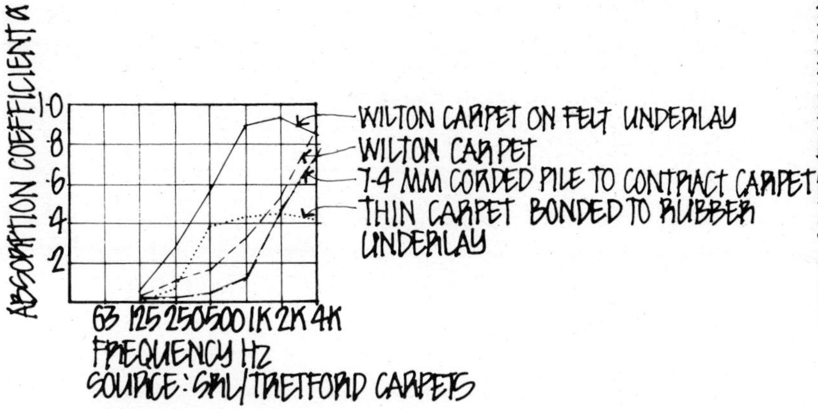

SOURCE: SRL/TRETFORD CARPETS

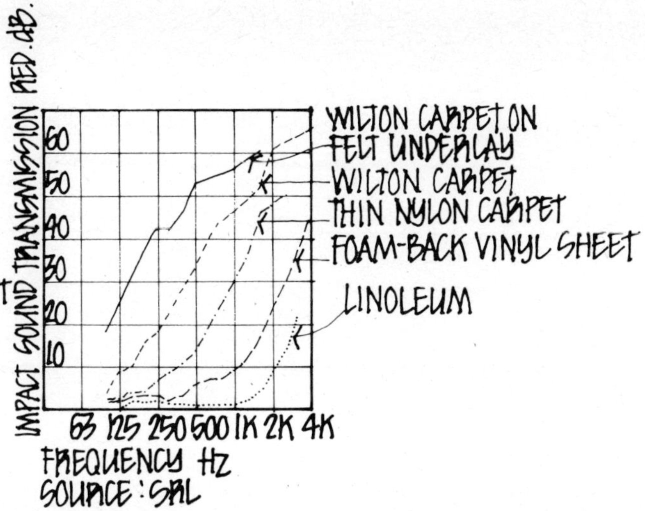

SOURCE: SRL

CARPET

CARPET IS USEFUL FOR BOTH IMPACT SOUND REDUCTION [FOOTFALL AND TO FLOOR BELOW] AND FOR SOUND ABSORPTION. VALUES ARE SHOWN IN APPENDIX BUT FOR THIN CARPET ON SOLID BACKING [TYPICAL OFFICE FOR LEASE] VALUES ARE LOW FOR LOW AND MID FREQUENCIES. CARPET IS SOMETIMES USED AS A HARDWEARING WALL FINISH, FOR EXAMPLE IN CINEMAS. CARPETS ON THIN WALL PANELS [SEE DETAIL ASSEMBLY (4)] GIVE A BROADER FREQUENCY PERFORMANCE.

CUSHIONED VINYL SHEET.

CUSHIONED VINYL SHEET ONLY 3MM THICK ('POLYTREAD' OR SIM.) ON A CONCRETE SLAB CAN GIVE IMPACT SOUND TRANSMISSION PERFORMANCE, EXCEEDING GRADE I BUILDING REGS. THIS IS A USEFUL MATERIAL FOR HOSPITALS, SCHOOLS AND OLD PEOPLES HOMES WHERE CARPET CANNOT BE USED

FINISHES [BRS 103 DEFINITION]

SOFT: CUSHIONED VINYL SHEET THIN CARPET WITH UNDERLAY. THICK CARPET WITH OR WITHOUT UNDERLAY, CORK TILES OVER 8MM TILES.

HARD: CONCRETE, TERRAZZO, PVC TILES, WOOD BLOCK, MASTIC ASPHALT.

FINISHES

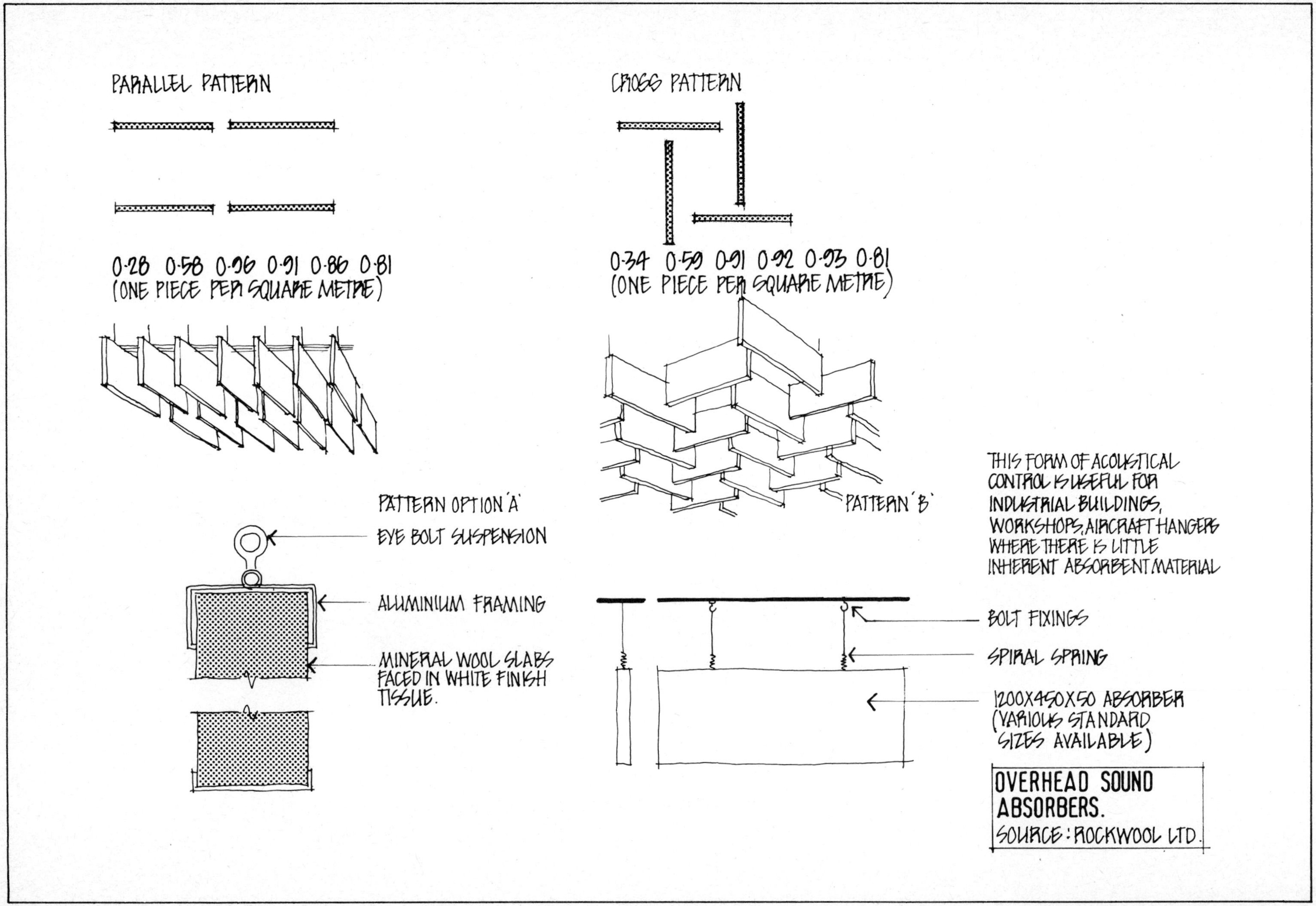
PARALLEL PATTERN
0·28 0·58 0·96 0·91 0·86 0·81
(ONE PIECE PER SQUARE METRE)
CROSS PATTERN
0·34 0·59 0·91 0·92 0·93 0·81
(ONE PIECE PER SQUARE METRE)
PATTERN OPTION 'A'
PATTERN 'B'
THIS FORM OF ACOUSTICAL CONTROL IS USEFUL FOR INDUSTRIAL BUILDINGS, WORKSHOPS, AIRCRAFT HANGERS WHERE THERE IS LITTLE INHERENT ABSORBENT MATERIAL
EYE BOLT SUSPENSION
ALUMINIUM FRAMING
MINERAL WOOL SLABS FACED IN WHITE FINISH TISSUE.
BOLT FIXINGS
SPIRAL SPRING
1200X450X50 ABSORBER (VARIOUS STANDARD SIZES AVAILABLE)
OVERHEAD SOUND ABSORBERS.
SOURCE: ROCKWOOL LTD.

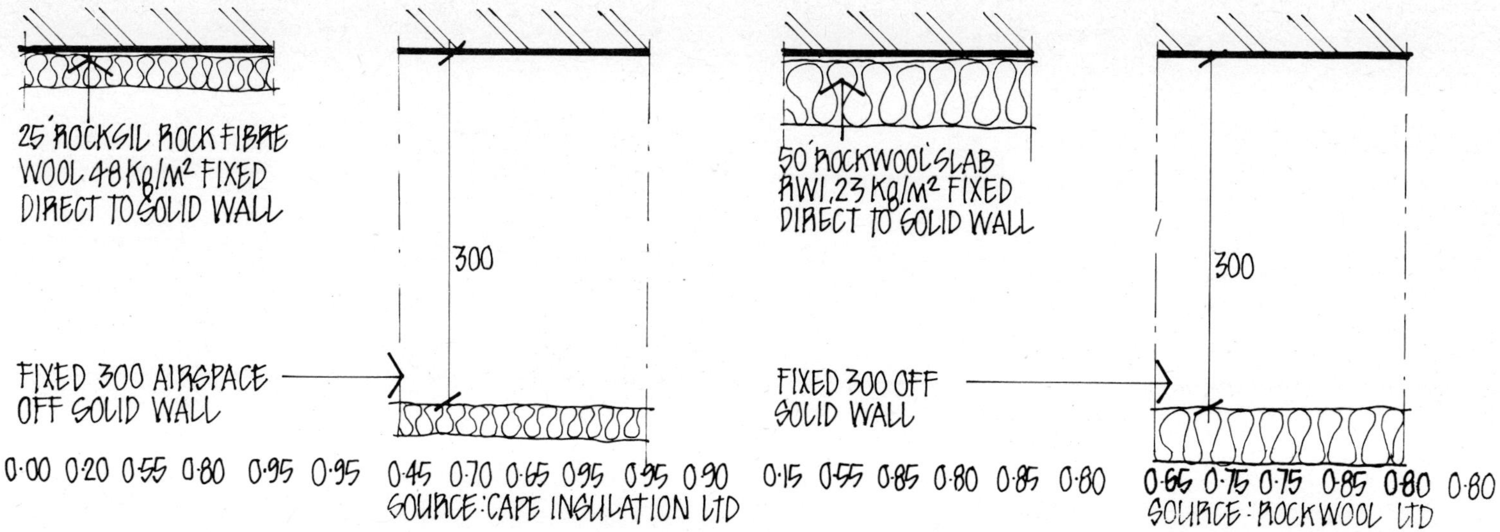

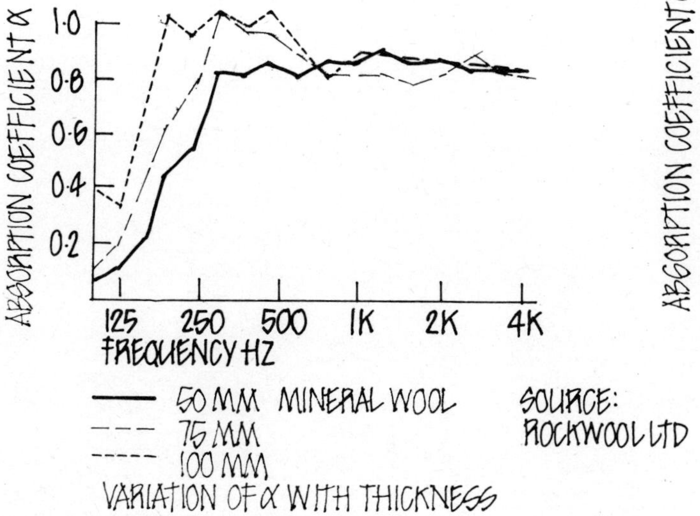

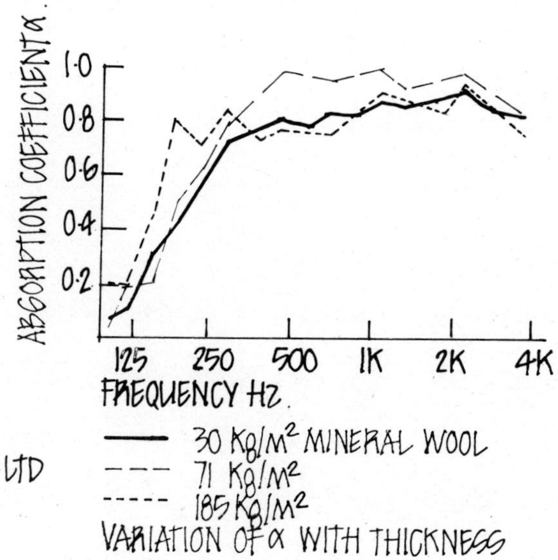

ALTHOUGH SOUND ABSORPTION RANGES FROM $\alpha = 0$ TO $\alpha = 1{\cdot}0$ (FULL REFLECTION TO COMPLETE SOUND ABSORPTION) FIGURES IN EXCESS OF 1·0 CAN OCCUR. THIS APPARENT ANOMALY IS DUE TO THE METHOD OF MEASUREMENT.

SOUND ABSORPTIVE LINING

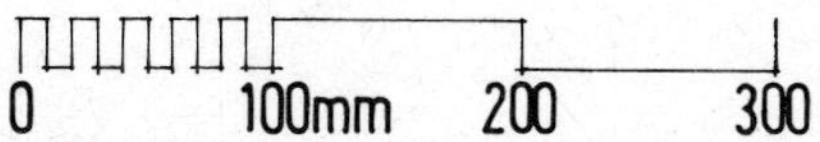

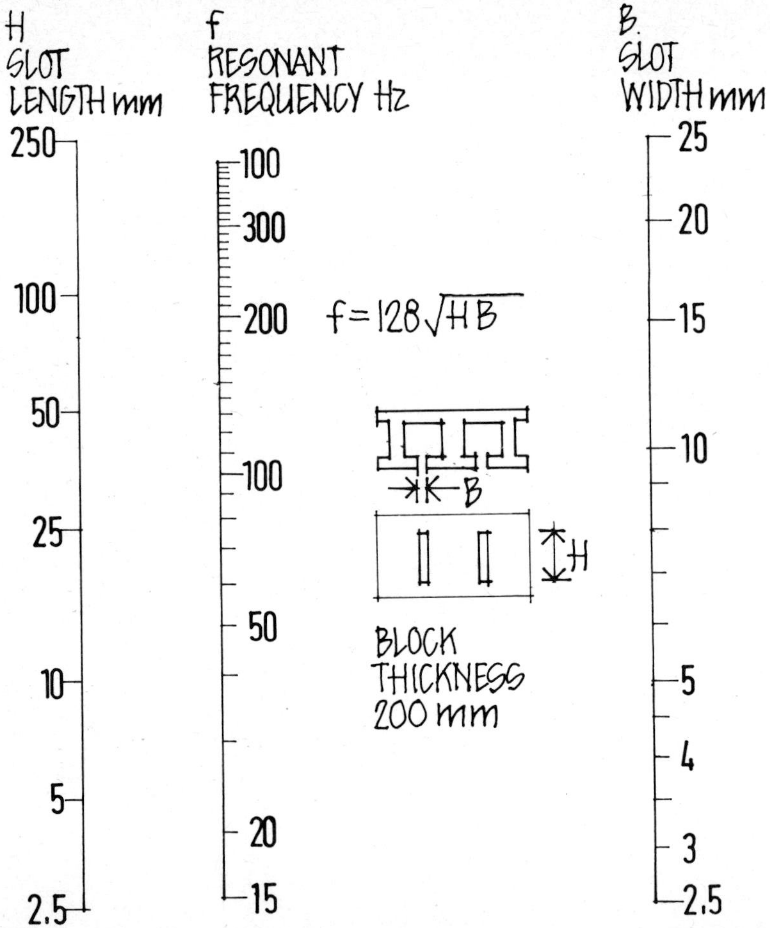

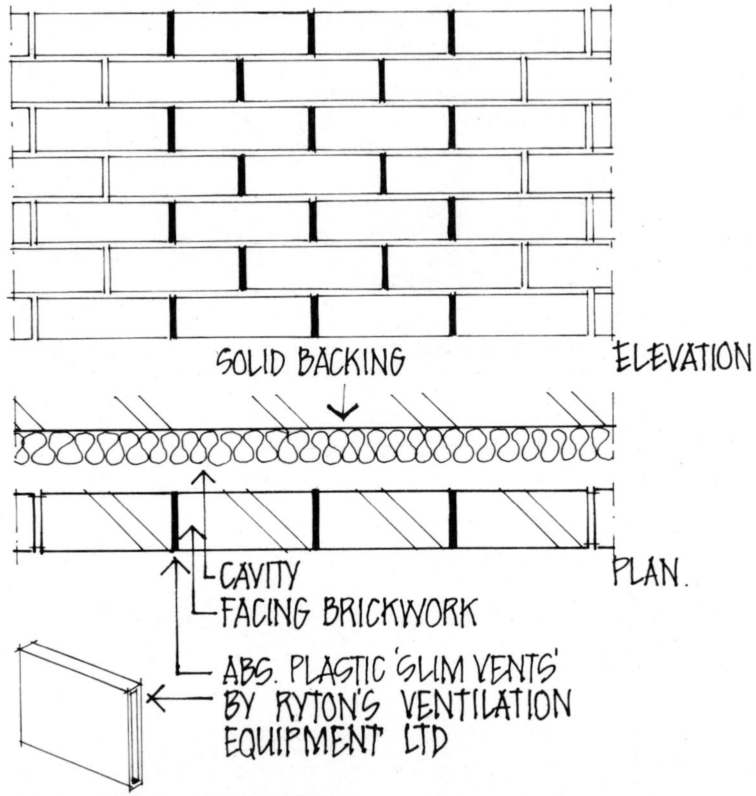

NOMOGRAM FOR COMPUTING THE HELMHOLTZ FREQUENCY FOR SLOTTED HOLLOW CONCRETE BLOCK. (SOURCE: HERRICK LABORATORIES, PURDUE UNIVERSITY. INDIANA U.S.)

ABSORPTION COEFFICIENTS FOR SLOTTED 200 BLOCK FILLED WITH INCOMBUSTIBLE FIBROUS MATERIAL

0·72 0·55 0·42 0·34 0·35 0·34

not to scale

A SIMILAR EFFECT IS OBTAINED FROM SLOTS IN BRICKWORK WITH CAVITY BEHIND (NOTE: LEAF INEFFECTIVE FOR SOUND REDUCTION BETWEEN SPACES) TO ASSESS ABSORPTION CHARACTERISTICS LABORATORY TESTS ARE REQUIRED FOR THE BRICK, SPACING AND SLOT PATTERN PROPOSED. APPLICATION: SELECTIVE ABSORPTION AT LOW FREQUENCIES.

SOUND ABSORPTIVE WALLING

'ACOUSTIC BRICKS' BY BLOCKLEYS LTD., TELFORD

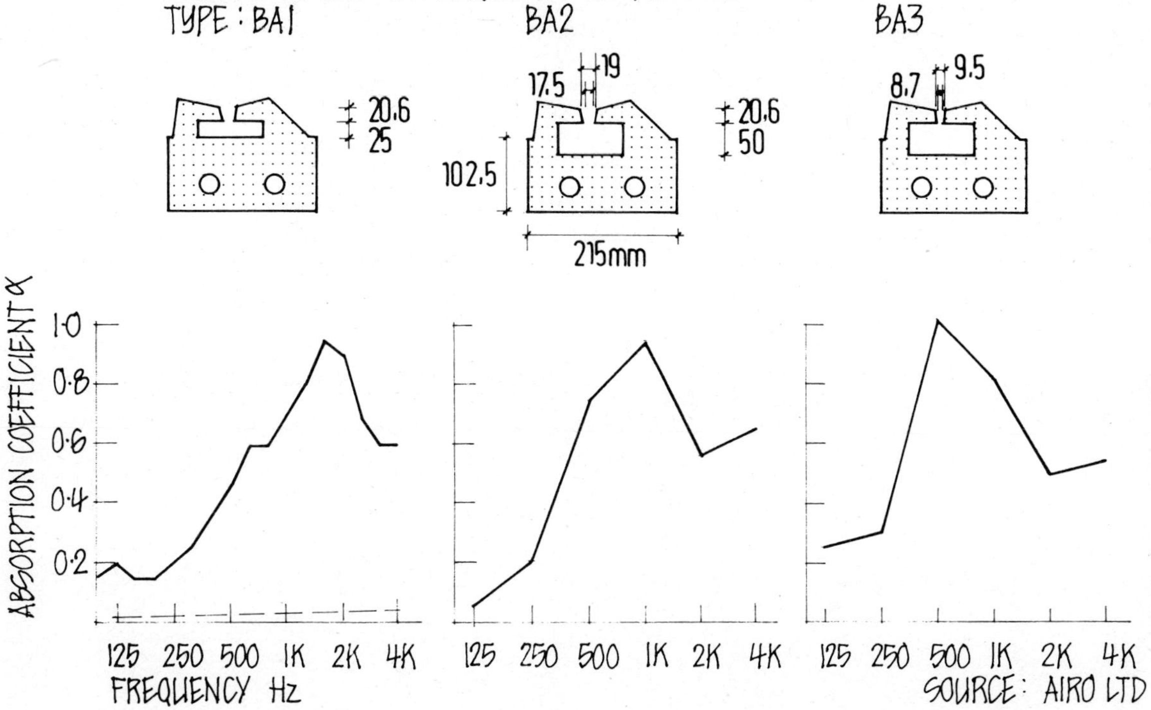

ELEVATION

0 10mm

ABSORPTION COEFFICIENT OF STANDARD FAIRFACED BRICKWORK

SPECIAL DESIGN BRICKS MAY BE CONSIDERED TO 'SOFTEN' THE SOUND QUALITY OF A SPACE WHERE HARD FINISHES ARE REQUIRED. A MIX OF THE 3 TYPES WILL ACHIEVE GOOD SOUND ABSORPTION IN THE RANGE 500–2000Hz. ANGLED FACES HELP DIFFUSE INCIDENT SOUND. MORE COSTLY THAN SLOTS BETWEEN STANDARD BRICKS, BUT ADVANTAGEOUS IN THAT WALL RETAINS SOME SOUND INSULATION VALUE.

SOUND ABSORPTIVE WALLING

not to scale

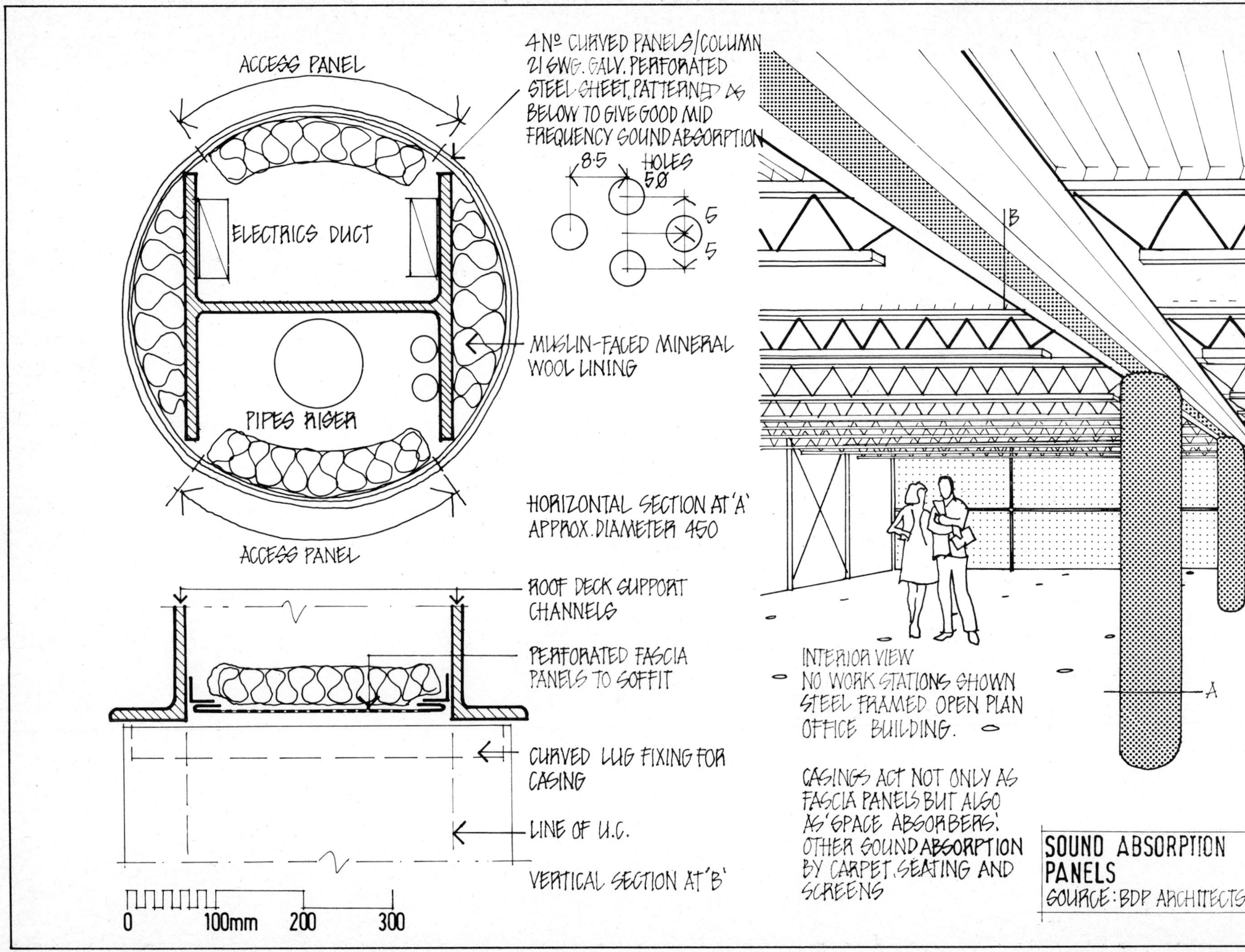
ACCESS PANEL
4Nº CURVED PANELS/COLUMN
21 SWG. GALV. PERFORATED
STEEL SHEET, PATTERNED AS
BELOW TO GIVE GOOD MID
FREQUENCY SOUND ABSORPTION
8·5
HOLES
5Ø
5
5
ELECTRICS DUCT
MUSLIN-FACED MINERAL
WOOL LINING
PIPES RISER
B
HORIZONTAL SECTION AT 'A'
APPROX. DIAMETER 450
ACCESS PANEL
ROOF DECK SUPPORT
CHANNELS
PERFORATED FASCIA
PANELS TO SOFFIT
CURVED LUG FIXING FOR
CASING
LINE OF U.C.
VERTICAL SECTION AT 'B'
0
100mm
200
300
INTERIOR VIEW
NO WORK STATIONS SHOWN
STEEL FRAMED OPEN PLAN
OFFICE BUILDING.
A
CASINGS ACT NOT ONLY AS
FASCIA PANELS BUT ALSO
AS 'SPACE ABSORBERS'.
OTHER SOUND ABSORPTION
BY CARPET, SEATING AND
SCREENS
SOUND ABSORPTION
PANELS
SOURCE: BDP ARCHITECTS

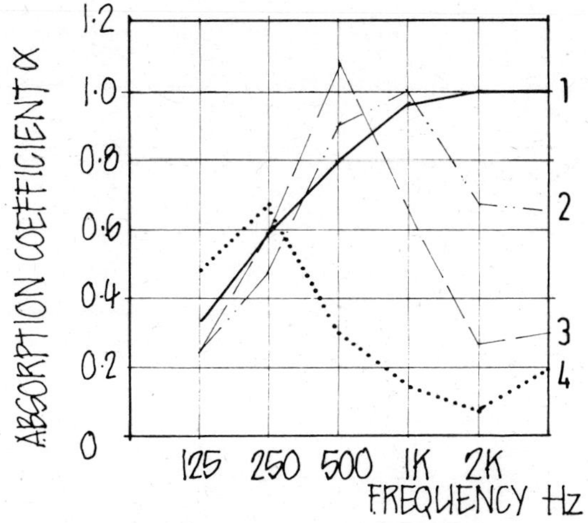

WALL PANELS CAN BE DESIGNED TO PROVIDE ABSORPTION OVER A NARROW RANGE OF FREQUENCIES BY ADJUSTING THE GAP BETWEEN PLANKS. THE PRINCIPLE OF THE SLIT RESONATOR IS THE SAME AS FOR A HELMHOLTZ RESONATOR, RESONANT FREQUENCY BEING DEPENDENT ON SLIT WIDTH DEPTH AND CROSS SECTIONAL AREA OF THE SPACE BEHIND THE SLITS FORMED BY THE PLANKS.

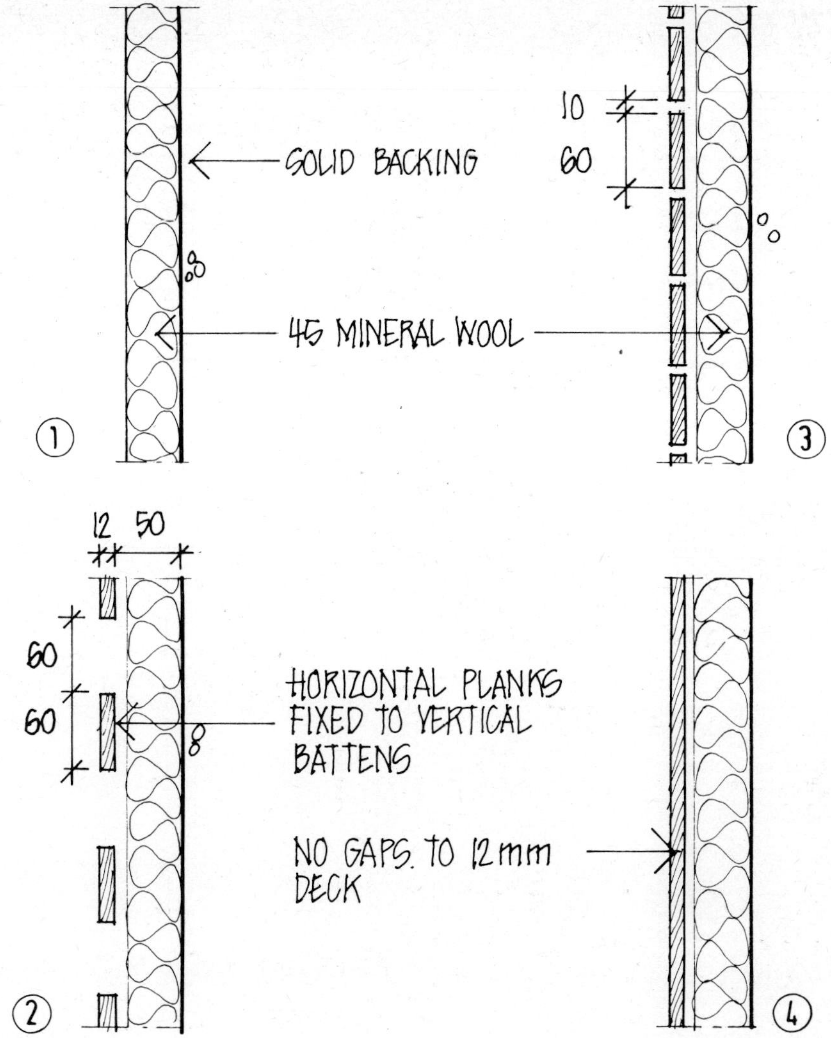

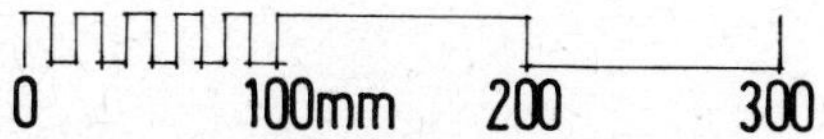

WALL PANELS
SOURCE: FURRER

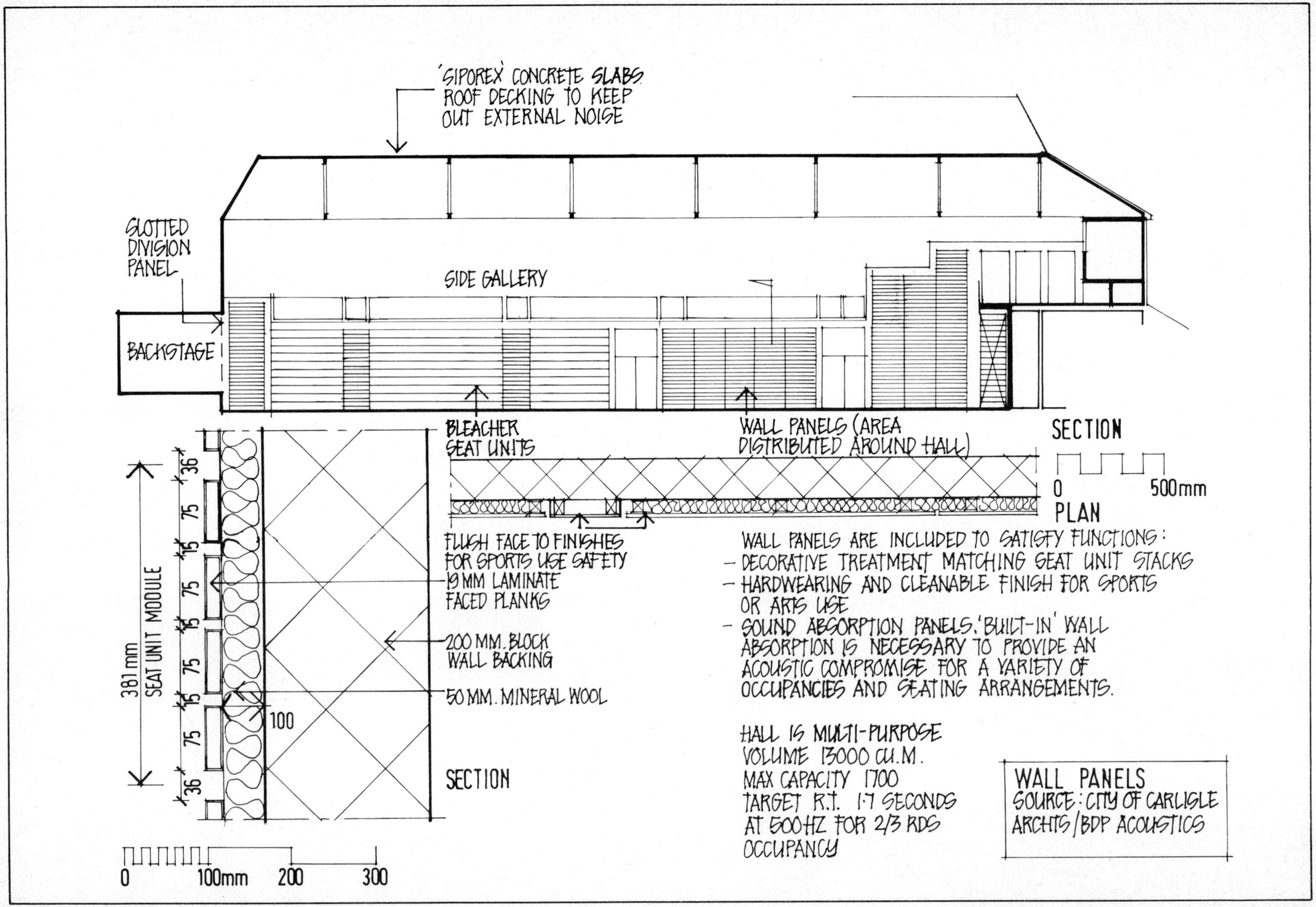
'SIPOREX' CONCRETE SLABS ROOF DECKING TO KEEP OUT EXTERNAL NOISE
SLOTTED DIVISION PANEL
SIDE GALLERY
BACKSTAGE
BLEACHER SEAT UNITS
WALL PANELS (AREA DISTRIBUTED AROUND HALL)
SECTION
0
500mm
PLAN
FLUSH FACE TO FINISHES FOR SPORTS USE SAFETY
19 MM LAMINATE FACED PLANKS
200 MM. BLOCK WALL BACKING
50 MM. MINERAL WOOL
100
381 mm SEAT UNIT MODULE
36
75
15
SECTION
0
100mm
200
300
WALL PANELS ARE INCLUDED TO SATISFY FUNCTIONS:
– DECORATIVE TREATMENT MATCHING SEAT UNIT STACKS
– HARDWEARING AND CLEANABLE FINISH FOR SPORTS OR ARTS USE
– SOUND ABSORPTION PANELS. 'BUILT-IN' WALL ABSORPTION IS NECESSARY TO PROVIDE AN ACOUSTIC COMPROMISE FOR A VARIETY OF OCCUPANCIES AND SEATING ARRANGEMENTS.
HALL IS MULTI-PURPOSE
VOLUME 13000 CU.M.
MAX CAPACITY 1700
TARGET R.T. 1·7 SECONDS AT 500HZ FOR 2/3 RDS OCCUPANCY
WALL PANELS
SOURCE: CITY OF CARLISLE ARCHTS/BDP ACOUSTICS

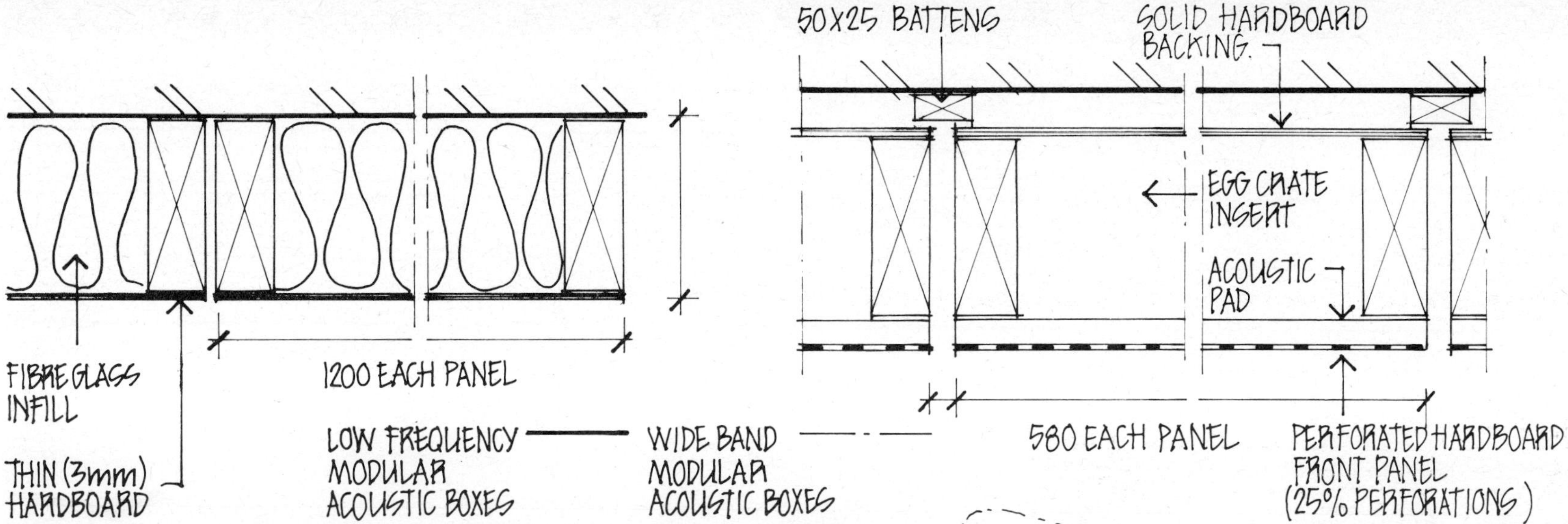

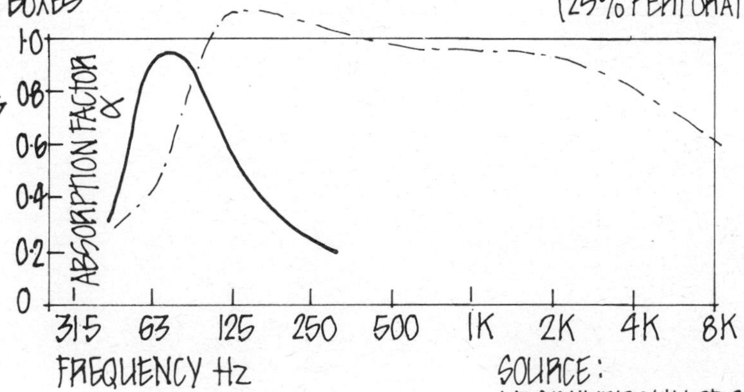

LOW FREQUENCY BOXES PROPOSED WITH WIDE BAND BOXES IN RATIO 1:2, TOTAL AREA 150 M². FOR A STUDIO REMODELLING TO ACHIEVE REVERBERATION TIME <0·5 SECONDS FOR FREQUENCIES BETWEEN 63-4K Hz. VOLUME 550 M³. FLOOR FINISH: CARPET.

SOURCE:
L.F. UNIVERSITY OF SALFORD
W.B. SHONE SOUND LTD

MODULAR ACOUSTIC BOXES ARE OF PARTICULAR USE IN BROADCASTING AND RADIO STUDIOS: SEEK SPECIALIST ADVICE CONCERNING USE. SOME USE OF SUCH PANELS MAY BE MADE IN eg. MUSIC ROOMS, DISCOTHEQUES. DECORATIVE OPEN WEAVE FINISH MAY BE STRETCHED OVER PANEL FRONTS. CHECK ANY REQUIREMENT FOR FIRE RESISTANCE.

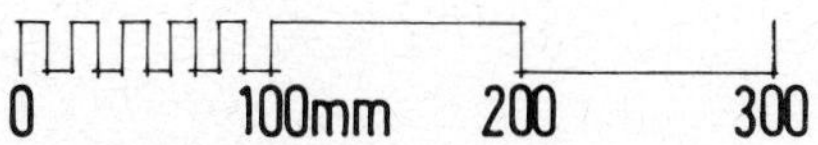

MODULAR ACOUSTIC BOXES

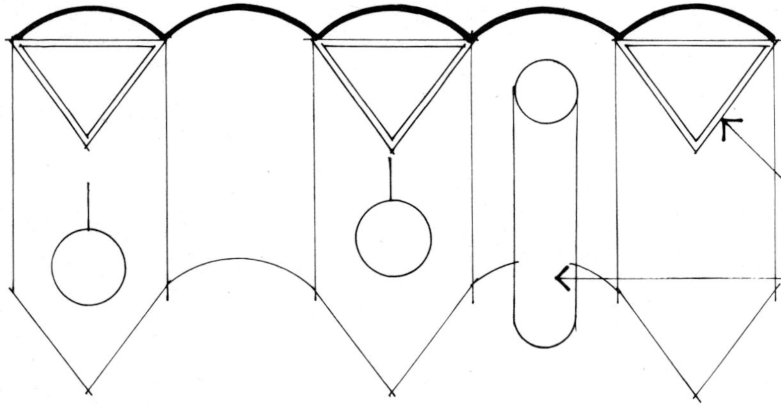

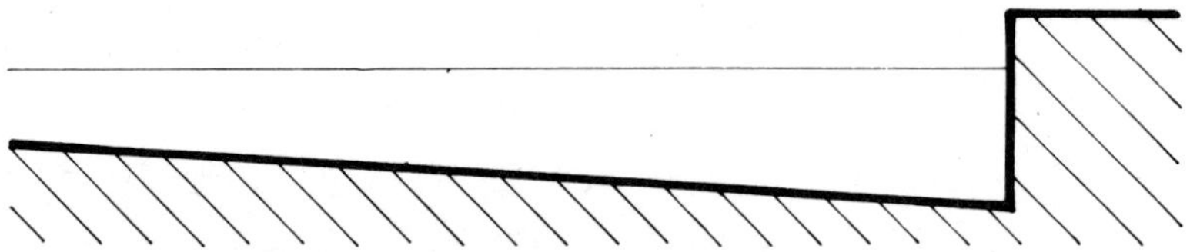

SECTION

SKETCH OF INTERIOR.

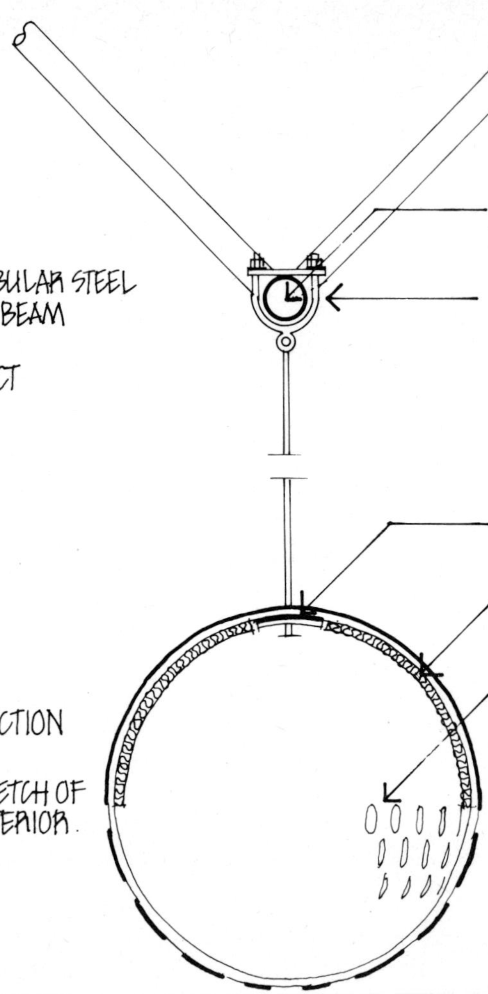

THESE GRP ORBS ARE USED IN SEVERAL SWIMMING AND LEISURE POOLS TO HELP ABSORB THE SOUND OF SHOUTS AND SPLASHES. POOL INTERIORS ARE PROBLEMATIC BECAUSE OF THE NEED FOR HARD DAMP RESISTANT MATERIALS; AVOID THE USE OF NORMAL ACOUSTIC CEILING PLASTER WHICH ABSORBS MOISTURE TO MANY TIMES ITS OWN WEIGHT AND FALLS OFF CEILING SOFFIT.

OVERHEAD SOUND ABSORBERS

not to scale

SERVICES

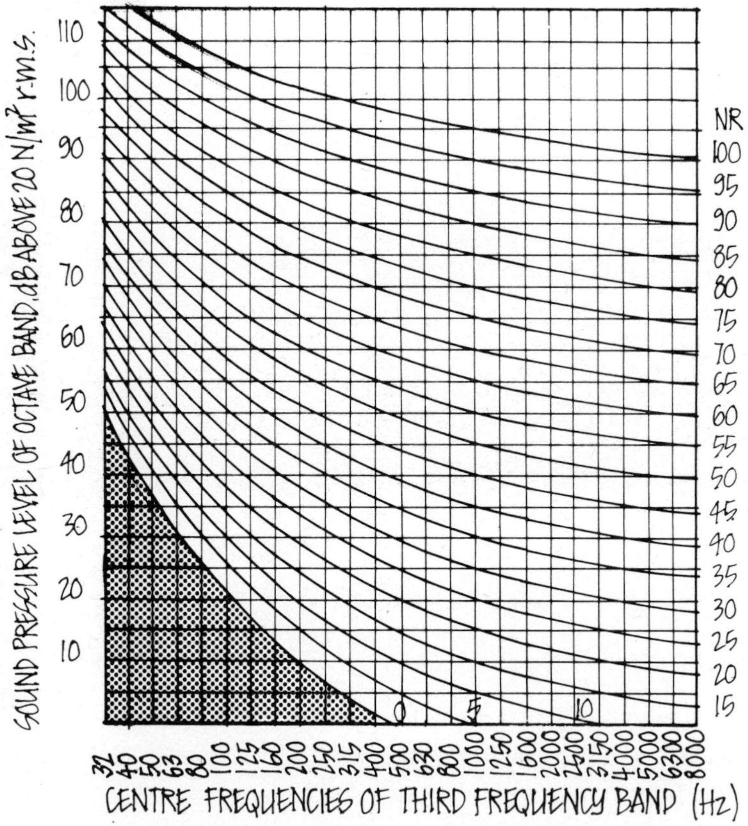

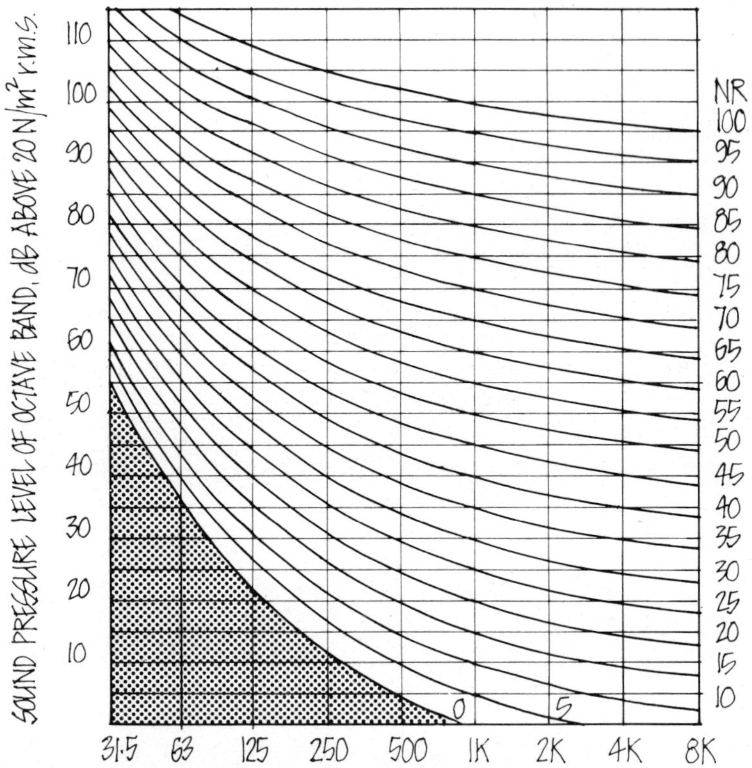

CENTRE FREQUENCIES OF OCTAVE BANDS
NOISE RATING CURVES – OCTAVE ANALYSIS

IT IS CONVENIENT TO SPECIFY THE LEVEL OF A PARTICULAR NOISE BY EQUAL LOUDNESS CONTOURS WHICH RECOGNISE THAT THE SUBJECTIVE RESPONSE BY THE EAR IS TO FIND LOW-FREQUENCY NOISE LESS NOTICEABLE THAN HIGH FREQUENCY FOR THE SAME SOUND POWER LEVEL. THE CONTOURS ABOVE ARE NOISE RATING (NR) CURVES, ANOTHER FORM IS PNC CURVES WHICH ARE VERY SIMILAR AT MOST FREQUENCIES. CURVES CAN BE USED TO MONITOR THAT THE AMBIENT SOUND LEVEL IN A SPACE IS NOT AT AN UNSUITABLY HIGH LEVEL DUE TO EXTERNAL, VENTILATION, MACHINERY OR ADJACENT ROOM NOISES. THE MAXIMUM BAND LEVEL OF THE NOISE SPECTRAL DISTRIBUTION DETERMINES ITS NR CURVE RATING. SOUND INSULATION VALUES CAN BE USED TO CHECK SOUND PENETRATING AT EACH FREQUENCY BAND.

AMBIENT NOISE CONTROL (1)

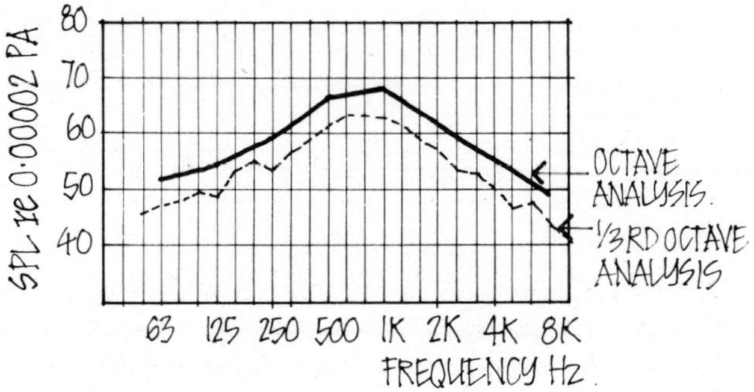

NOISE ANALYSIS IN A CANTEEN
SOURCE: PARKIN, HUMPHREYS AND COWELL.

CARE SHOULD BE TAKEN THAT FIGURES USED ARE FOR OCTAVE BAND ANALYSIS AS ONE-THIRD OCTAVE BAND REQUIRES USE OF THE LHS. SET OF NR CURVES. BECAUSE OF THE REDUCED BAND-WIDTH COMPONENT INDIVIDUAL LEVELS ARE ABOUT 5dB DOWN AND THE 1/3RD OCTAVE NR CURVES ALLOW FOR THIS. WHEN 1/3RD OCTAVE BAND LEVELS ARE ADDED TOGETHER, A HIGHER FIGURE RESULTS AS THE LEVEL OVER THE WHOLE FREQUENCY RANGE.

RECOMMENDED NR. LEVELS FOR AMBIENT NOISE CONTROL TO SUIT USE	NR
STUDIOS SPACES FOR RECORDING OR BROADCAST	15
CONCERT HALLS	15-20
THEATRES, MULTI-PURPOSE HALLS CONFERENCE ROOMS, COURT ROOMS.	25
HOSPITAL WARDS, HOTEL BEDROOMS LIBRARIES	25-30
CLASSROOMS, MEETING ROOMS FOR SMALL GROUPS, EXECUTIVE OFFICES, HOMES.	30
RESTAURANTS, OPEN PLAN AND GENERAL OFFICES	35
CAFETERIAS, CIRCULATION	40
KITCHENS, TOILETS AND WORKSHOPS	45
CAR PARKS, SHOPPING MALLS, BUS, RAIL OR AIRPORT CONCOURSES.	50
OFFICE (TYPING, WORD PROCESS MACHINES)	55
WORKSHOPS, INDUSTRIAL PROCESS	65

DESIRABLE dB(A) CRITERIA 5-10 ABOVE RECOMMENDED NR LEVEL.

AMBIENT NOISE CONTROL (2)

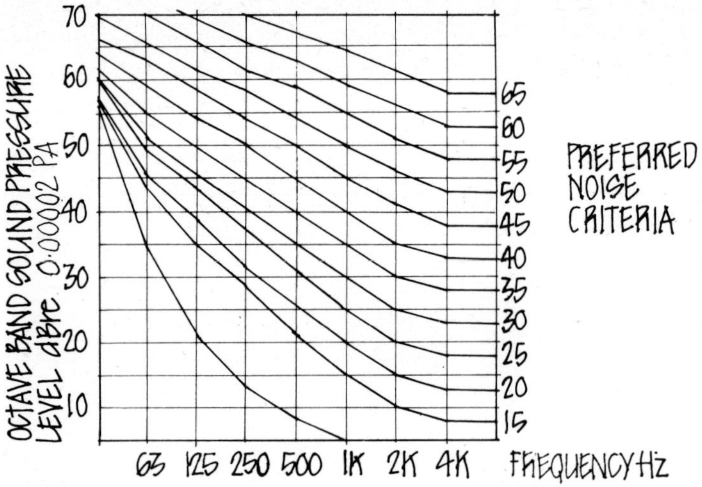

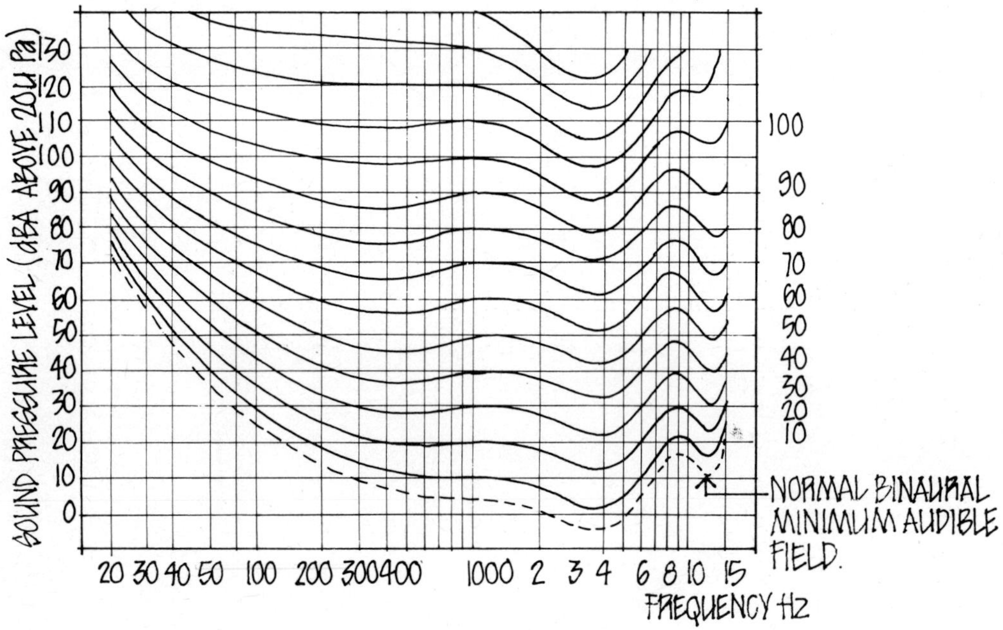

NR CURVES TEND TO BE USED IN EUROPE, WHERE AS IN THE USA THE 1971 PREFERRED NOISE CRITERIA (PNC) CURVES HAVE REPLACED THE EARLIER NOISE CRITERION (NC) CURVES. THEY HAVE, HOWEVER, NOT RECEIVED UNIVERSAL ACCEPTANCE.
COMPARISON WITH
THE EQUAL LOUDNESS CONTOURS SHOWS THE INTENTION OF EITHER SET OF CURVES TO REFLECT THE DIFFERENT SENSITIVITY OF EAR AT DIFFERENT PARTS OF THE FREQUENCY RANGE.

AMBIENT NOISE CONTROL (3)

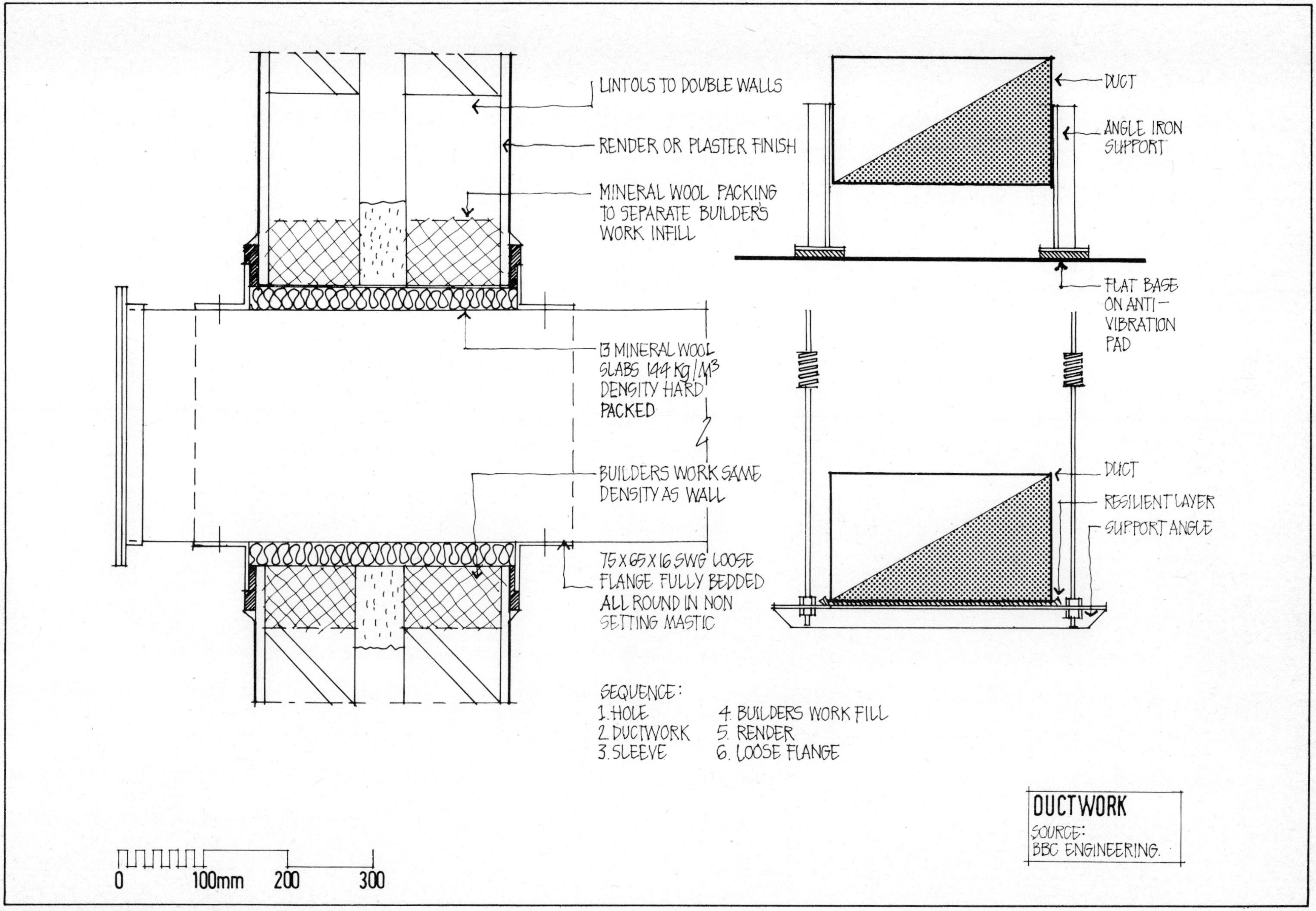
LINTOLS TO DOUBLE WALLS
RENDER OR PLASTER FINISH
MINERAL WOOL PACKING TO SEPARATE BUILDER'S WORK INFILL
13 MINERAL WOOL SLABS 144 Kg/M3 DENSITY HARD PACKED
BUILDERS WORK SAME DENSITY AS WALL
75 X 65 X 16 SWG LOOSE FLANGE FULLY BEDDED ALL ROUND IN NON SETTING MASTIC
DUCT
ANGLE IRON SUPPORT
FLAT BASE ON ANTI-VIBRATION PAD
DUCT
RESILIENT LAYER
SUPPORT ANGLE
SEQUENCE:
1. HOLE
2. DUCTWORK
3. SLEEVE
4. BUILDERS WORK FILL
5. RENDER
6. LOOSE FLANGE
DUCTWORK
SOURCE:
BBC ENGINEERING.
0
100mm
200
300

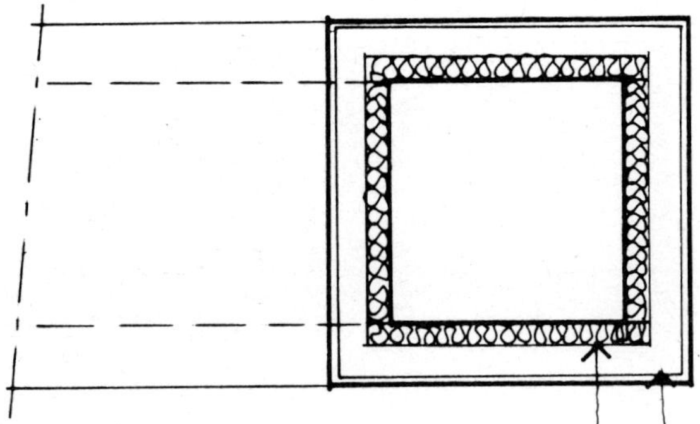

DUCT LINING.

WHERE LARGE LOW AIR VELOCITY DUCTS ARE NECESSARY TO ACHIEVE LOW AMBIENT NOISE LEVELS, BUILDERS WORK DUCTS MAY BE AN ECONOMIC SOLUTION FOR BOTH SUPPLY AND EXTRACT.

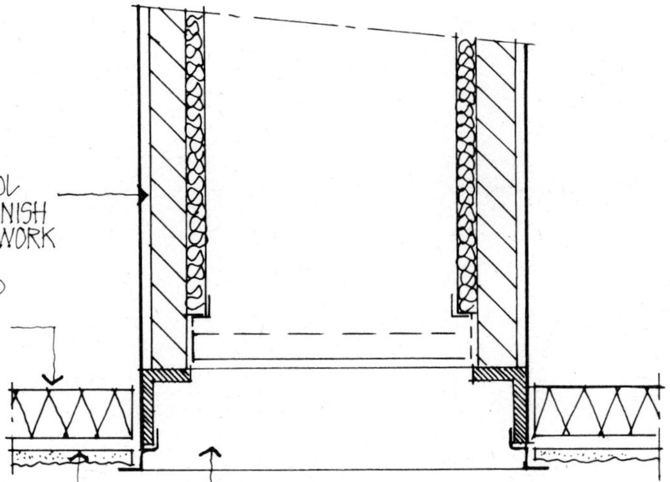

SUPPLY AIR/RETURN DIFFUSERS IN STUDIO CEILINGS.

ANALYSIS SEQUENCE FOR DUCT NOISE CONTROL:

1. FAN NOISE-SUPPLY SIDE
2. FAN NOISE -EXTRACT SIDE
3. AIR THROUGH SYSTEM/ REGENERATED NOISE
4. BREAKOUT THROUGH DUCT WALLS
5. TRANSFER TO ATMOSPHERE
6. AIRBORNE TRANSMISSION OUT OF PLANT ROOMS.

DUCTWORK
SOURCE:
BBC ENGINEERING

not to scale

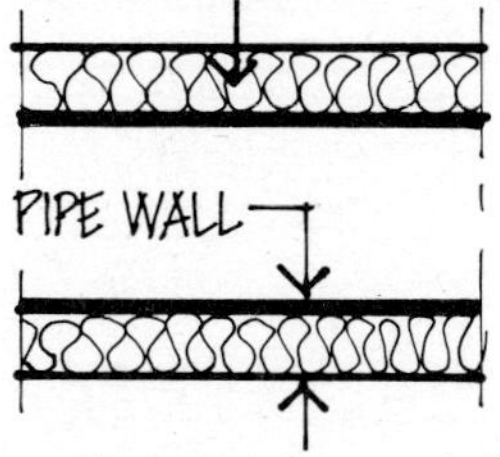

MASS LAGGING TO REDUCE PIPE NOISE

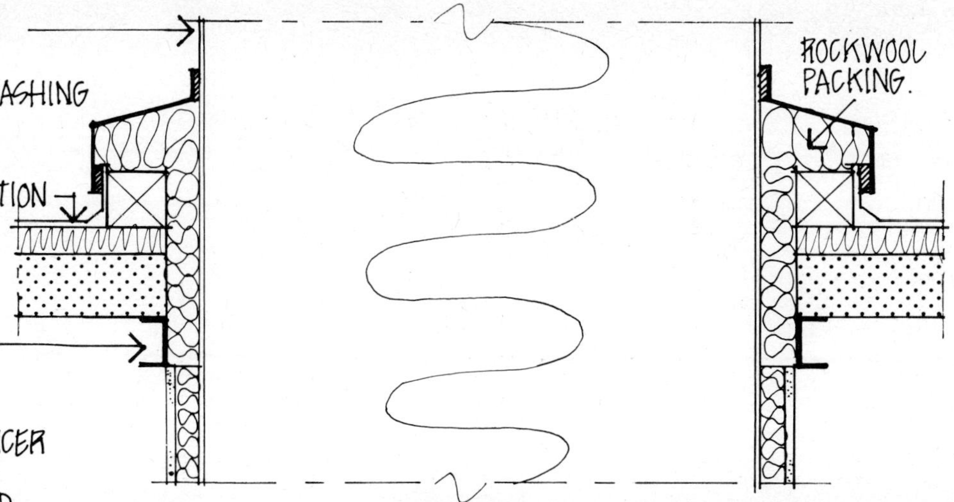

DUCT THROUGH ROOF

SOURCE: BDP

BELOW SILENCER, DUCT CASED ONLY WITH ACOUSTIC DAMPING MATERIAL, REVAC DD 2010 SA

DETAIL USED FOR A ROOF TOP H&V PLANT ON A SPORTS HALL. IT IS ADVISABLE TO AVOID OVERHEAD ROOF PLANT ON LONG SPAN AUDITORIA BECAUSE VIBRATION IS DIFFICULT TO DAMP AND IT IS DIFFICULT TO PREVENT NOISE BREAK-IN.

OPTIONS ON LAGGING

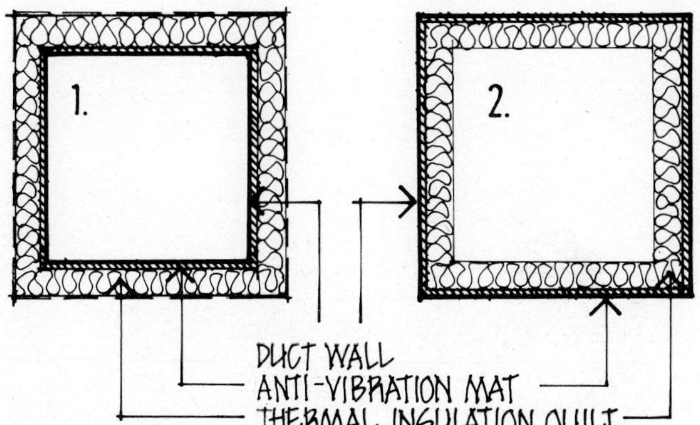

1. EXTERNAL CASING TO DUCT: EASIER TO APPLY AFTER INSTALLATION AND IF A LARGE Nº OF DUCTS ADDS SOUND ABSORPTION TO CEILING VOID OR ROOM.
2. INTERNAL - GIVES GOOD ATTENUATION TO DUCT-BORNE NOISE, BETTER APPEARANCE IF IN EXPOSED LOCATION

1 OR 2mm REDUC OR SIMILAR SOUND REDUCING MAT IS PARTICULARLY USEFUL IN LARGE DUCT SECTIONS WHERE DUCT WALLS CAN BE SET INTO EXCITATION BY AIR TURBULENCE (ESPECIALLY AT REDUCTION SECTIONS, T OR ELBOW JUNCTIONS.)

DUCTWORK

not to scale

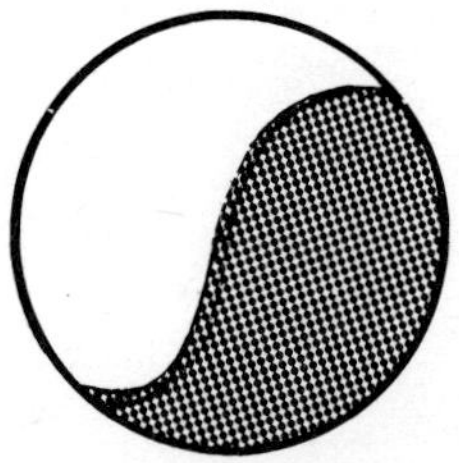

VS.

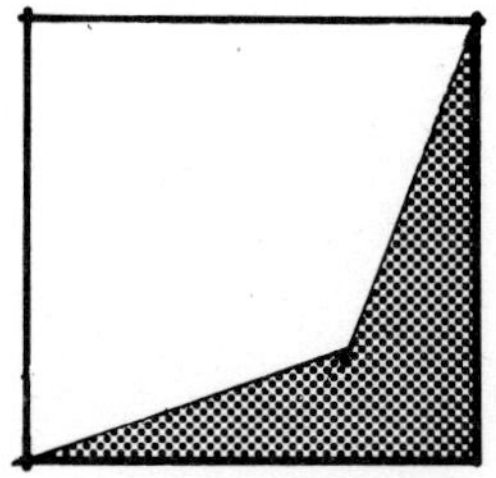

CIRCULAR DUCTS ARE INHERENTLY RIGID AND OF MINIMUM PERIMETER FOR SECTIONAL AREA: NOISE IS CONTAINED WITHIN RATHER THAN TRANSMITTED THROUGH DUCT TO SPACE. USE OF CIRCULAR DUCTWORK MAY THEREFORE BE PREFERABLE FOR EXPOSED INSTALLATIONS..

RECTANGULAR DUCTS HAVE LESS RIGID WALLS AND NOISE WITHIN EXCITES THE FLAT METAL SECTIONS. THIS GIVES USEFUL LOW FREQUENCY ATTENUATION ALONG DUCTS. THERE IS ALSO A LARGER INTERNAL AREA FOR LINING WITH MINERAL WOOL: USE OF SQUARE OR RECTANGULAR DUCTWORK IS BEST FOR INSTALLATIONS MOUNTED IN CEILINGS, SUPPLYING OR EXTRACTING AIR AT A LOW NOISE LEVEL, INHERENT ATTENUATION MAY ALLOW ECONOMY OF ATTENUATORS ALONG DUCTS. LINING DUCTS IS PARTICULARLY EFFECTIVE FOR SMALL DUCTS AND FOR HIGHER FREQUENCIES.

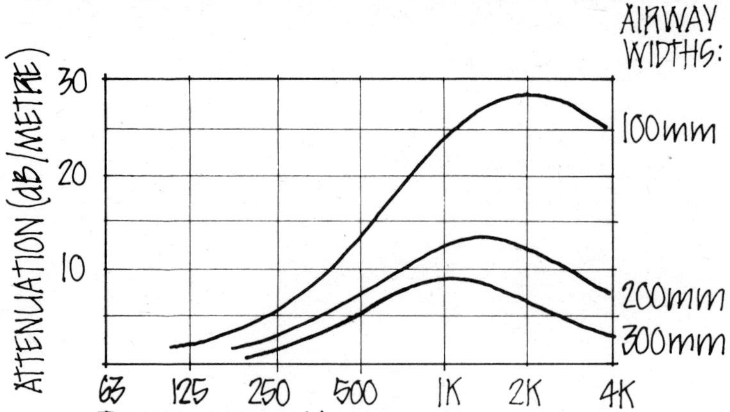

TYPICAL ATTENUATION FOR DUCTS LINED WITH 25 MINERAL WOOL

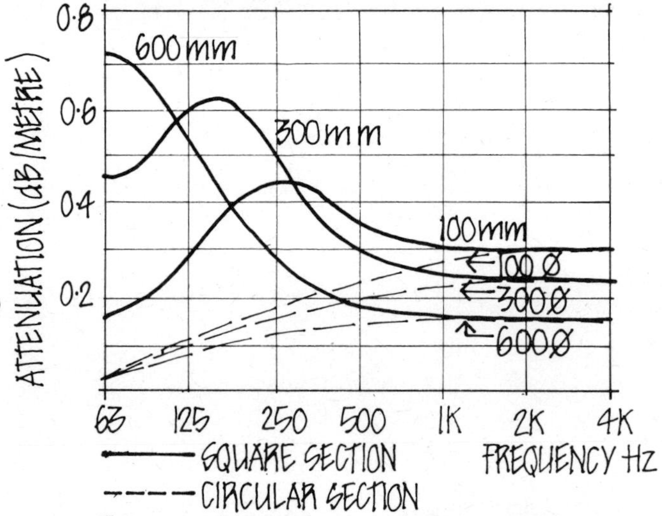

TYPICAL ATTENUATION ALONG UNLINED DUCTS.

AIR HANDLING DUCTWORK
SOURCE: PARKER HUMPHREYS & COWELL

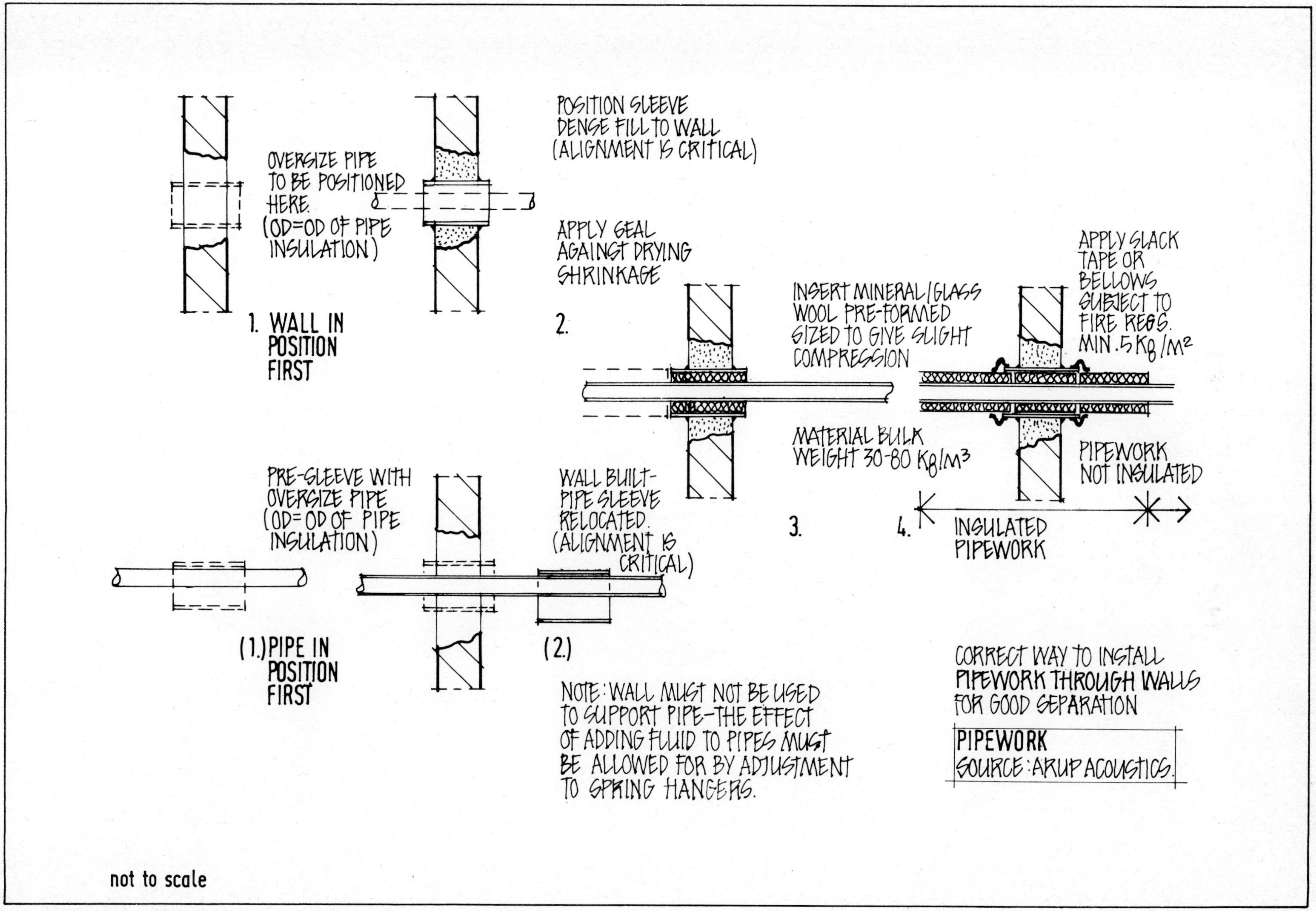

OVERSIZE PIPE TO BE POSITIONED HERE. (OD=OD OF PIPE INSULATION)
1. WALL IN POSITION FIRST
POSITION SLEEVE DENSE FILL TO WALL (ALIGNMENT IS CRITICAL)
APPLY SEAL AGAINST DRYING SHRINKAGE
2.
INSERT MINERAL/GLASS WOOL PRE-FORMED SIZED TO GIVE SLIGHT COMPRESSION
MATERIAL BULK WEIGHT 30-80 Kg/M3
3.
APPLY SLACK TAPE OR BELLOWS SUBJECT TO FIRE REGS. MIN. 5 Kg/M2
PIPEWORK NOT INSULATED
4.
INSULATED PIPEWORK
PRE-SLEEVE WITH OVERSIZE PIPE (OD=OD OF PIPE INSULATION)
(1.) PIPE IN POSITION FIRST
WALL BUILT-PIPE SLEEVE RELOCATED. (ALIGNMENT IS CRITICAL)
(2.)
NOTE: WALL MUST NOT BE USED TO SUPPORT PIPE—THE EFFECT OF ADDING FLUID TO PIPES MUST BE ALLOWED FOR BY ADJUSTMENT TO SPRING HANGERS.
CORRECT WAY TO INSTALL PIPEWORK THROUGH WALLS FOR GOOD SEPARATION
PIPEWORK
SOURCE: ARUP ACOUSTICS.
not to scale

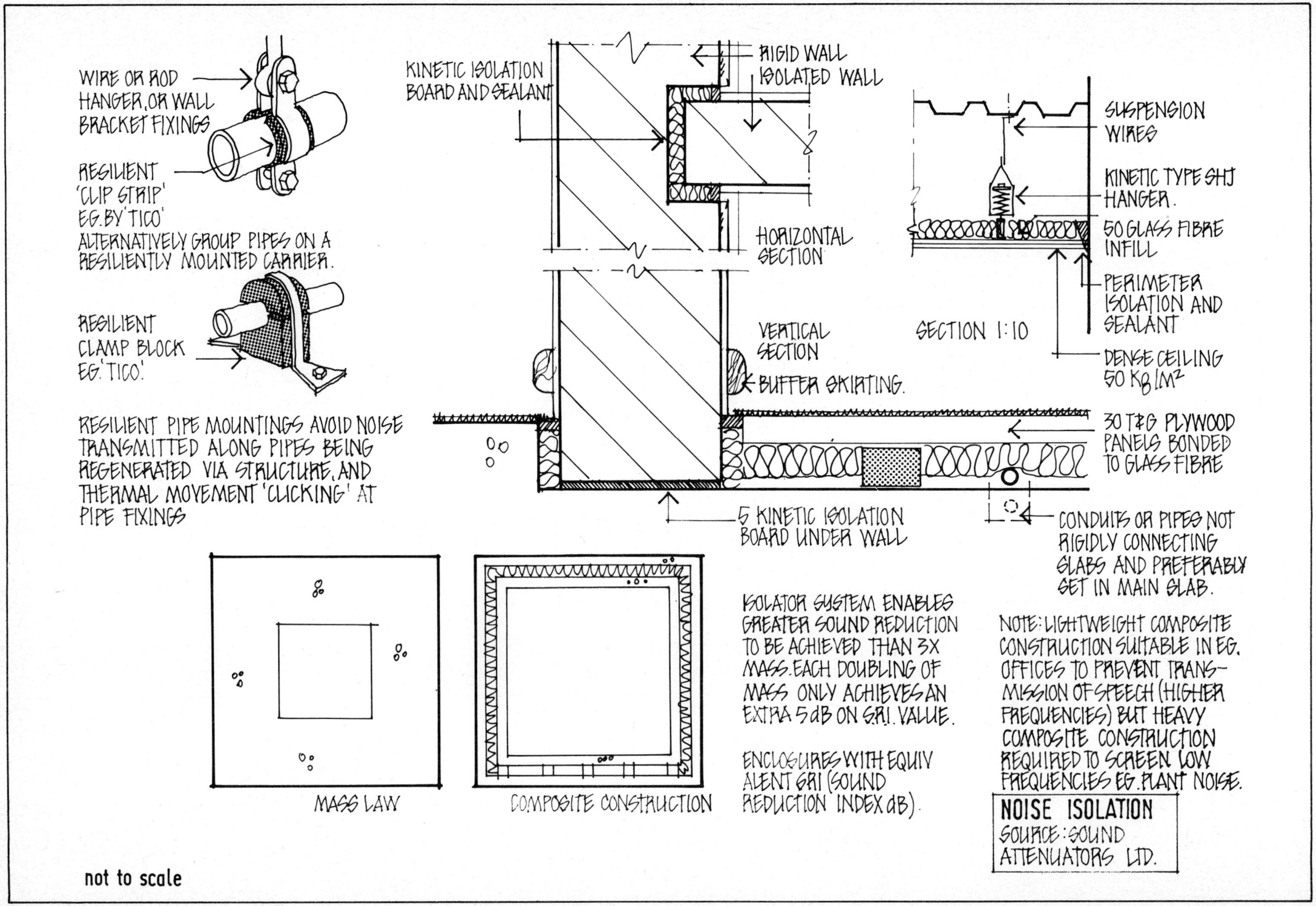
WIRE OR ROD HANGER, OR WALL BRACKET FIXINGS
RESILIENT 'CLIP STRIP' EG. BY 'TICO'
ALTERNATIVELY GROUP PIPES ON A RESILIENTLY MOUNTED CARRIER.
RESILIENT CLAMP BLOCK EG. 'TICO'.
RESILIENT PIPE MOUNTINGS AVOID NOISE TRANSMITTED ALONG PIPES BEING REGENERATED VIA STRUCTURE, AND THERMAL MOVEMENT 'CLICKING' AT PIPE FIXINGS
KINETIC ISOLATION BOARD AND SEALANT
RIGID WALL
ISOLATED WALL
HORIZONTAL SECTION
VERTICAL SECTION
BUFFER SKIRTING.
5 KINETIC ISOLATION BOARD UNDER WALL
SUSPENSION WIRES
KINETIC TYPE SHJ HANGER.
50 GLASS FIBRE INFILL
PERIMETER ISOLATION AND SEALANT
SECTION 1:10
DENSE CEILING 50 Kg/M²
30 T&G PLYWOOD PANELS BONDED TO GLASS FIBRE
CONDUITS OR PIPES NOT RIGIDLY CONNECTING SLABS AND PREFERABLY SET IN MAIN SLAB.
MASS LAW
COMPOSITE CONSTRUCTION
ISOLATOR SYSTEM ENABLES GREATER SOUND REDUCTION TO BE ACHIEVED THAN 3X MASS. EACH DOUBLING OF MASS ONLY ACHIEVES AN EXTRA 5dB ON S.R.I. VALUE.
ENCLOSURES WITH EQUIVALENT SRI (SOUND REDUCTION INDEX dB).
NOTE: LIGHTWEIGHT COMPOSITE CONSTRUCTION SUITABLE IN EG. OFFICES TO PREVENT TRANSMISSION OF SPEECH (HIGHER FREQUENCIES) BUT HEAVY COMPOSITE CONSTRUCTION REQUIRED TO SCREEN LOW FREQUENCIES EG. PLANT NOISE.
NOISE ISOLATION
SOURCE: SOUND ATTENUATORS LTD.
not to scale

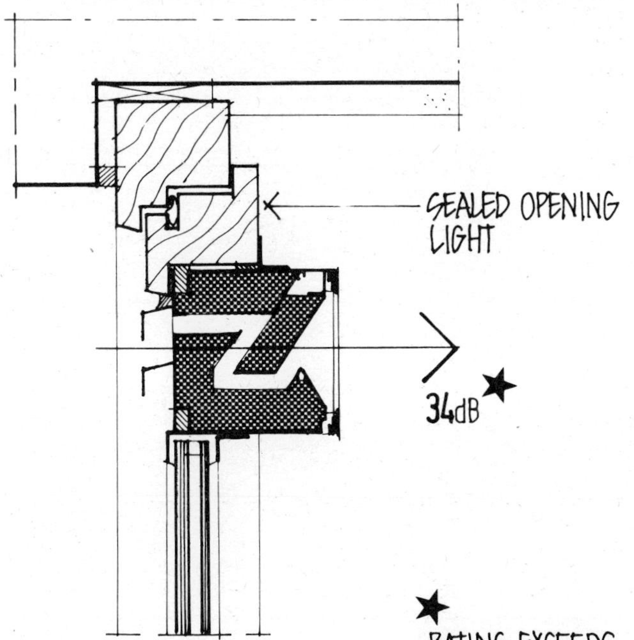

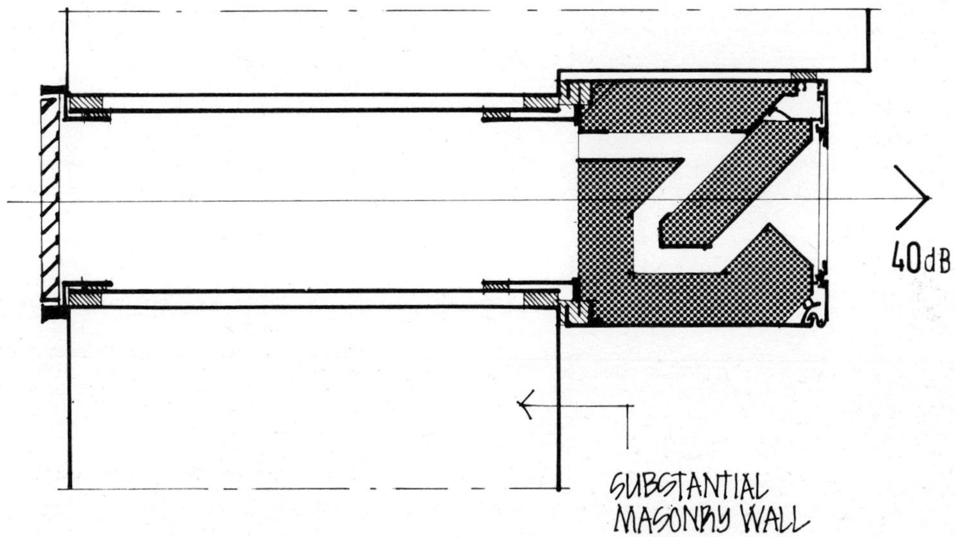

★ RATING EXCEEDS PATH VIA GLAZING

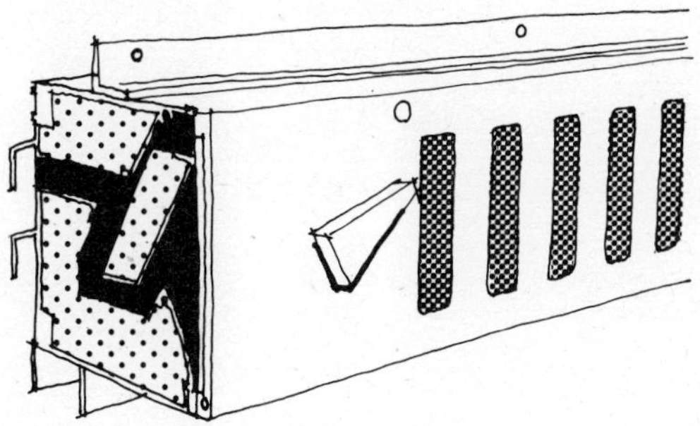

MODEL Z100F

LINEAR SOUND-ABSORBERS VENTILATORS ANODISED ALUMINIUM BODIES. VENTILATION POTENTIAL UP TO 200 M³ OF AIR/HOUR/METRE LENGTH. SEE ALSO VENTILATOR SHOWN UNDER 'HOUSING.' USE IN OFFICES, HOUSING.

MODEL ZR 150

SECTIONS
SECTIONAL SIZES
100 X 100, 150 X 150

VENTILATORS
SOURCE: GU. UNITAS.

not to scale

TUNGSTEN

TUNGSTEN LAMPS GIVE OFF NEGLIGIBLE NOISE EXCEPT WHEN ABOUT TO FAIL: THE FILAMENTS THEN 'RATTLE'. TUNGSTEN LIGHTING IS THEREFORE USED FOR ANECHOIC CHAMBERS AND OTHER NOISE SENSITIVE ROOMS. DIMMER CONTROLS ON LIGHTING MAY CAUSE SOME NOISE AT LOW ILLUMINATION SETTING.

MERCURY DISCHARGE.

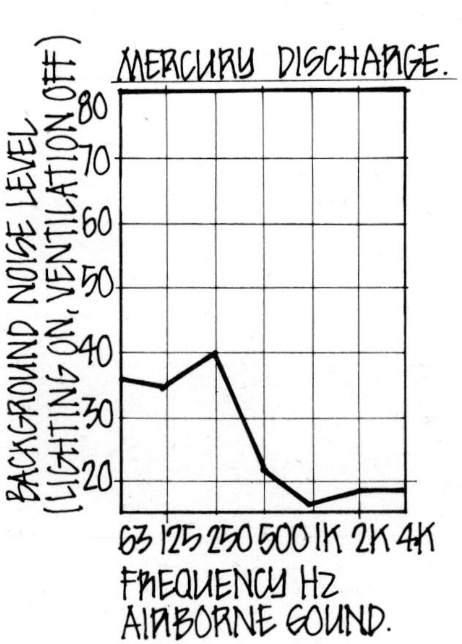

POWERFUL LIGHTING MAY CAUSE SIGNIFICANT NOISE AT SOME PARTICULAR FREQUENCY. THE DISTRIBUTION AT LHS IS A TOTAL OF 55 400 W LUMINAIRES (TYPE HPI MBI SPECIAL SPORTS LIGHTS) MOUNTED UNDER HIGH LEVEL CATWALKS. IN A MULTI-PURPOSE SPORTS HALL OF 11300 CU.M. AND RT=2 SECONDS AT MID FREQUENCY. AMBIENT SOUND LEVEL FOR SPORTS LIGHTING ON, HOUSE LIGHTS AND VENTILATION OFF CORRESPONDS TO NR=30.

SOURCE: SANDY BROWN ASSOCIATES.

FLUORESCENT.

THERE IS A GREAT VARIATION OF NOISE FROM THE MAIN SOURCE, THE CONTROL GEAR, EVEN FOR FITTINGS OF THE SAME TYPE FROM THE SAME MANUFACTURER. THE LAMINATED IRON CORE PRODUCES NOISE DUE TO MAGNETISATION OF THE LAMINATIONS, VARYING ACCORDING TO THE TYPE OF CIRCUIT AND WATTAGE RATING. THE METHOD OF MOUNTING FITTINGS AFFECTS NOISE OUTPUT: HANGING FITTINGS ARE PREFERABLE TO SOLID FIXING. IN SENSITIVE AREAS, REMOTE CONTROL GEAR MAY HAVE TO BE CONSIDERED. INDIVIDUAL FITTINGS MAY NOT BE NOISY. FOR INSTANCE A FITTING RECENTLY TESTED HAD A WORST CASE SWL OF 23.3 dB AT 630 Hz.

HOWEVER, 32 SUCH FITTINGS IN A ROOM AS A TYPICAL ARRANGEMENT WERE CALCULATED TO PRODUCE A SOUND LEVEL APPROXIMATING NR=25 AT 630 Hz

SOURCE: UNIVERSITY OF SALFORD

LIGHTING

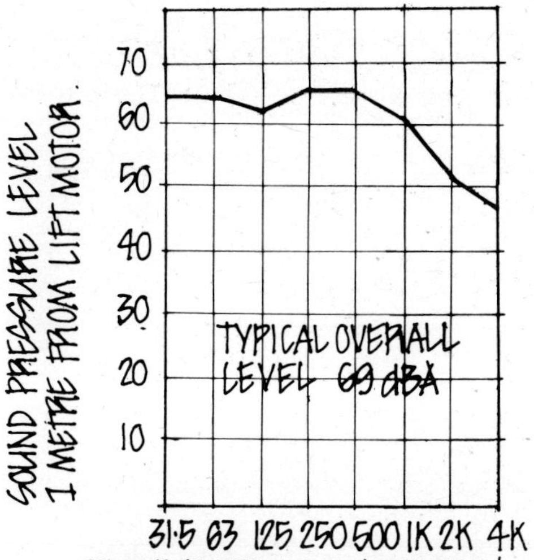

SINGLE HYDRAULIC LIFT MOTOR/PUMP UNIT.

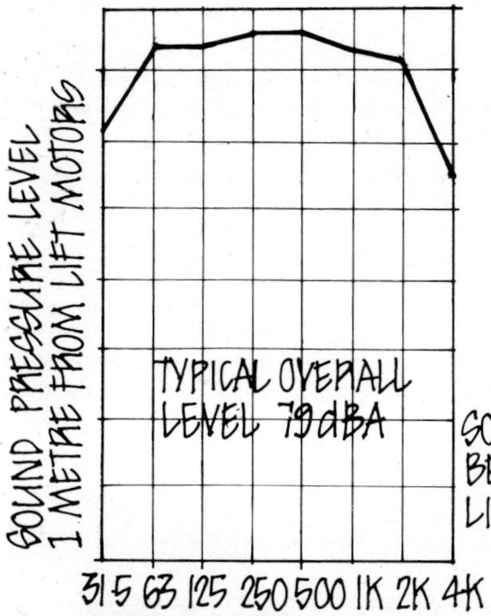

SOURCE: BENNIE LIFTS LTD

DOUBLE HYDRAULIC LIFT MOTOR/PUMP UNITS

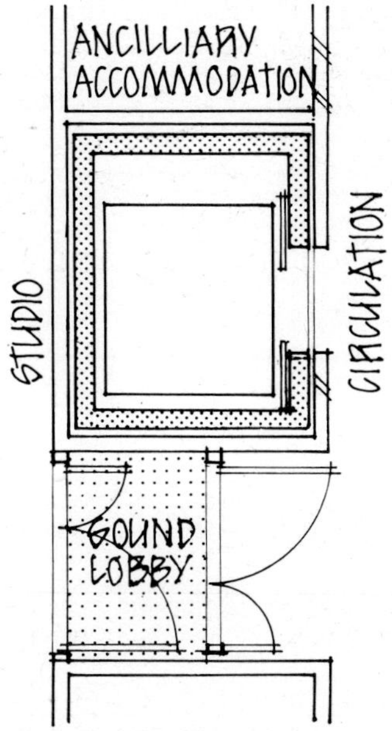

TYPICAL ISOLATED SHAFT PLAN ARRANGEMENT.

1. HYDRAULIC LIFT MOTORS ARE MORE NOISY THAN ELECTRIC MOTORS: MINIMISE BY SPECIFYING PUMP AND MOTOR TO BE SUBMERSIBLE TYPE AND TO HAVE CASINGS LINED WITH SOUND DEADENING MATERIAL.
2. SELECT LIFT DOORS FOR QUIET OPERATION
3. CHECK HIGH SPEED/HIGH RISE LIFTS FOR WIND NOISE UPSHAFT
4. ALL MACHINERY TO BE ON ANTI-VIBRATION MOUNTINGS
5. AVOID BUILDERS WORK HOLES BETWEEN LIFT SHAFTS AND OTHER AREAS.
6. CHOOSE ELECTRONIC PROXIMITY SWITCHES ETC. AS A WAY OF AVOIDING 'CLICKING' WORKING PARTS IN CONTROL GEAR
7. ENSURE GOOD MAINTENANCE OF LIFTS.
8. IN REFURBISHMENT SCHEMES, CHECK NOISE FROM EXISTING LIFT INSTALLATIONS (OLDER TYPES, ESPECIALLY OPEN SHAFT, WERE FAR NOISIER.)

LIFT NOISE IS OF PARTICULAR CONCERN IN HOTELS OR HOSPITALS WHERE THE NOISE LEVELS FROM MACHINERY MAY NOT BE HIGH, BUT THE DISCONTINUITY OF OPERATION DRAWS ATTENTION TO THE NOISE. TYPICAL SOUND LEVELS WITHIN LIFT CAR ARE 65 dB FOR HYDRAULIC, 60 dB FOR ELECTRIC OPERATION.

LIFTS

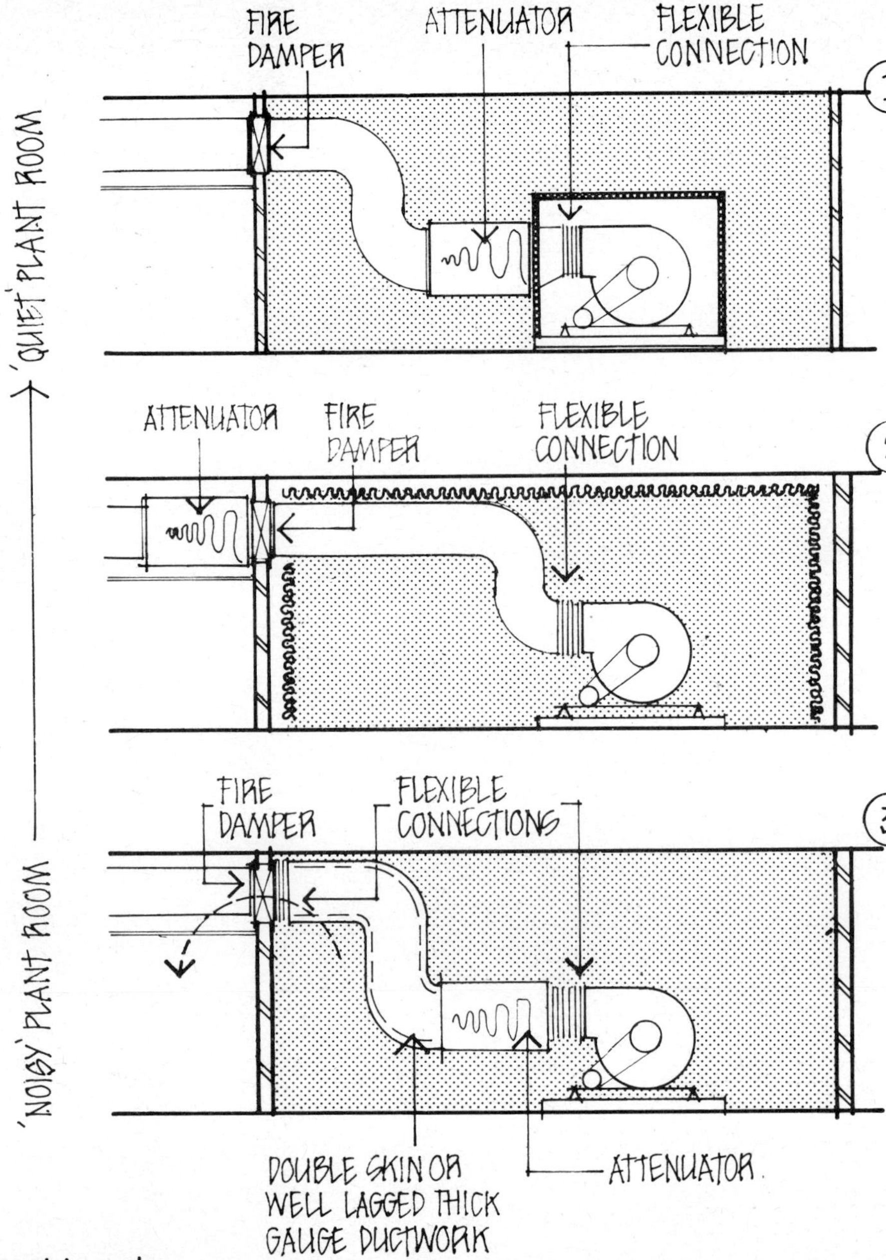

1. NOISE ENCLOSED AT SOURCE

ACOUSTIC ENCLOSURES TO NOISY ITEMS OF PLANT WITHIN PLANT ROOM

A.V. MOUNTS TO ISOLATED BASE.

GOOD TO ENCLOSE NOISE AT SOURCE BUT THIS MAY NOT BE POSSIBLE IF THERE ARE MANY NOISE SOURCES, OR PLANT ITEMS REQUIRING COMPLETE ACCESS AND VENTILATION TO ALL SIDES.

2. PLANT ROOM LINED WITH SOUND ABSORBENT QUILT SPACED OFF WALLS.

ENCLOSURE OF INDIVIDUAL ITEMS OF PLANT MAY NOT BE PRACTICAL. LINING OF PLANT ROOM MAY REDUCE REVERBERANT SOUND.

3. SUBSTANTIAL ROOM ENCLOSURE ENSURES ISOLATION. GOOD INSULATION VALUE TO WALLS, ROOF, FLOOR: SEPARATE STRUCTURE IF POSSIBLE. AIRBORNE SOUND LEVEL INSIDE PLANTROOM HIGH, PARTICULARLY WITH AMPLIFICATION FACTOR OF HARD FINISHES WHICH INCREASE NOISE LEVEL WITHIN BY UP TO 9 dB.

PARTICULAR CARE RE. NOISE BREAKOUT VIA DUCTWORK NECESSARY. FOR VERY NOISY PLANT AND SENSITIVE ROOMS AROUND ALL MEASURES MAY BE NECESSARY: ENCLOSURE OF PLANT, LININGS TO PLANT ROOM, SUBSTANTIAL FABRIC.

PLANTROOMS

not to scale

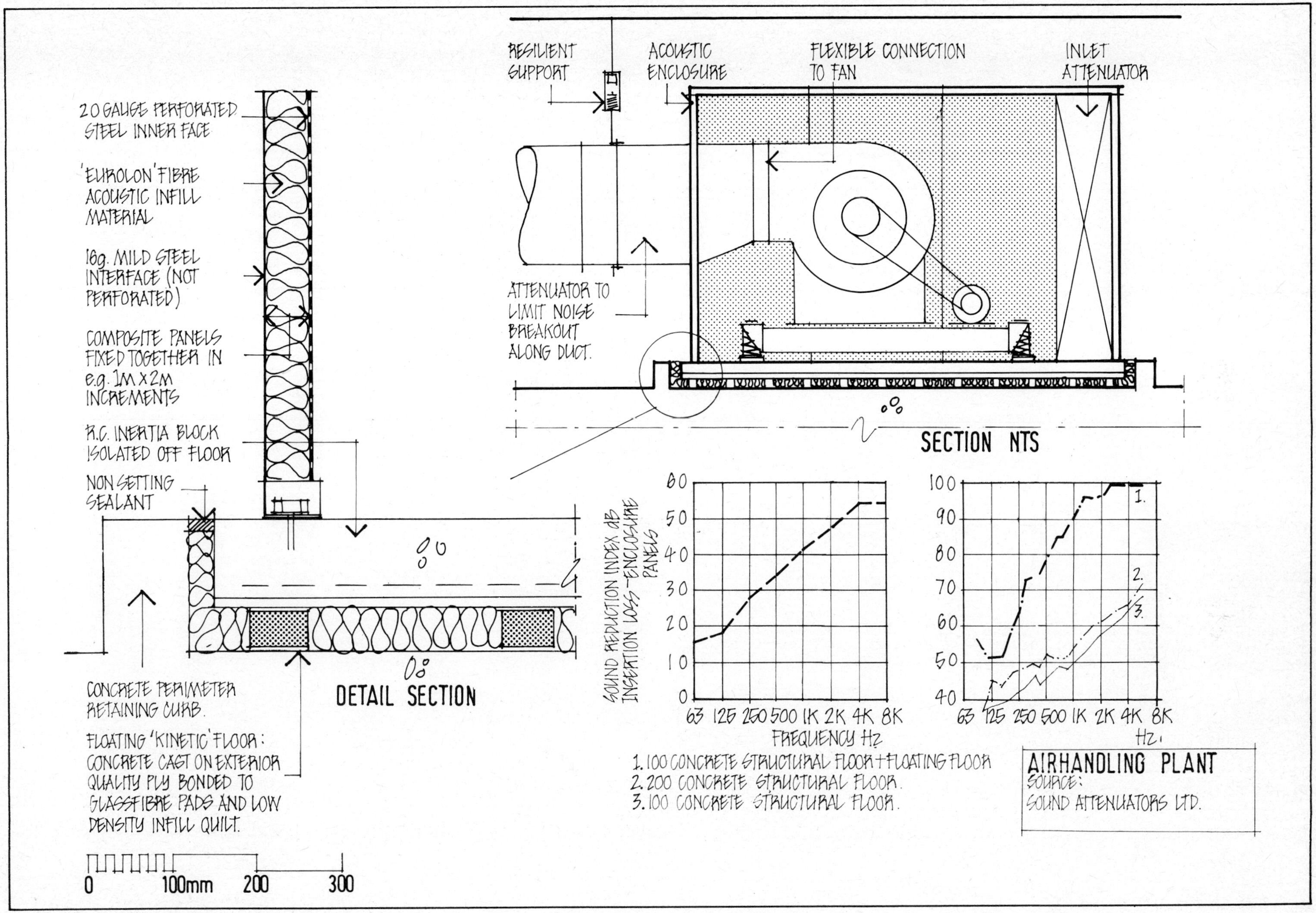
RESILIENT SUPPORT
ACOUSTIC ENCLOSURE
FLEXIBLE CONNECTION TO FAN
INLET ATTENUATOR
ATTENUATOR TO LIMIT NOISE BREAKOUT ALONG DUCT.
SECTION NTS
20 GAUGE PERFORATED STEEL INNER FACE
'EUROLON' FIBRE ACOUSTIC INFILL MATERIAL
16g. MILD STEEL INTERFACE (NOT PERFORATED)
COMPOSITE PANELS FIXED TOGETHER IN e.g. 1M x 2M INCREMENTS
R.C. INERTIA BLOCK ISOLATED OFF FLOOR
NON SETTING SEALANT
CONCRETE PERIMETER RETAINING CURB.
FLOATING 'KINETIC' FLOOR: CONCRETE CAST ON EXTERIOR QUALITY PLY BONDED TO GLASSFIBRE PADS AND LOW DENSITY INFILL QUILT.
DETAIL SECTION
0 100mm 200 300
SOUND REDUCTION INDEX dB INSERTION LOSS - ENCLOSURE PANELS
0 10 20 30 40 50 60
63 125 250 500 1K 2K 4K 8K
FREQUENCY Hz
40 50 60 70 80 90 100
1. 2. 3.
63 125 250 500 1K 2K 4K 8K
Hz.
1. 100 CONCRETE STRUCTURAL FLOOR + FLOATING FLOOR
2. 200 CONCRETE STRUCTURAL FLOOR.
3. 100 CONCRETE STRUCTURAL FLOOR.
AIRHANDLING PLANT
SOURCE:
SOUND ATTENUATORS LTD.

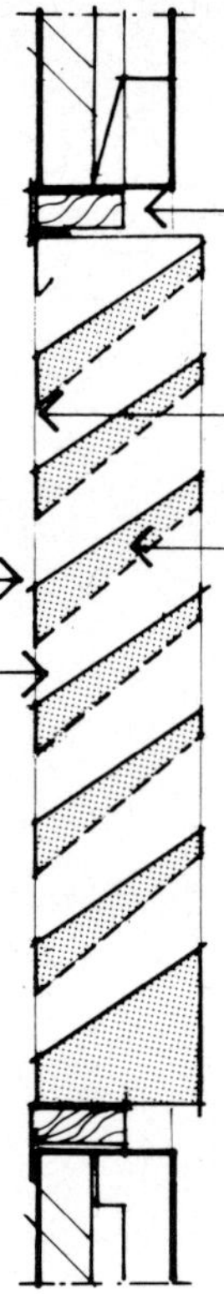

300 DEEP ACOUSTIC LOUVRE

LOUVRES CAN ALSO BE USED IN DOORS OR AS ENCLOSING PANELS

SOURCE:
BESTOBELL ACOUSTICS LTD (150 AND 300 DEEP)
SOUND SOLUTIONS LTD (600 DEEP).
PERFORMANCE DEPENDS TO A DEGREE ON PANEL SIZE AND AIR VELOCITY: OVER 4 M/SEC NOT RECOMMENDED.

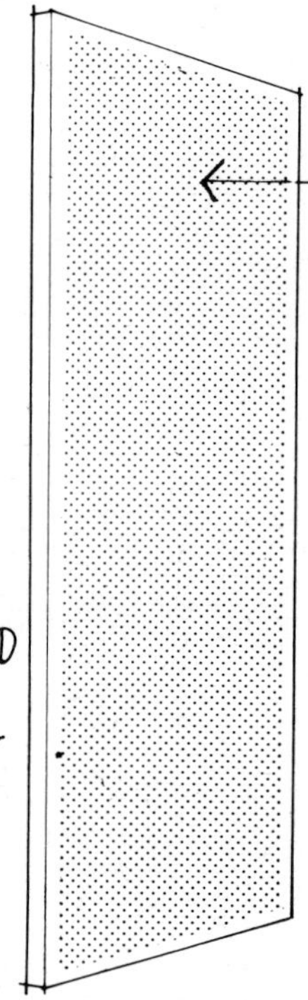

'GULLFIBER' MODULAR ACOUSTIC SCREEN 2m X 1m X 40
FINISH: YELLOW PAINTED STEEL.
PERFORATIONS BOTH SIDES (ONE SIDE PERFORATED OPTIONAL)

·15 ·40 ·80 ·90 ·90 ·85

SOURCE: LUND INSTITUTE OF TECHNOLOGY

(INSERTION LOSS TEST BY ROOM METHOD TO ISO 354)

INSERTION LOSS IS DEFINED AS THE SOUND LEVEL DIFFERENCE BEFORE AND AFTER INSERTING A SOUND ABSORBER OR BARRIER BETWEEN NOISE SOURCE AND REFERENCE POINT.

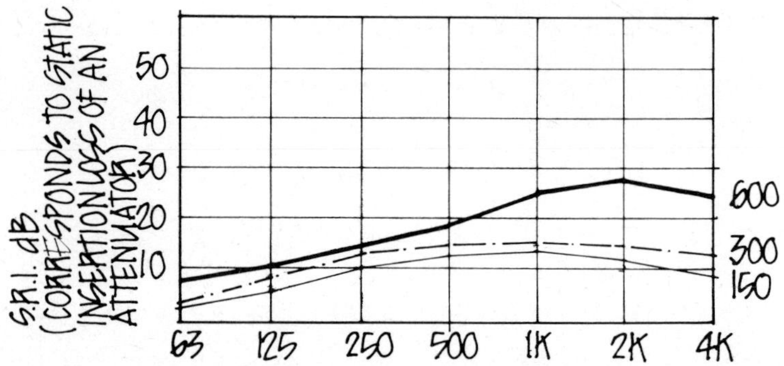

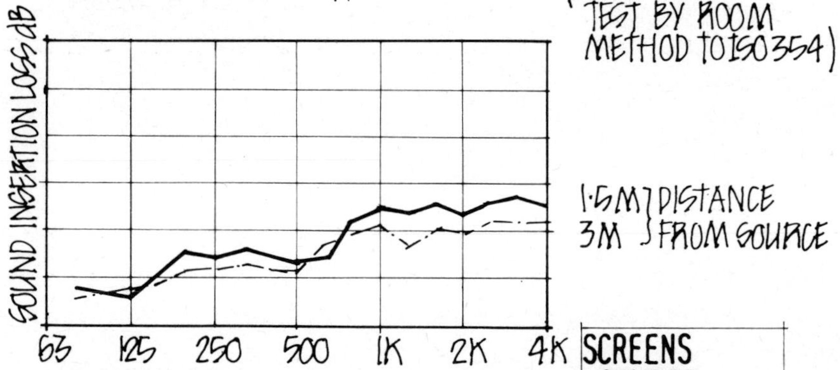

SCREENS
LOUVRES

BUILDING TYPES

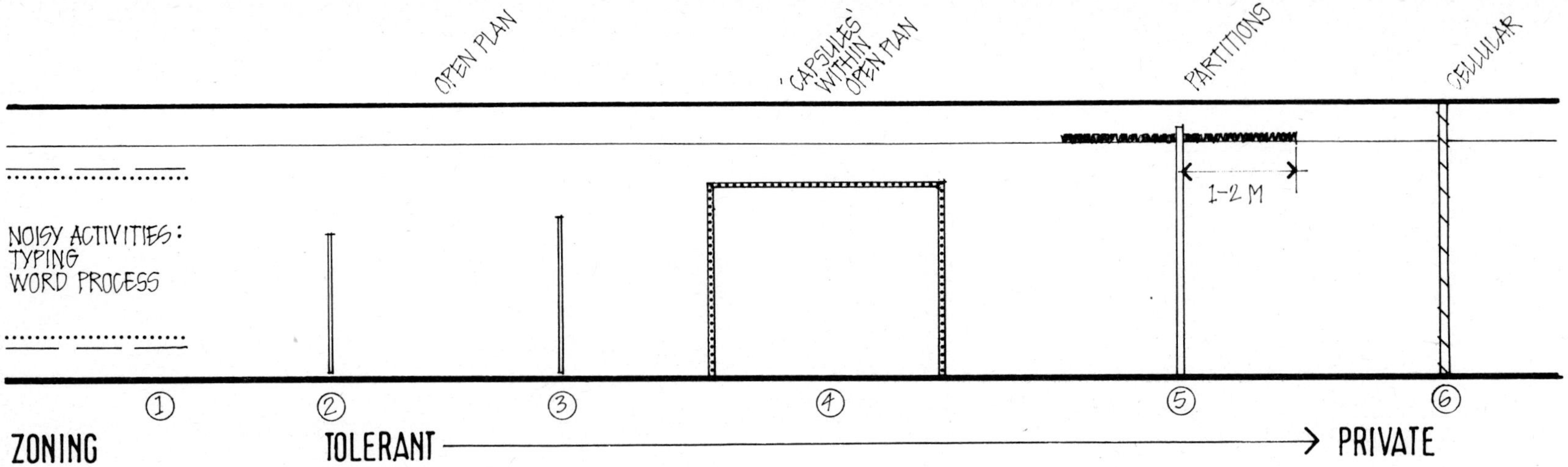

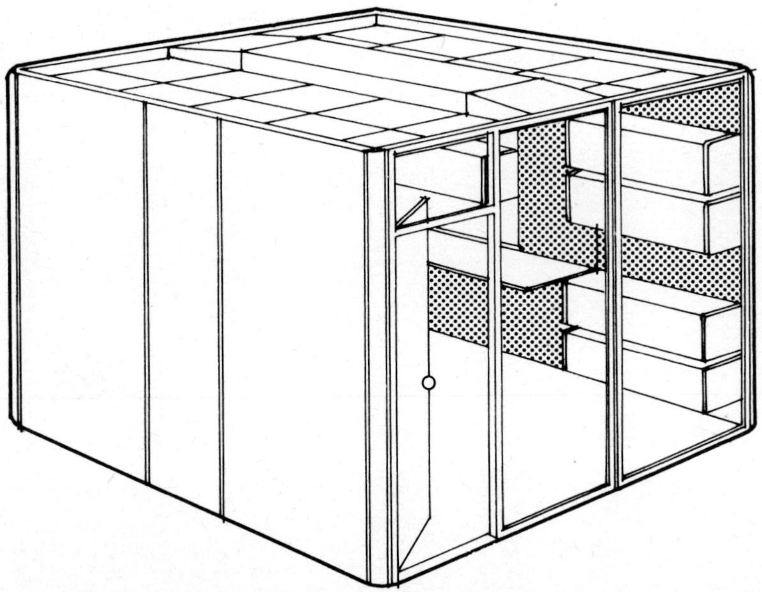

HERMAN MILLER EXECUTIVE OFFICE LOCATED WITHIN AN OPEN PLAN OFFICE

OPTIONS IN SELECTING COMPONENT ASSEMBLIES

① GROUP NOISY ACTIVITIES AND ISOLATE OR SCREEN. INCLUDE LOTS OF ABSORPTION.

② SCREENS TO HEAD HEIGHT, 1800 mm.

③ SCREENS TO GREATER HEIGHT, 2100 OR 2400 mm. SEPARATION BY SCREENING < 20 dBA BETWEEN WORK STATIONS

④ SENIOR MANAGEMENT OR CONFIDENTIAL MEETING ROOMS WITHIN OVERALL SPACE. CAN BE MOVED. SEPARATE VENTILATION OR TRANSFER VENTILATION.

⑤ LIGHT PARTITIONS STOPPED OFF AT CEILING. DENSE QUILT OR LEAD FOIL IN CEILING BY PARTITION REDUCES CEILING VOID FLANKING.

OPTIONS IN SELECTING COMPONENT ASSEMBLIES.

OFFICES

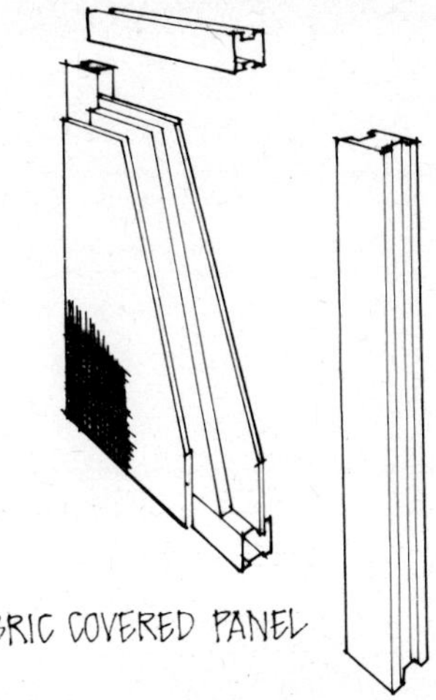

FABRIC COVERED PANEL

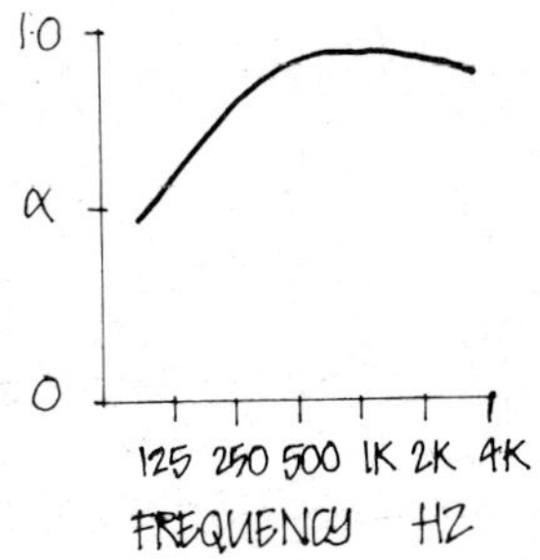

①

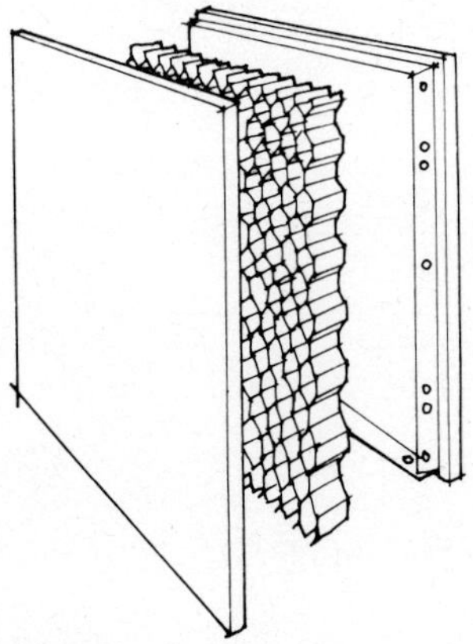

METAL PANEL

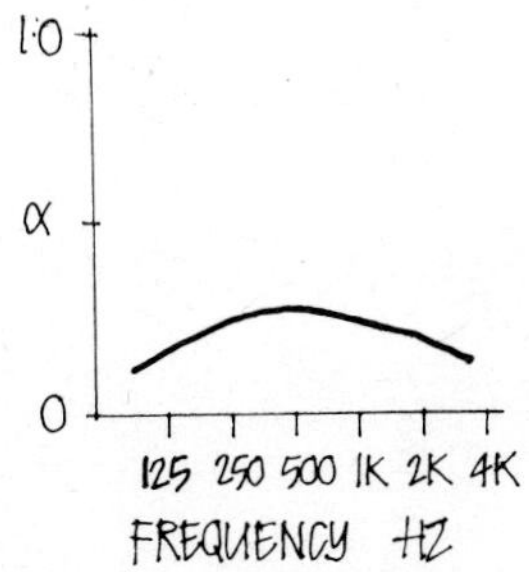

②

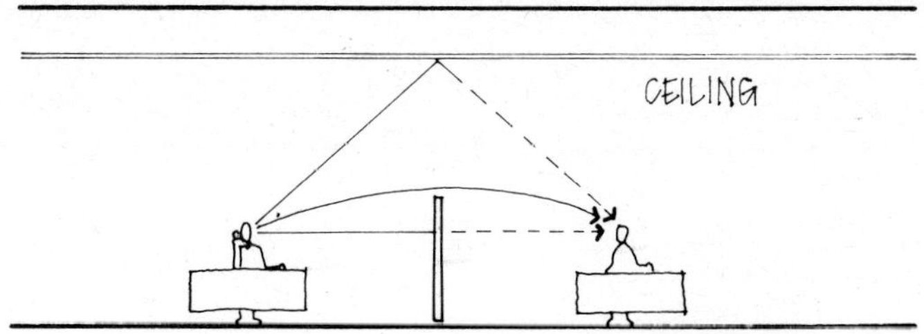

SECTION

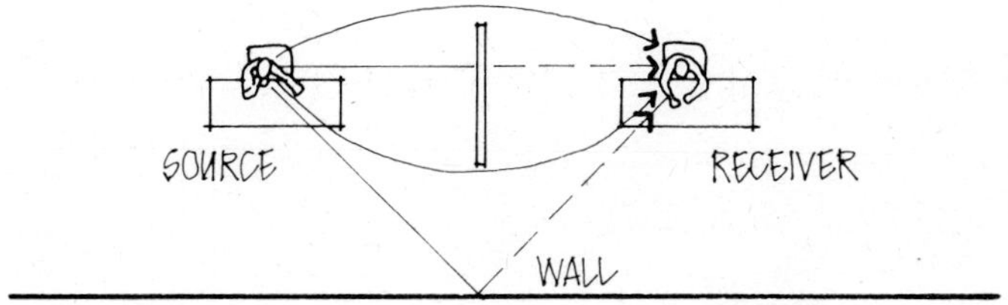

PLAN

FOR A REASONABLE SPACING OF OCCUPANTS EG. 12-14 M^2 / PERSON SOUND ATTENUATION BETWEEN WORKPLACES ONLY 20-25 dBA MAX, EMPIRICAL FORMULA TO CHECK ADEQUATE PRIVACY: BACKGROUND NOISE (dBA) + SOUND ATTENUATION $\geqslant$ 75. BY THIS CRITERIA MANY OFFICES HAVE AN AMBIENT LEVEL TOO LOW AND MASKING SOUND MAY BE CONSIDERED.

IN CHOOSING PARTITION SYSTEMS OR DIVIDER PANELS, LOOK FOR SANDWICH OF ABSORPTION-SOLID CORE-ABSORPTION IF A GOOD ABSORPTION RATING (OVER 0·6) IS REQUIRED. BOTH EXAMPLE 1 AND 2 ABOVE ARE EFFECTIVE AS SCREENS. GOOD ABSORPTION AT 1-2 K IMPORTANT TO MASK SPEECH.

OPEN PLAN OFFICES

1. 'ARENA' EFFECT ALLOWS PEOPLE TO WORK ON THE PERIMETER OF A SPACE BUT TURN INWARDS FOR DISCUSSION.

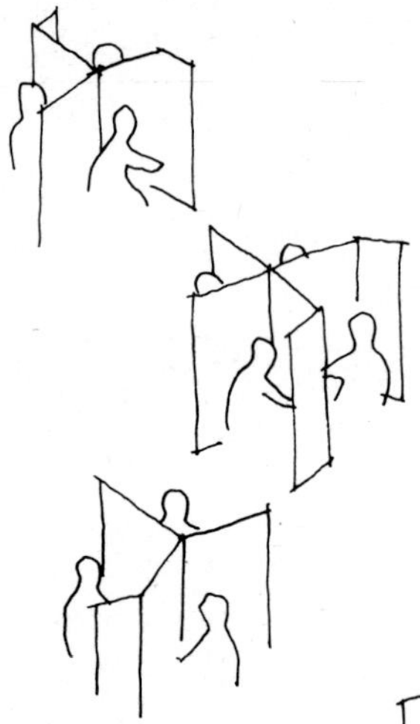

2. 'SPOKE' LAYOUTS CAUSE UNCOMFORTABLE BACK EXPOSURE AND NOISE INTRUSION FROM NEAR BUT HIDDEN NEIGHBOURS

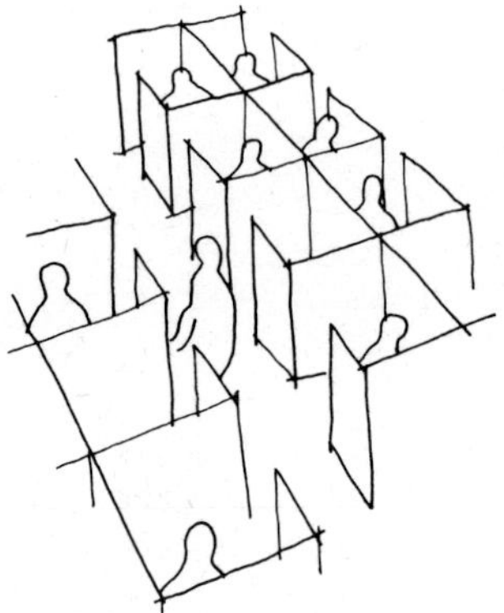

3. 'CUBICLES' GIVE HIGH DENSITY AND INVISIBILITY BUT LITTLE ACOUSTIC PRIVACY

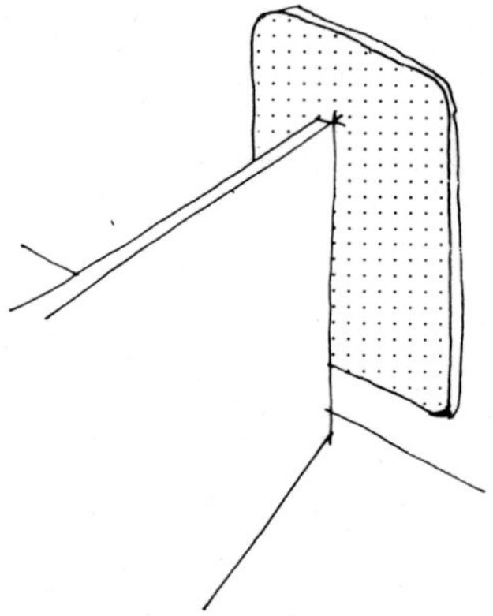

TO AVOID REFLECTED SOUND AT PERIMETER CURTAIN WALL, TRAP SOUND BY ENCLOSING ABSORPTIVE END PANELS.

LAYOUT TYPE	
FREEFLOW	12%
PLANNED TO WORKFLOW AND RELATIONSHIPS	60%
DIVISION BY ACOUSTIC SCREENS	44%
DIVISION BY FURNITURE	32%
DIVISION BY HALF HEIGHT PARTITIONS	16%

STAFF ATTITUDES	
NATURE OF COMMON COMPLAINTS AND REQUESTS:	
MORE PRIVACY	42%
INTERNAL NOISE	42%
TEMPERATURE	42%
VENTILATION	38%
TOO MUCH DISTRACTION	25%
DRAUGHTS	17%
BETTER FURNITURE	12%
BETTER LAYOUT	8%

RESULTS OF A LARGE U.K. OFFICE USE SURVEY
SOURCE: SPACE PLANNING SERVICES

ABILITY TO CONVERSE AS A FUNCTION OF BACKGROUND NOISE

NOISE LEVEL (NR)	MAX DISTANCE FOR NORMAL CONVERSATION (M) ABSORBENT	REFLECTING	Nº PEOPLE IN CONFERENCE
30	8	50	50
35	6	25	20
40	4	10	10
50	2	2	4
55	1	1	2

SOURCE: R.A. WALLER. APPLIED ACOUSTICS (2) 1961 PP 121–130

OPEN-PLAN OFFICES

SOURCE: HERMAN MILLER RESEARCH CORPORATION.

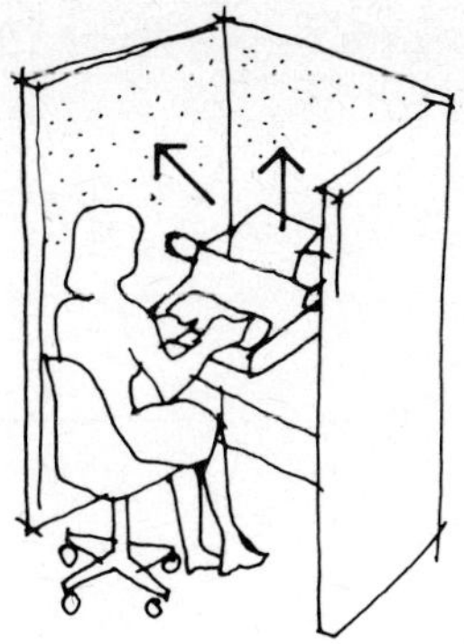

ENCLOSE NOISY ACTIVITIES
USE SCREENS EXTENDING
TO FLOOR.

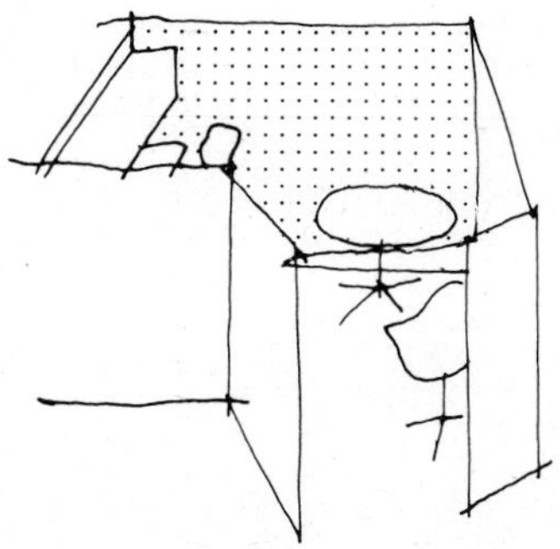

L-SHAPED WORKSPACES
AFFORD PRIVACY

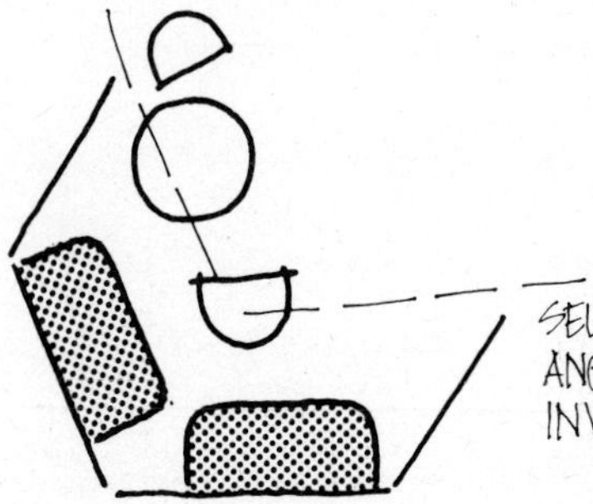

SELECTION OF WORK
ANGLES ALLOWS VARIETY.
INWARD FACE FOR PRIVACY.

DESIGN GUIDE

1. DESIGN IN A SPACE OVER 11 METRES WIDE.
2. PLAN 12 M^2/PERSON INCLUDING CIRCULATION
3. ZONE OFFICE MACHINES APART AND SCREEN.
4. DEVELOP 'ARENA' ORIENTATION LAYOUTS
5. CONTAIN BACKGROUND SOUND LEVEL WITHIN RANGE 40-50dBA
6. DEVELOP GOOD SCREEN LAYOUTS AVOID 'SPOKES' AND 'CUBICLES'.
7. LOCATE ACOUSTIC PANELS AROUND SOUND SOURCES TO BE CONTROLLED. WITHIN 1200.
8. DESIGNATE 1/3 PANELS TO BE ACOUSTIC RATHER THAN HARD SURFACE.
9. CHOOSE HIGH (2100) PANELS WHERE WORKERS STAND OR CIRCULATE
10. CHOOSE A HIGH ABSORPTION CEILING
11. CHOOSE A THICK CARPET ON A DENSE UNDERLAY.
12. CONSIDER WHETHER MASKING BACKGROUND SOUND MAY BE NECESSARY
13. MINIMISE NOISE FROM IMPULSE OR INTERMITTENT SOURCES-TELEPHONES, TYPEWRITERS, VDUS.
14. CHECK FOR SPEECH PRIVACY: GOOD IF A LISTENER 5 METRES FROM A PERSON READING AT NORMAL CONVERSATIONAL LEVEL CAN ONLY UNDERSTAND 10% OF SPEECH. (STANDARD SPEECH ARTICULATION TEST). INCREASE ABSORPTIVE SCREENS IF NECESSARY.
15. CHECK RT: GENERAL OFFICE DESIRED MID FREQUENCY 0·70 SECONDS. EXECUTIVE OFFICES 0·5 SEC, OPEN PLAN 0·45.

OPEN-PLAN OFFICES
SOURCE: HERMAN MILLER
RESEARCH CORPORATION.

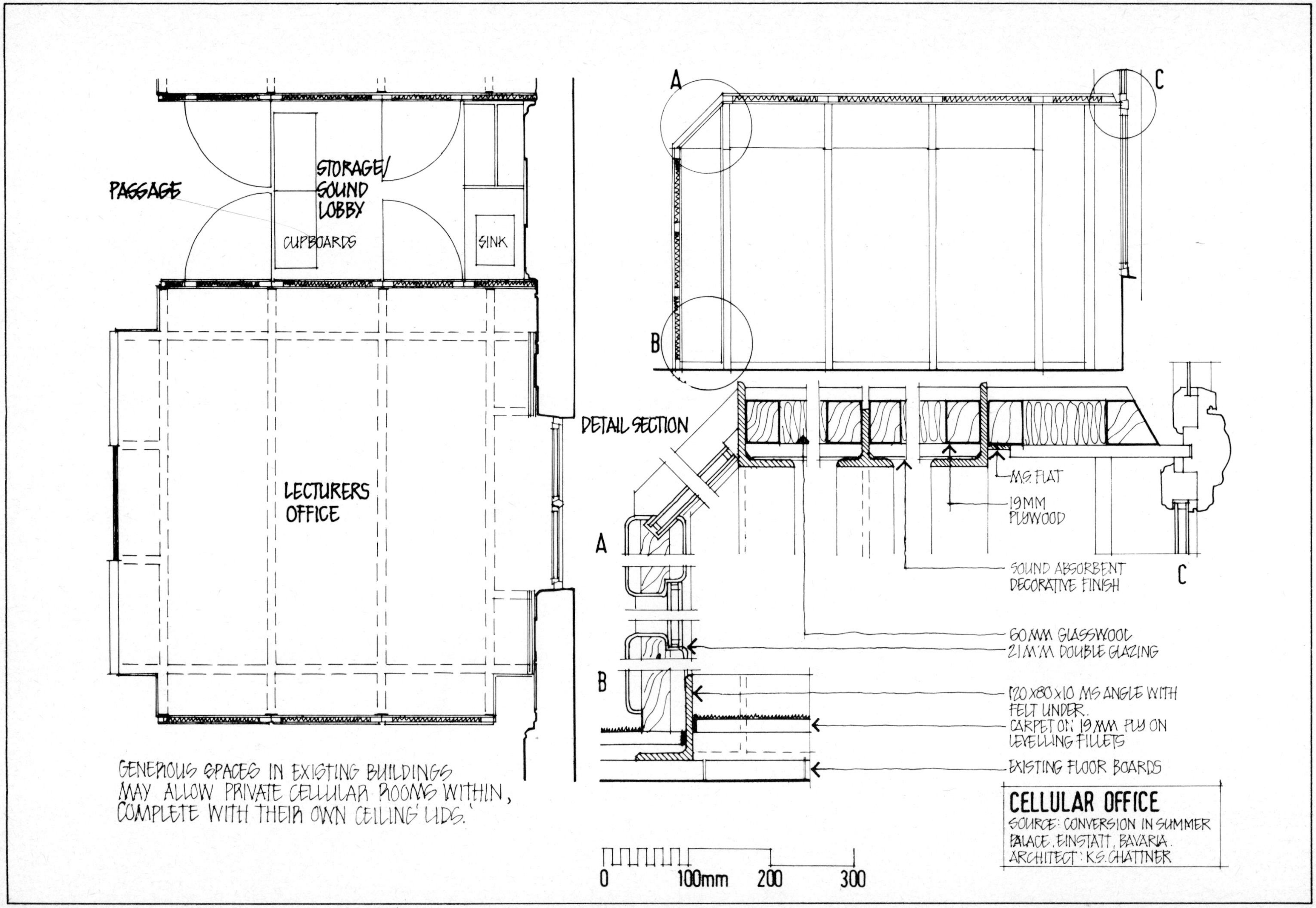
PASSAGE
STORAGE/ SOUND LOBBY
CUPBOARDS
SINK
LECTURERS OFFICE
DETAIL SECTION
A
B
C
M.S. FLAT
19MM PLYWOOD
SOUND ABSORBENT DECORATIVE FINISH
60MM GLASSWOOL
21MM DOUBLE GLAZING
120 X80 X10 MS ANGLE WITH FELT UNDER.
CARPET ON 19MM PLY ON LEVELLING FILLETS
EXISTING FLOOR BOARDS
GENEROUS SPACES IN EXISTING BUILDINGS MAY ALLOW PRIVATE CELLULAR ROOMS WITHIN, COMPLETE WITH THEIR OWN CEILING 'LIDS.'
0
100mm
200
300
CELLULAR OFFICE
SOURCE: CONVERSION IN SUMMER PALACE, EINSTATT, BAVARIA.
ARCHITECT: K.S. CHATTNER

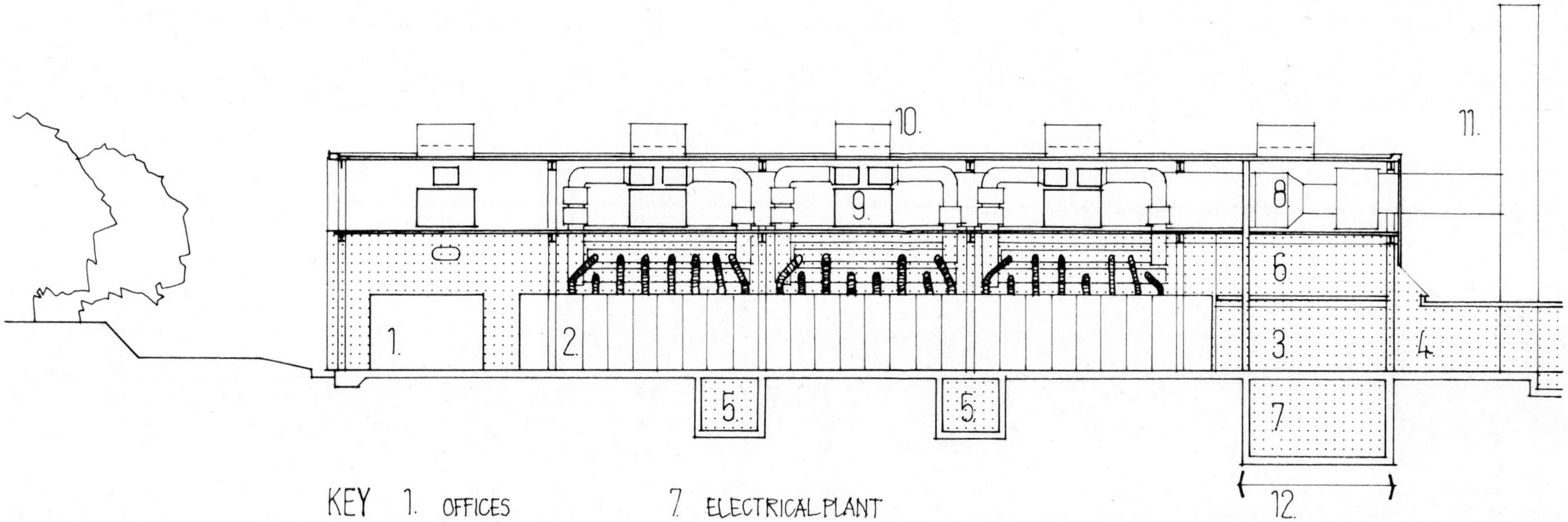

KEY

1. OFFICES
2. LABORATORY BOX
3. TOILETS
4. LINK
5. DRAINAGE
6. MECHANICAL PLANT
7. ELECTRICAL PLANT
8. EXTRACT PLANT
9. SUPPLY PLANT
10. AIR INTAKES
11. EXHAUST SHAFT
12. SERVICE CORE

NOISE ZONING:
SMALLER ACTIVITIES CARRIED ON WITHIN LARGE SPACES BY ENCLOSURE WITHIN THE OVERALL SPACE.

LABORATORY (1)
SOURCE: BDP.

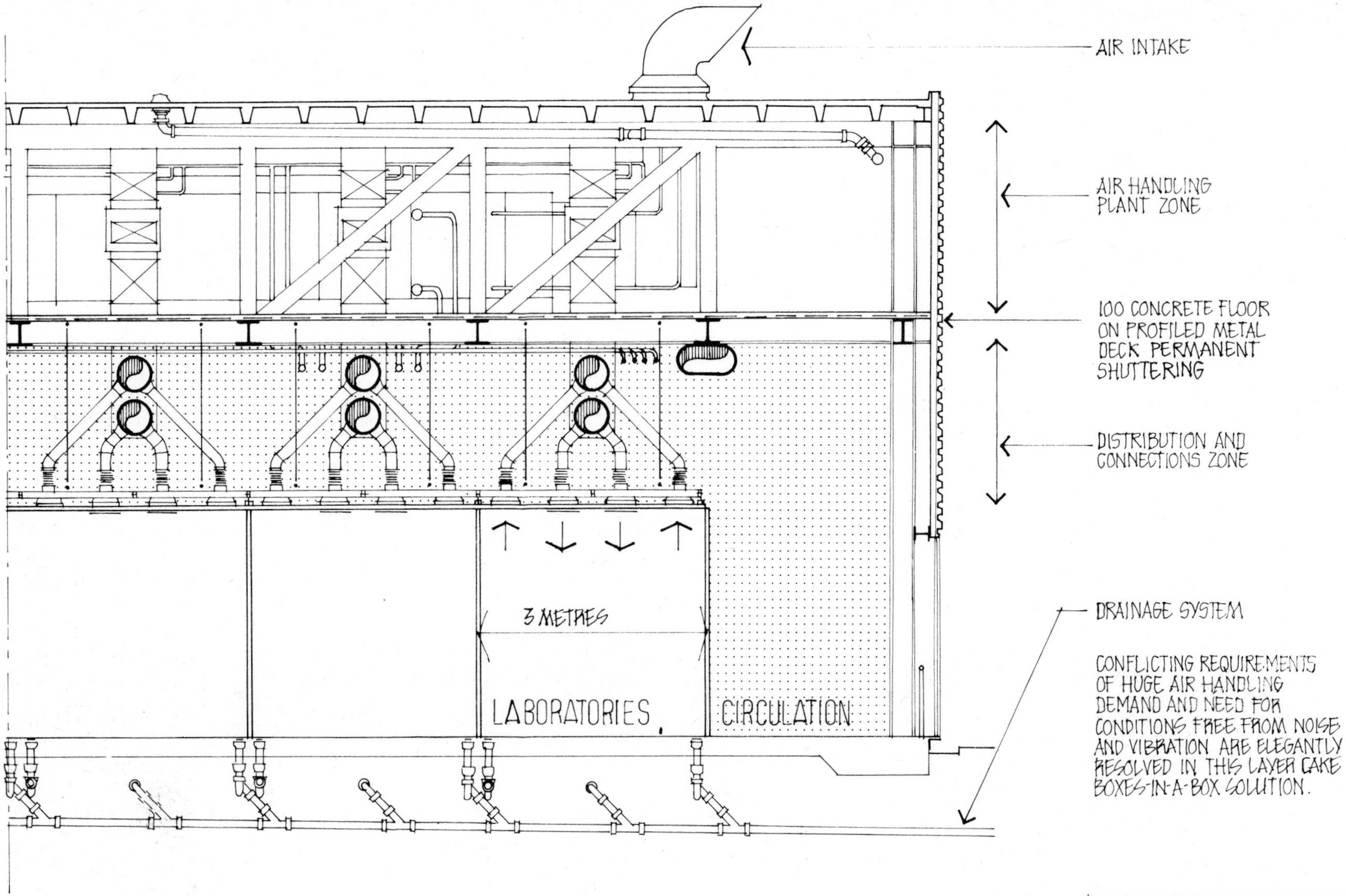

LABORATORY (2)
SOURCE: BDP

not to scale

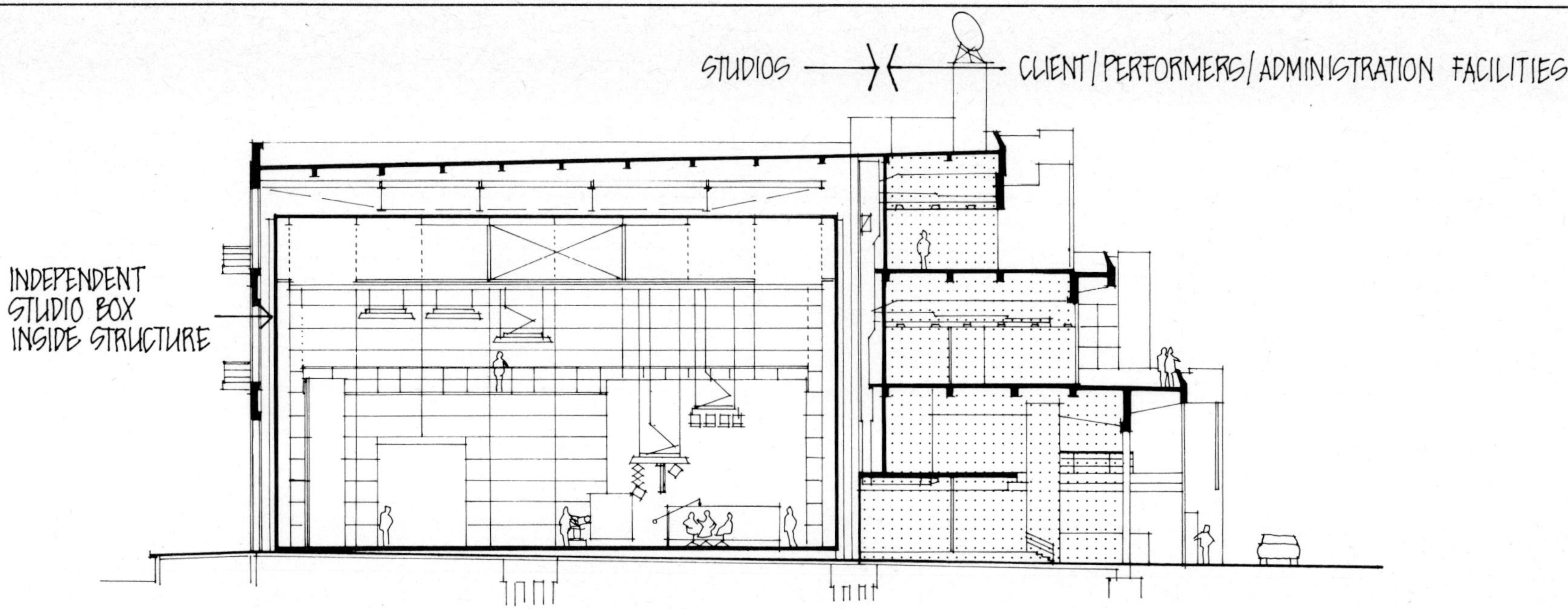

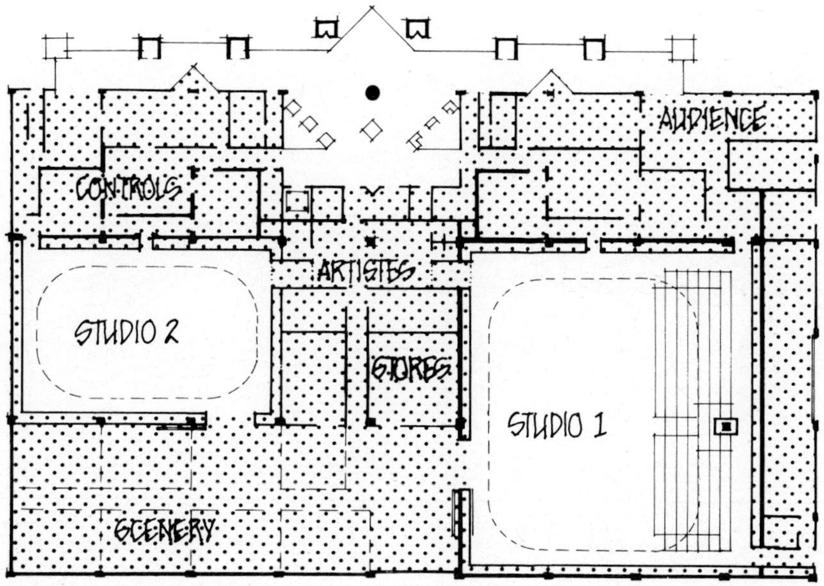

NEW STUDIOS, LONDON, INCORPORATE AN ELECTRO-ACOUSTIC SYSTEM INTEGRATED INTO THE INDEPENDENT STRUCTURE. THIS IS DESIGNED TO ADJUST REVERBERATION TIME BETWEEN 0·8 SECONDS AND 1·4 SECONDS TO SUIT PERFORMERS

CELLULAR ACCOMMODATION BUFFERS STUDIOS FROM PUBLIC AREAS.

TV STUDIOS
SOURCE: TERRY FARRELL PARTNERSHIP ARCHITECTS., SANDY BROWN ASSOC. ACOUS.

not to scale

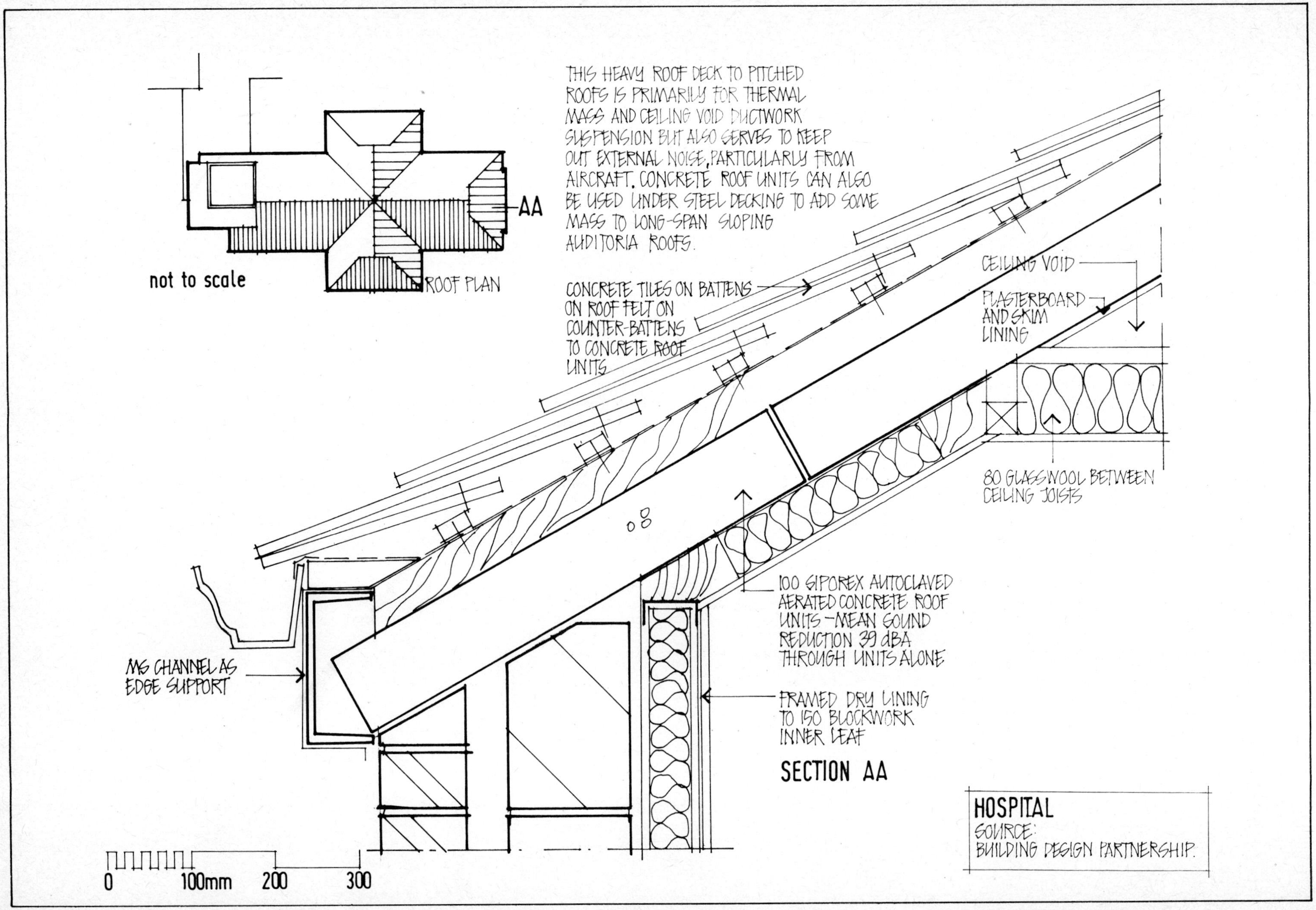

THIS HEAVY ROOF DECK TO PITCHED ROOFS IS PRIMARILY FOR THERMAL MASS AND CEILING VOID DUCTWORK SUSPENSION BUT ALSO SERVES TO KEEP OUT EXTERNAL NOISE, PARTICULARLY FROM AIRCRAFT. CONCRETE ROOF UNITS CAN ALSO BE USED UNDER STEEL DECKING TO ADD SOME MASS TO LONG-SPAN SLOPING AUDITORIA ROOFS.
AA
not to scale
ROOF PLAN
CONCRETE TILES ON BATTENS ON ROOF FELT ON COUNTER-BATTENS TO CONCRETE ROOF UNITS
CEILING VOID
PLASTERBOARD AND SKIM LINING
80 GLASSWOOL BETWEEN CEILING JOISTS
100 SIPOREX AUTOCLAVED AERATED CONCRETE ROOF UNITS - MEAN SOUND REDUCTION 39 dBA THROUGH UNITS ALONE
MS CHANNEL AS EDGE SUPPORT
FRAMED DRY LINING TO 150 BLOCKWORK INNER LEAF
SECTION AA
0
100mm
200
300
HOSPITAL
SOURCE:
BUILDING DESIGN PARTNERSHIP.

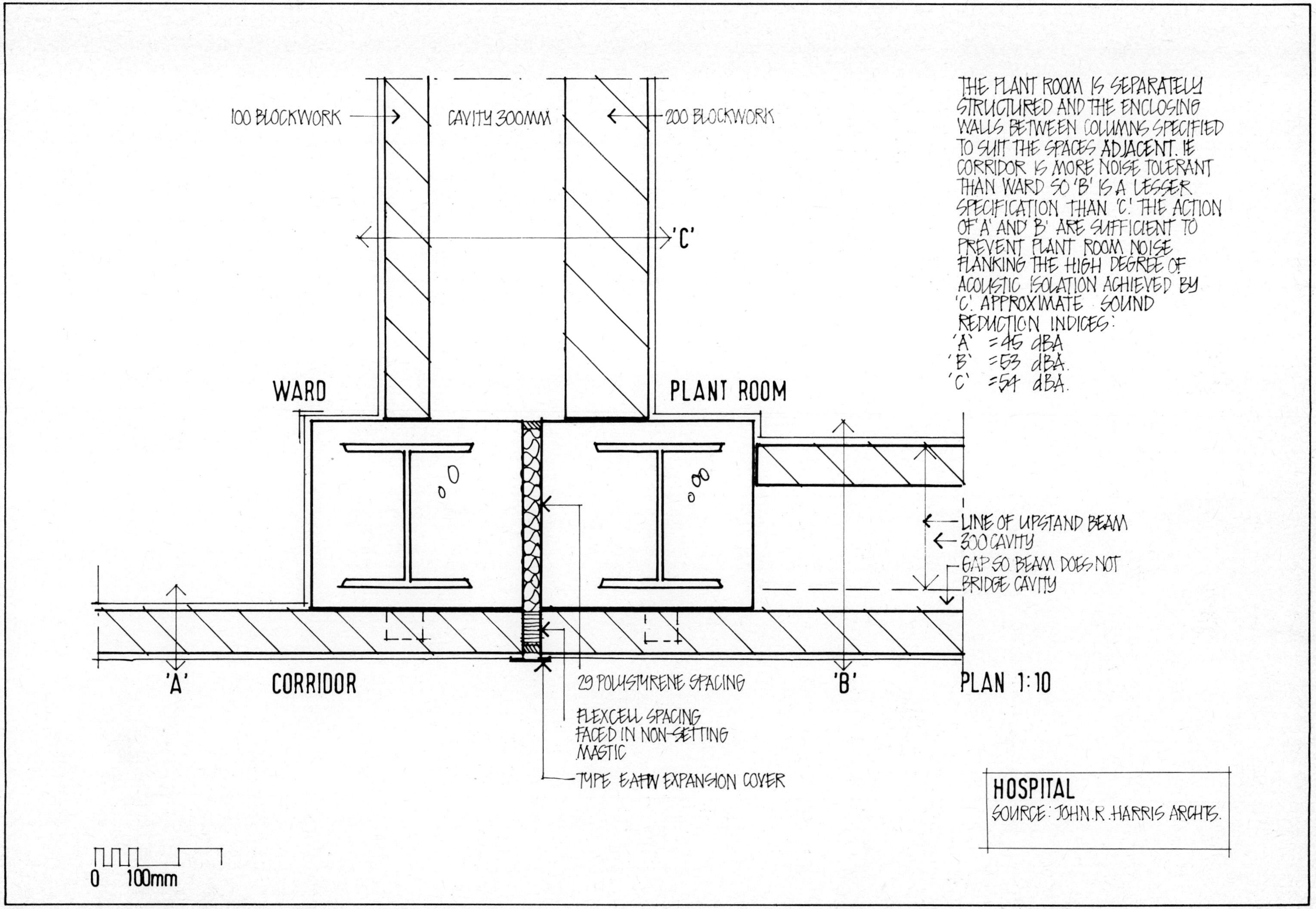
100 BLOCKWORK
CAVITY 300MM
200 BLOCKWORK
'C'
THE PLANT ROOM IS SEPARATELY STRUCTURED AND THE ENCLOSING WALLS BETWEEN COLUMNS SPECIFIED TO SUIT THE SPACES ADJACENT. IE CORRIDOR IS MORE NOISE TOLERANT THAN WARD SO 'B' IS A LESSER SPECIFICATION THAN 'C'. THE ACTION OF 'A' AND 'B' ARE SUFFICIENT TO PREVENT PLANT ROOM NOISE FLANKING THE HIGH DEGREE OF ACOUSTIC ISOLATION ACHIEVED BY 'C'. APPROXIMATE SOUND REDUCTION INDICES:
'A' =45 dBA
'B' =53 dBA.
'C' =54 dBA.
WARD
PLANT ROOM
LINE OF UPSTAND BEAM
300 CAVITY
GAP SO BEAM DOES NOT BRIDGE CAVITY
'A'
CORRIDOR
29 POLYSTYRENE SPACING
'B'
PLAN 1:10
FLEXCELL SPACING FACED IN NON-SETTING MASTIC
TYPE EAFW EXPANSION COVER
HOSPITAL
SOURCE: JOHN.R.HARRIS ARCHTS.
0 100mm

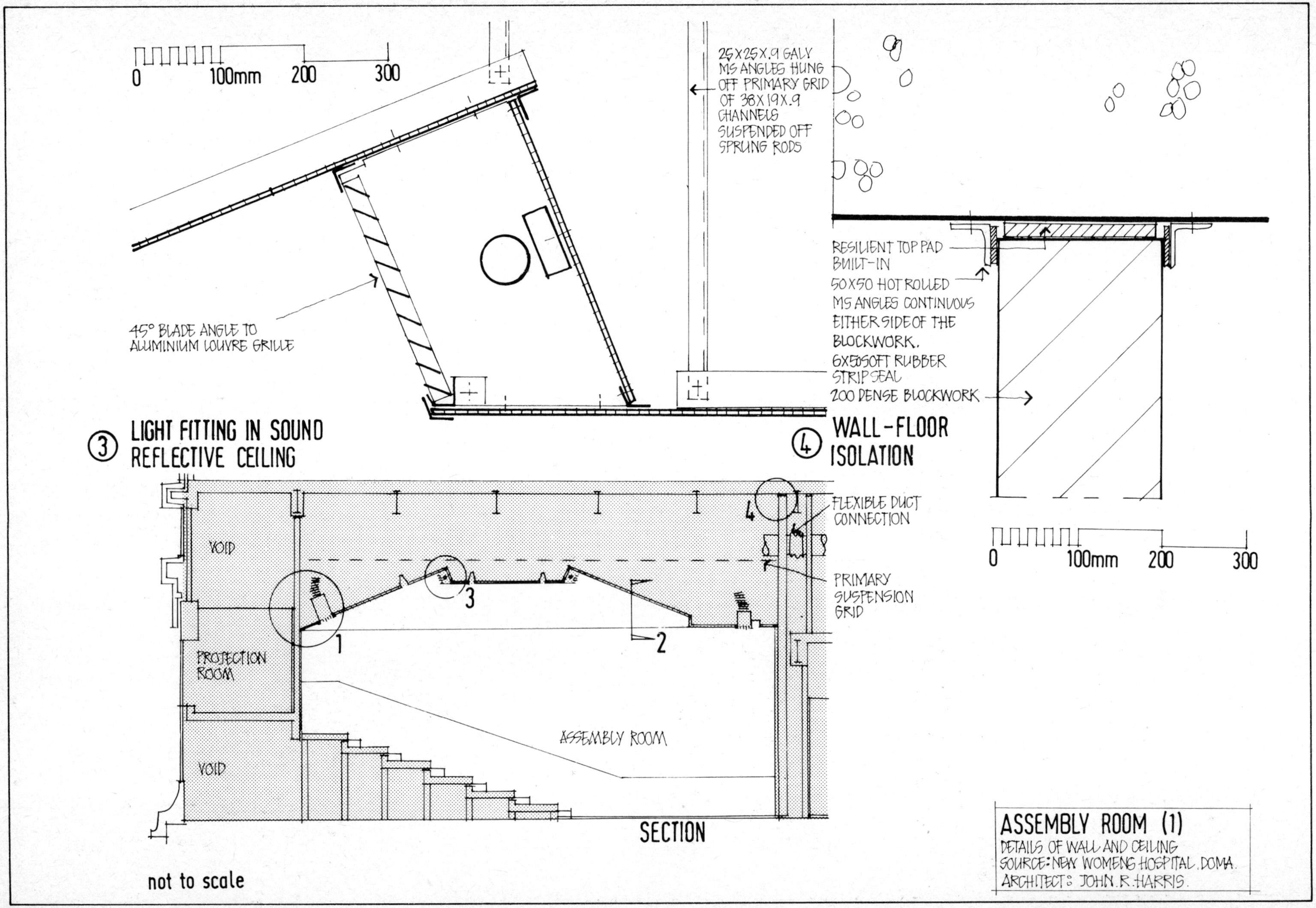
0 100mm 200 300
25X25X.9 GALV MS ANGLES HUNG OFF PRIMARY GRID OF 38X19X.9 CHANNELS SUSPENDED OFF SPRUNG RODS
45° BLADE ANGLE TO ALUMINIUM LOUVRE GRILLE
3 LIGHT FITTING IN SOUND REFLECTIVE CEILING
RESILIENT TOP PAD BUILT-IN
50X50 HOT ROLLED MS ANGLES CONTINUOUS EITHER SIDE OF THE BLOCKWORK.
6X50 SOFT RUBBER STRIP SEAL
200 DENSE BLOCKWORK
4 WALL-FLOOR ISOLATION
0 100mm 200 300
VOID
PROJECTION ROOM
VOID
ASSEMBLY ROOM
1
2
3
4
FLEXIBLE DUCT CONNECTION
PRIMARY SUSPENSION GRID
SECTION
not to scale
ASSEMBLY ROOM (1)
DETAILS OF WALL AND CEILING
SOURCE: NEW WOMENS HOSPITAL, DOMA.
ARCHITECT: JOHN. R. HARRIS.

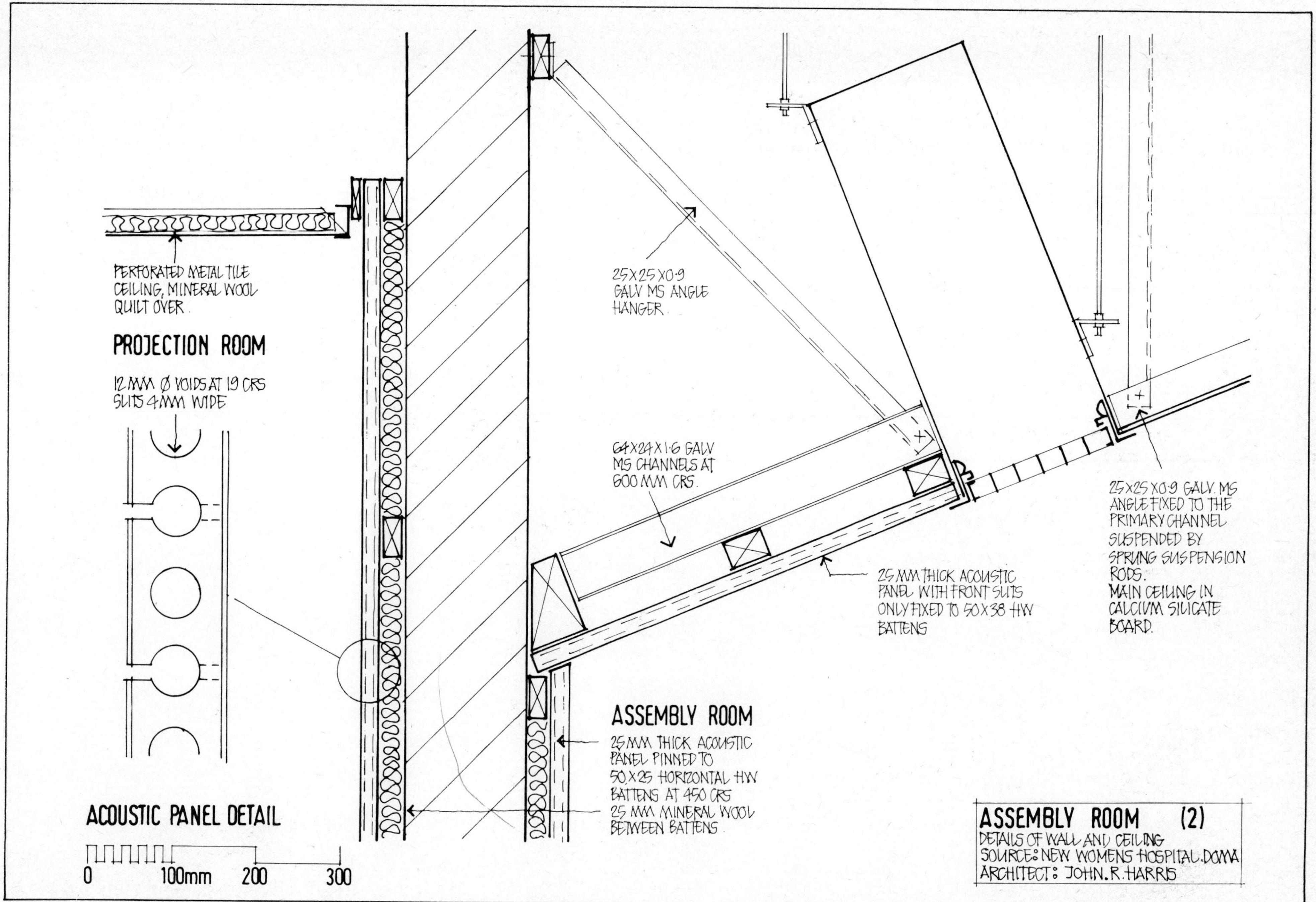

PERFORATED METAL TILE CEILING, MINERAL WOOL QUILT OVER.
PROJECTION ROOM
12 MM Ø VOIDS AT 19 CRS SLITS 4MM WIDE
25X25X0·9 GALV MS ANGLE HANGER
64X24X1·6 GALV MS CHANNELS AT 600 MM CRS.
25 MM THICK ACOUSTIC PANEL WITH FRONT SLITS ONLY FIXED TO 50X38 HW BATTENS
25X25X0·9 GALV. MS ANGLE FIXED TO THE PRIMARY CHANNEL SUSPENDED BY SPRUNG SUSPENSION RODS. MAIN CEILING IN CALCIUM SILICATE BOARD.
ASSEMBLY ROOM
25 MM THICK ACOUSTIC PANEL PINNED TO 50X25 HORIZONTAL HW BATTENS AT 450 CRS
25 MM MINERAL WOOL BETWEEN BATTENS.
ACOUSTIC PANEL DETAIL
0
100mm
200
300
ASSEMBLY ROOM (2)
DETAILS OF WALL AND CEILING
SOURCE: NEW WOMENS HOSPITAL.DOHA
ARCHITECT: JOHN.R.HARRIS

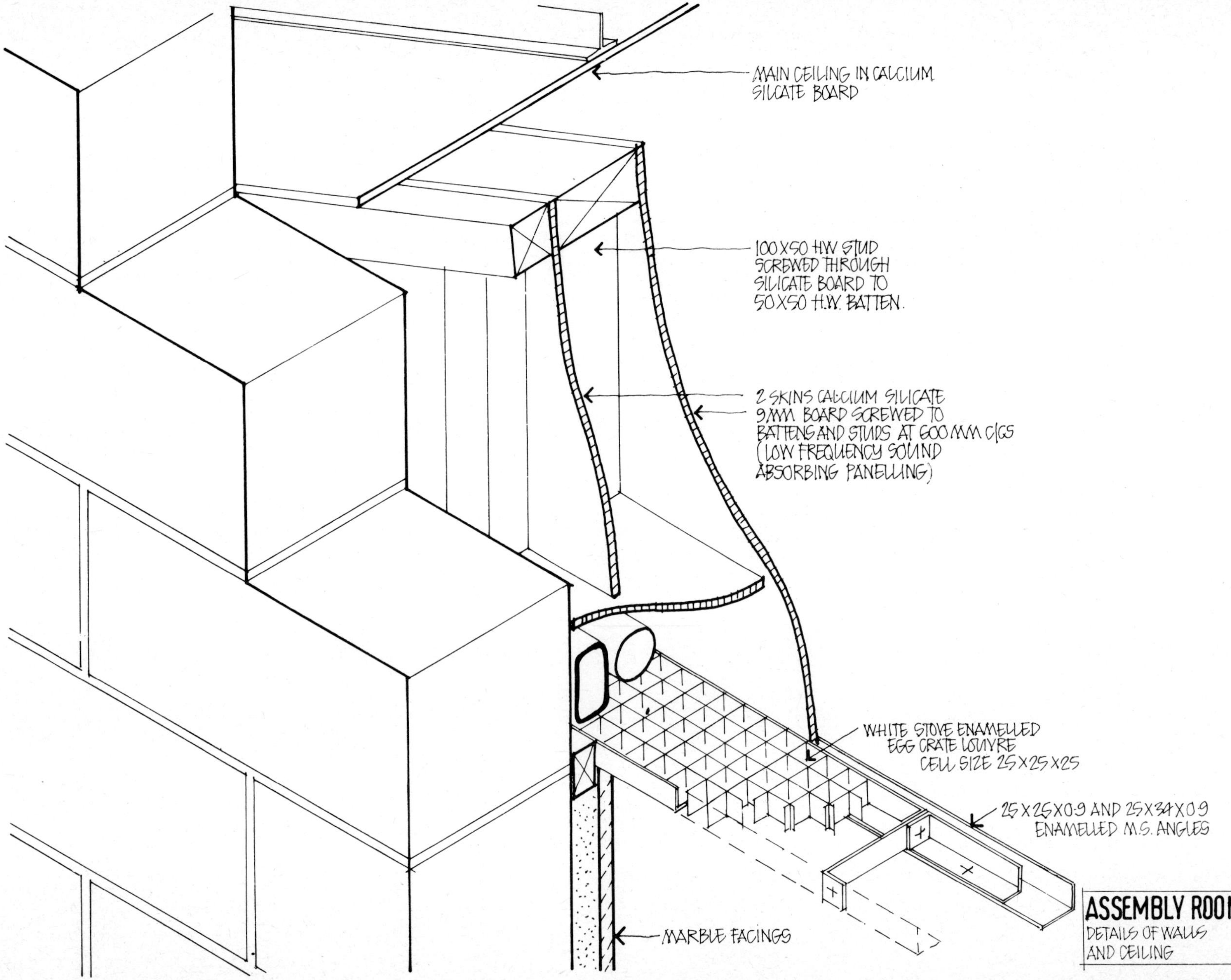

ASSEMBLY ROOM (3)
DETAILS OF WALLS AND CEILING

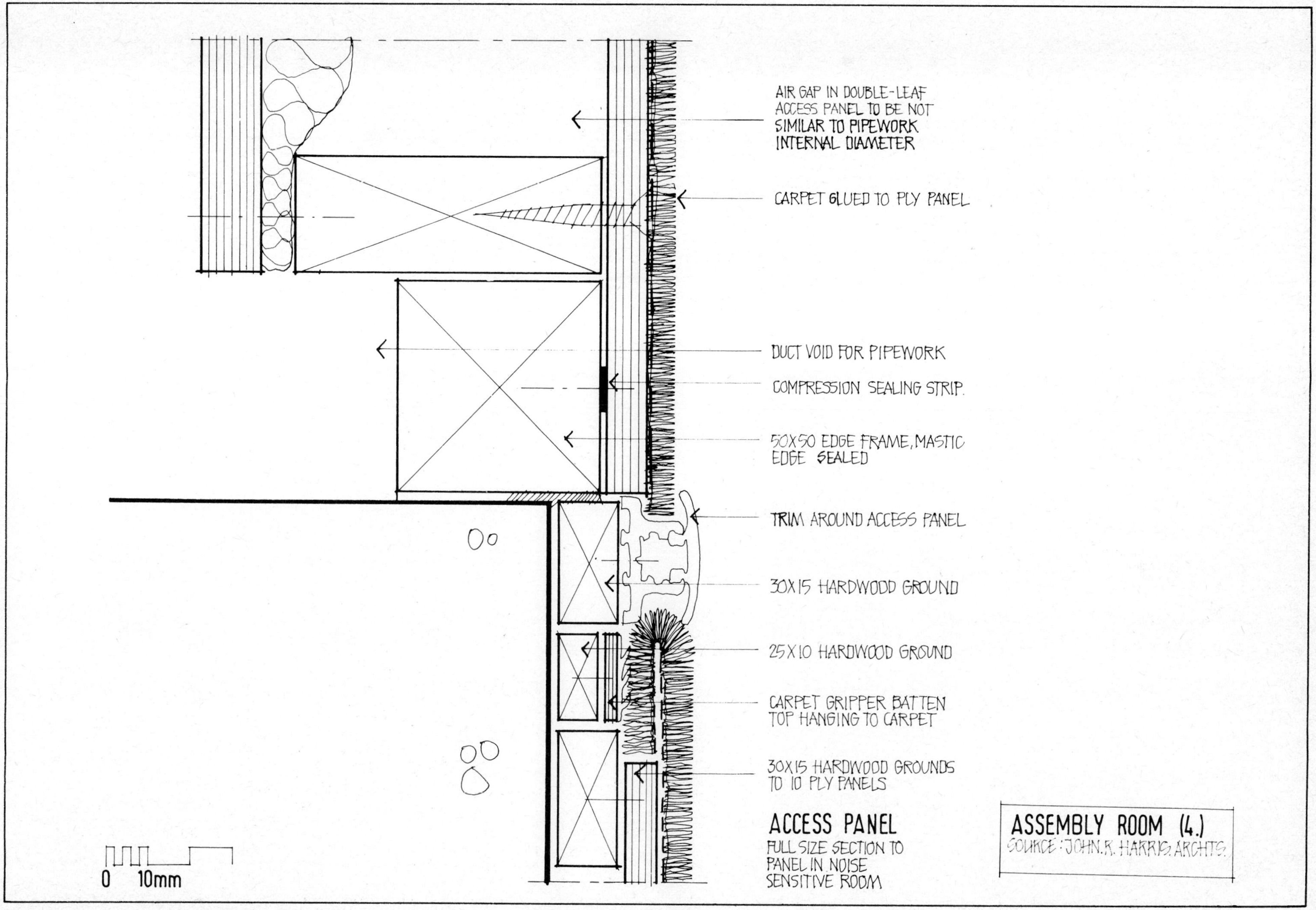
AIR GAP IN DOUBLE-LEAF ACCESS PANEL TO BE NOT SIMILAR TO PIPEWORK INTERNAL DIAMETER
CARPET GLUED TO PLY PANEL
DUCT VOID FOR PIPEWORK
COMPRESSION SEALING STRIP.
50X50 EDGE FRAME, MASTIC EDGE SEALED
TRIM AROUND ACCESS PANEL
30X15 HARDWOOD GROUND
25X10 HARDWOOD GROUND
CARPET GRIPPER BATTEN TOP HANGING TO CARPET
30X15 HARDWOOD GROUNDS TO 10 PLY PANELS
0 10mm
ACCESS PANEL
FULL SIZE SECTION TO PANEL IN NOISE SENSITIVE ROOM
ASSEMBLY ROOM (4.)
SOURCE: JOHN.R. HARRIS. ARCHTS.

HORIZONTAL SECTION

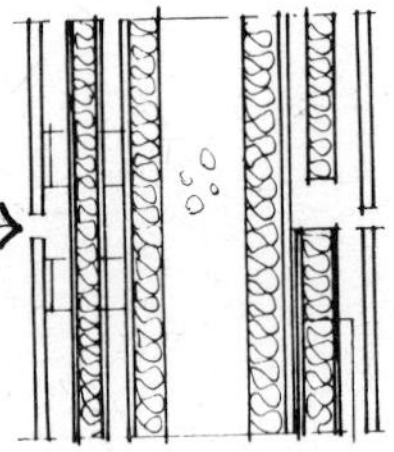

SECTION BETWEEN AUDITORIA

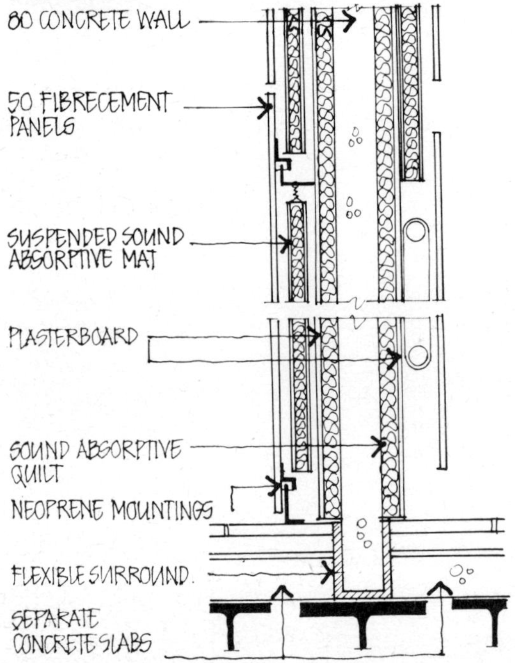

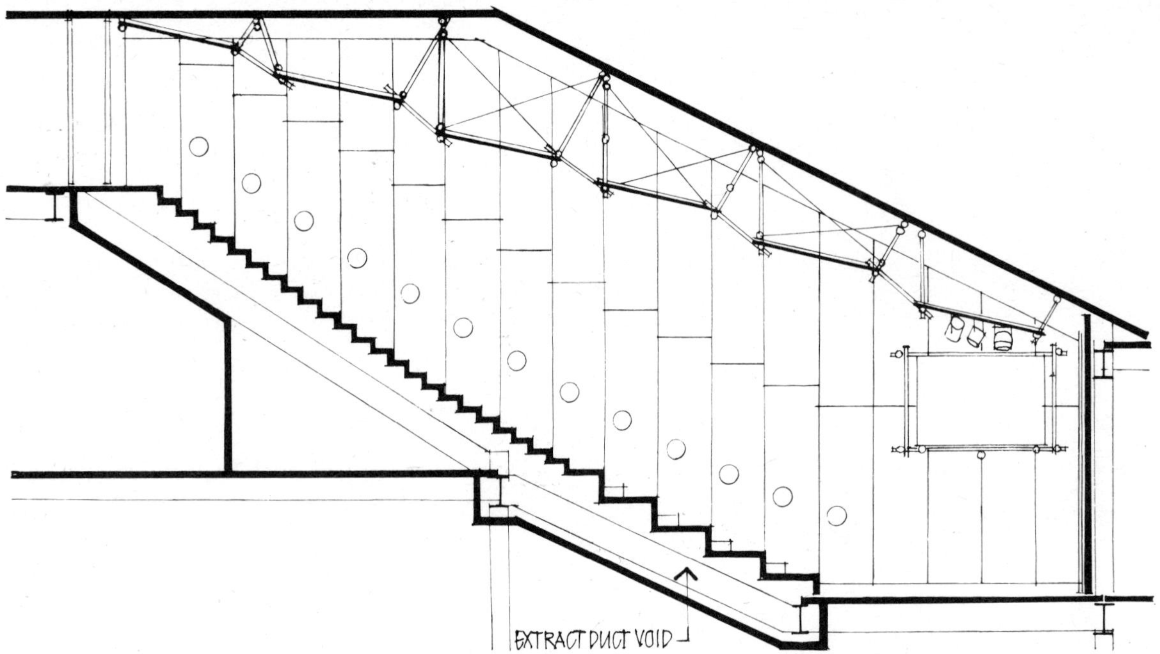

SECTION

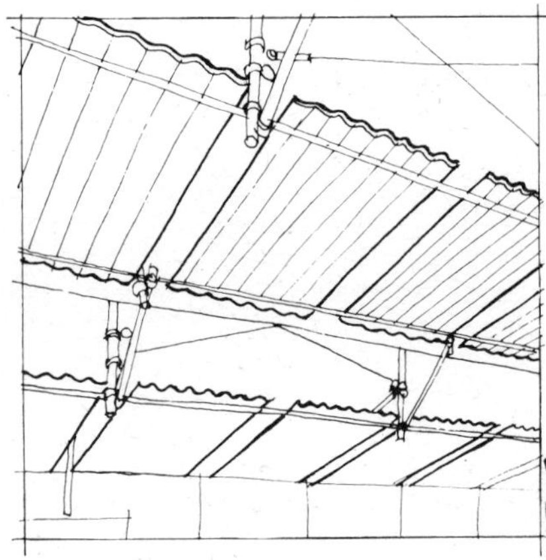

VIEW OF OVERHEAD PANELS

LECTURE THEATRE

[LOCATION: DEPT OF BUILDING TECHNICAL COLLEGE, BERLIN]

not to scale

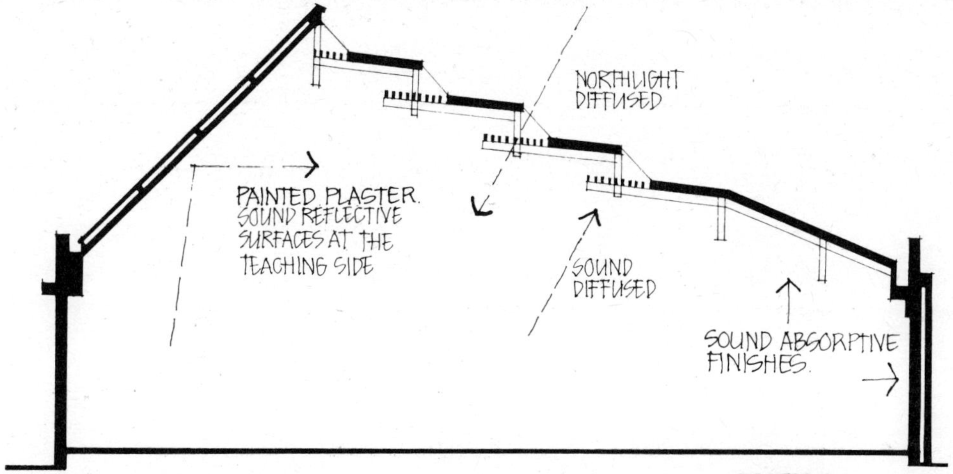

SECTION

REFLECTIVE FINISHES AT FRONT ALLOW VOICE PROJECTION TO REAR OF ROOM
SOUND ABSORPTIVE FINISHES TO BE USED AT REAR OF ROOM
DESIGN RT AT APPROX 0·75 SECONDS

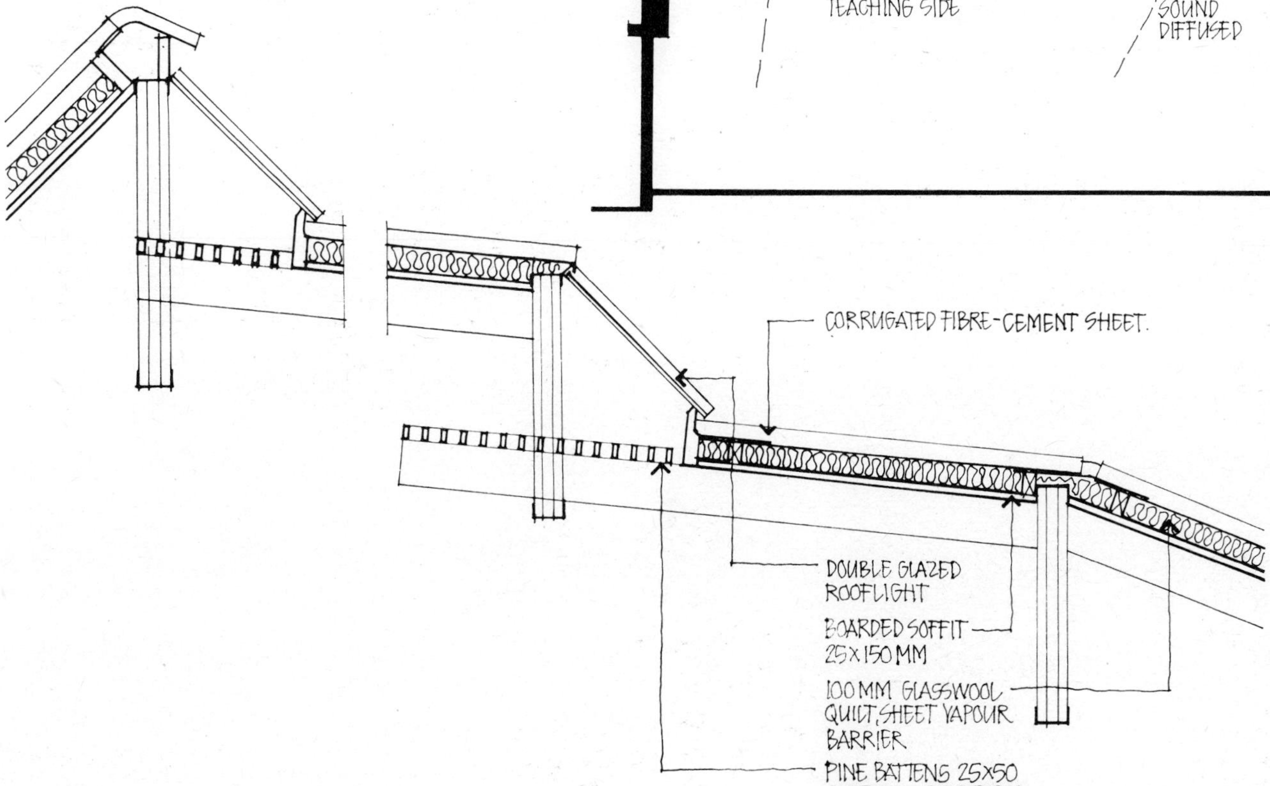

DETAIL SECTION

CLASSROOM

not to scale

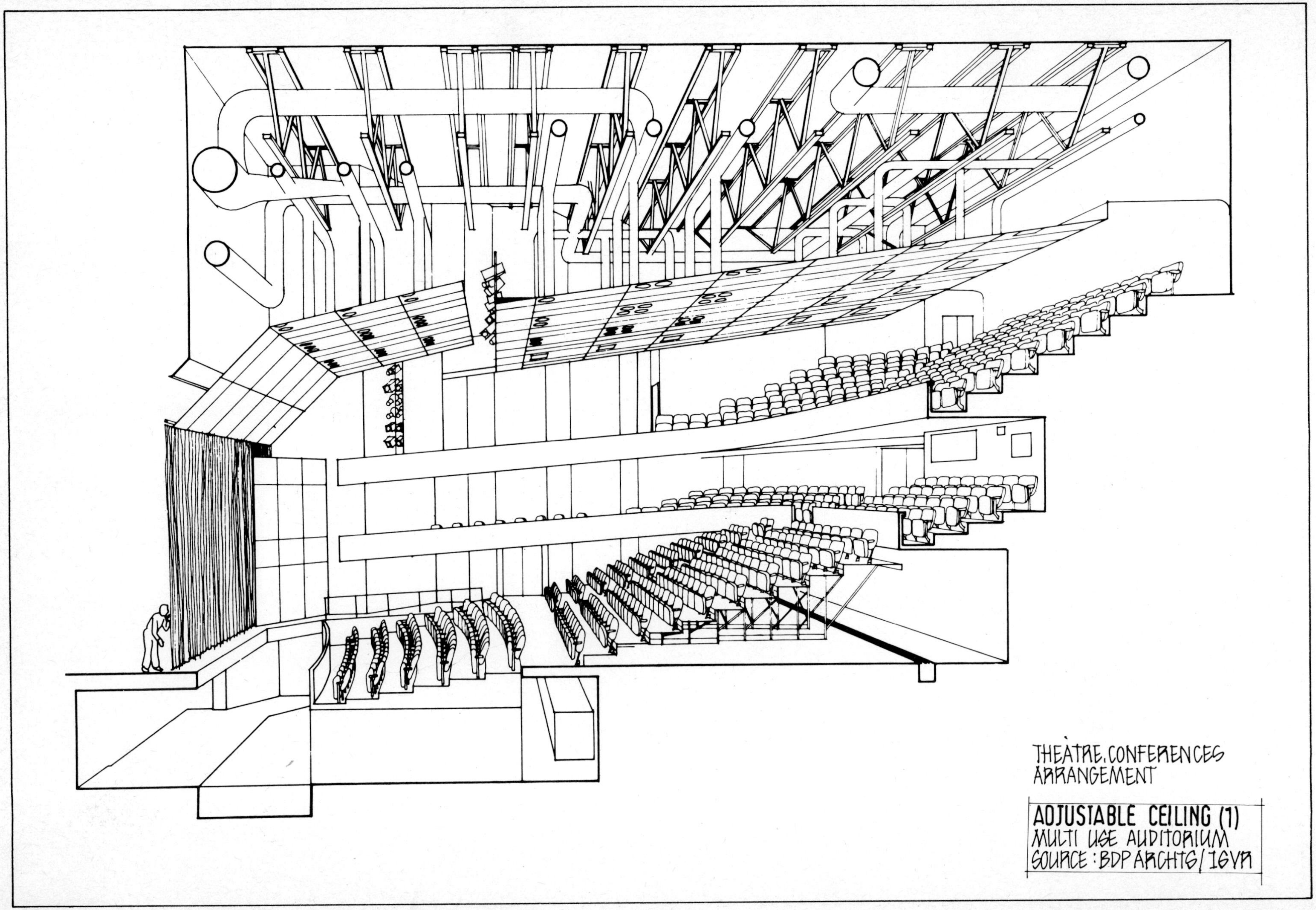
THEÀTRE, CONFERENCES
ARRANGEMENT
ADJUSTABLE CEILING (1)
MULTI USE AUDITORIUM
SOURCE : BDP ARCHTS / IGVR

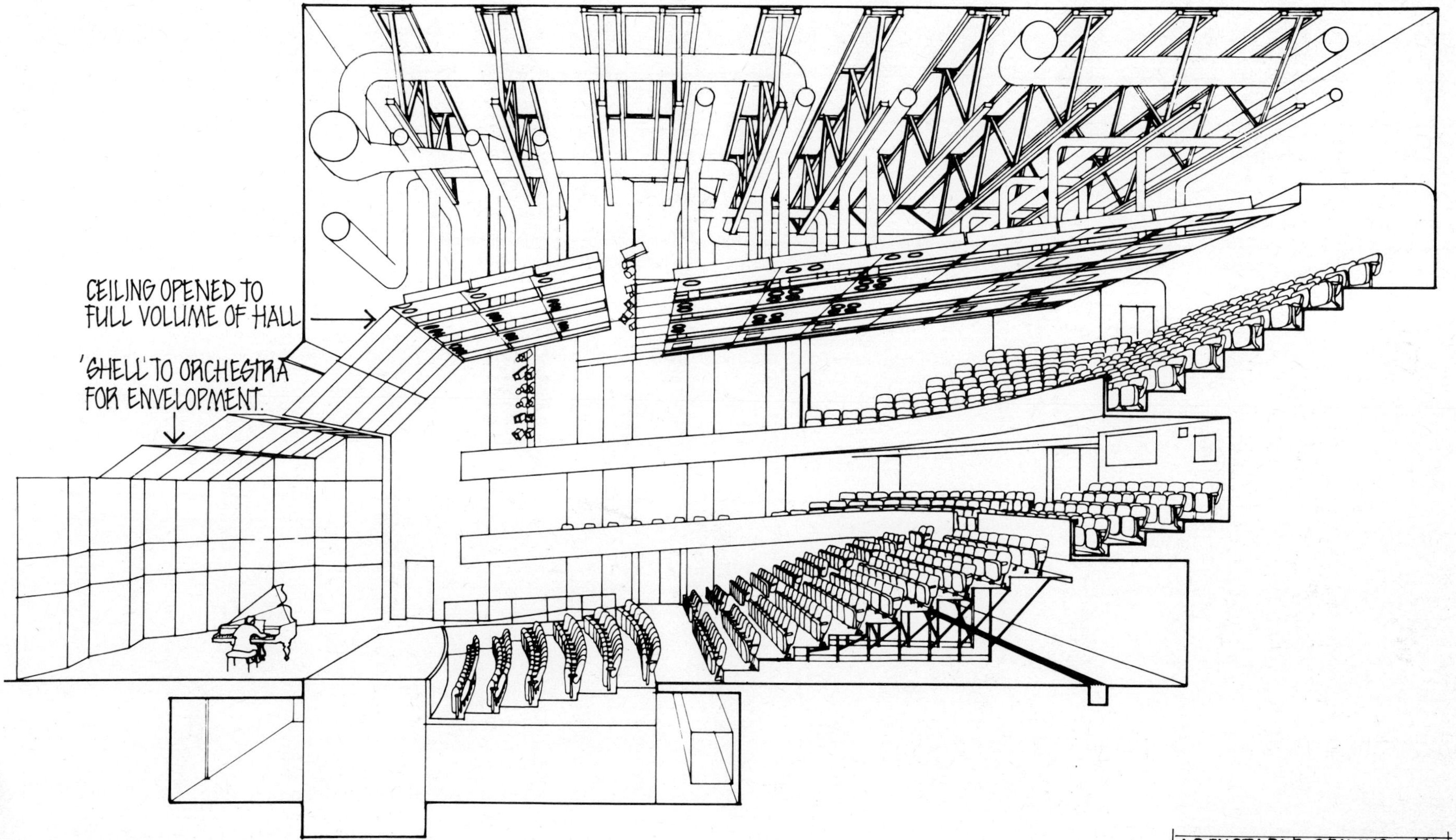

ADJUSTABLE CEILING (2)
MULTI USE AUDITORIUM
SOURCE: BDP ARCHTS/ISVR

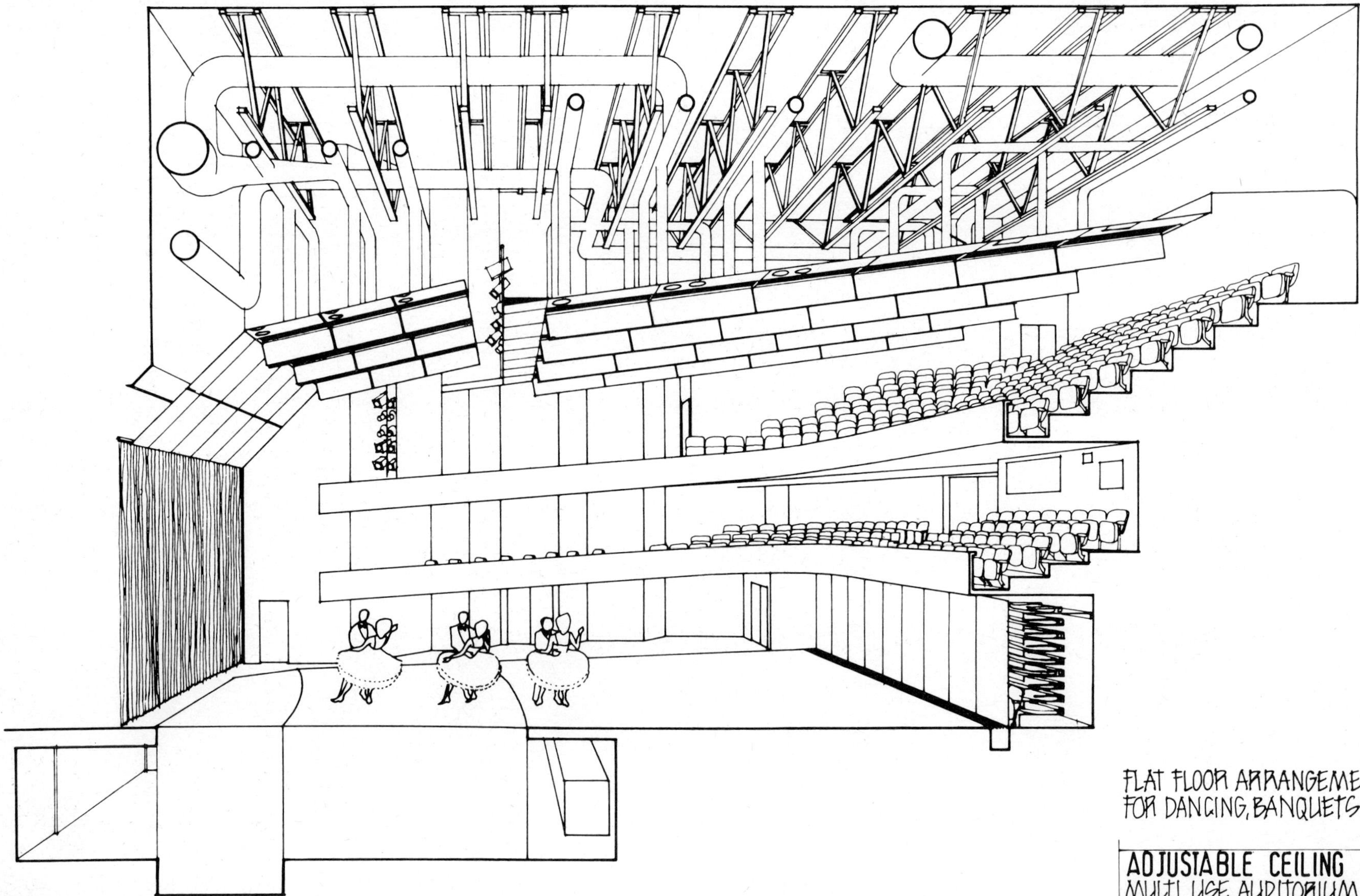

FLAT FLOOR ARRANGEMENT
FOR DANCING, BANQUETS ETC.

ADJUSTABLE CEILING (3)
MULTI USE AUDITORIUM
SOURCE: BDP ARCHTS / I.S.V.R.

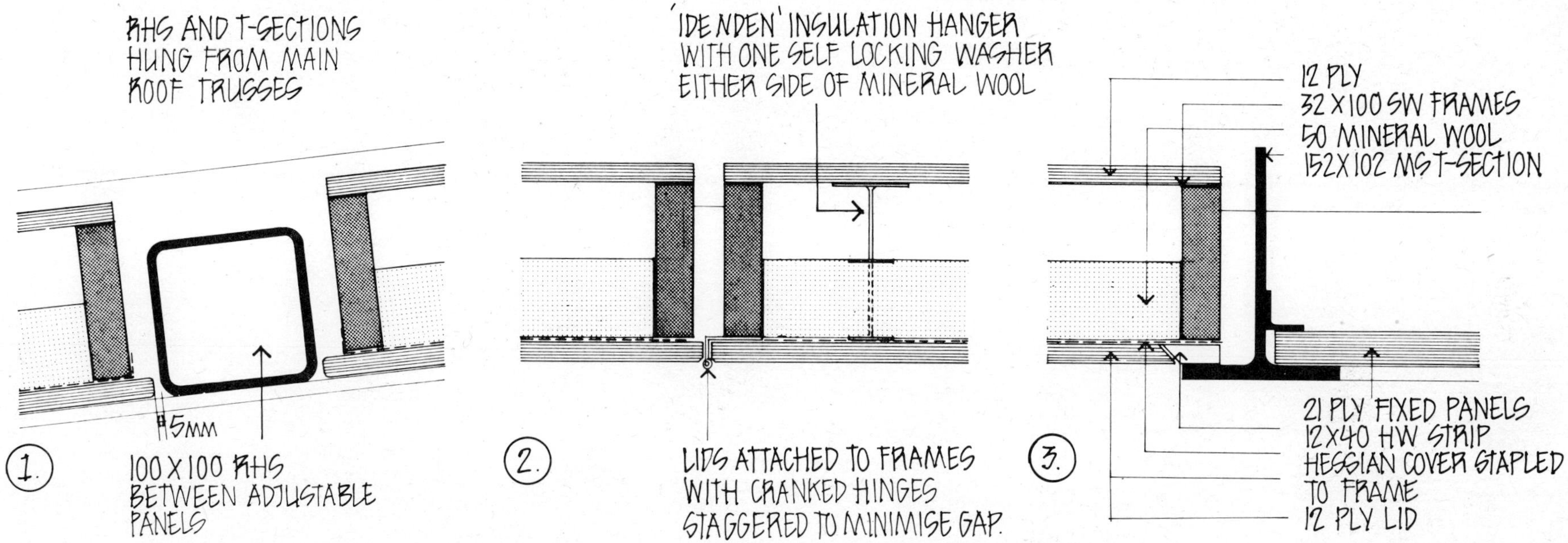

ADJUSTABLE CEILING (4)
MULTI-USE AUDITORIUM
SOURCE: BDP ARCHTS/ISVR.

not to scale

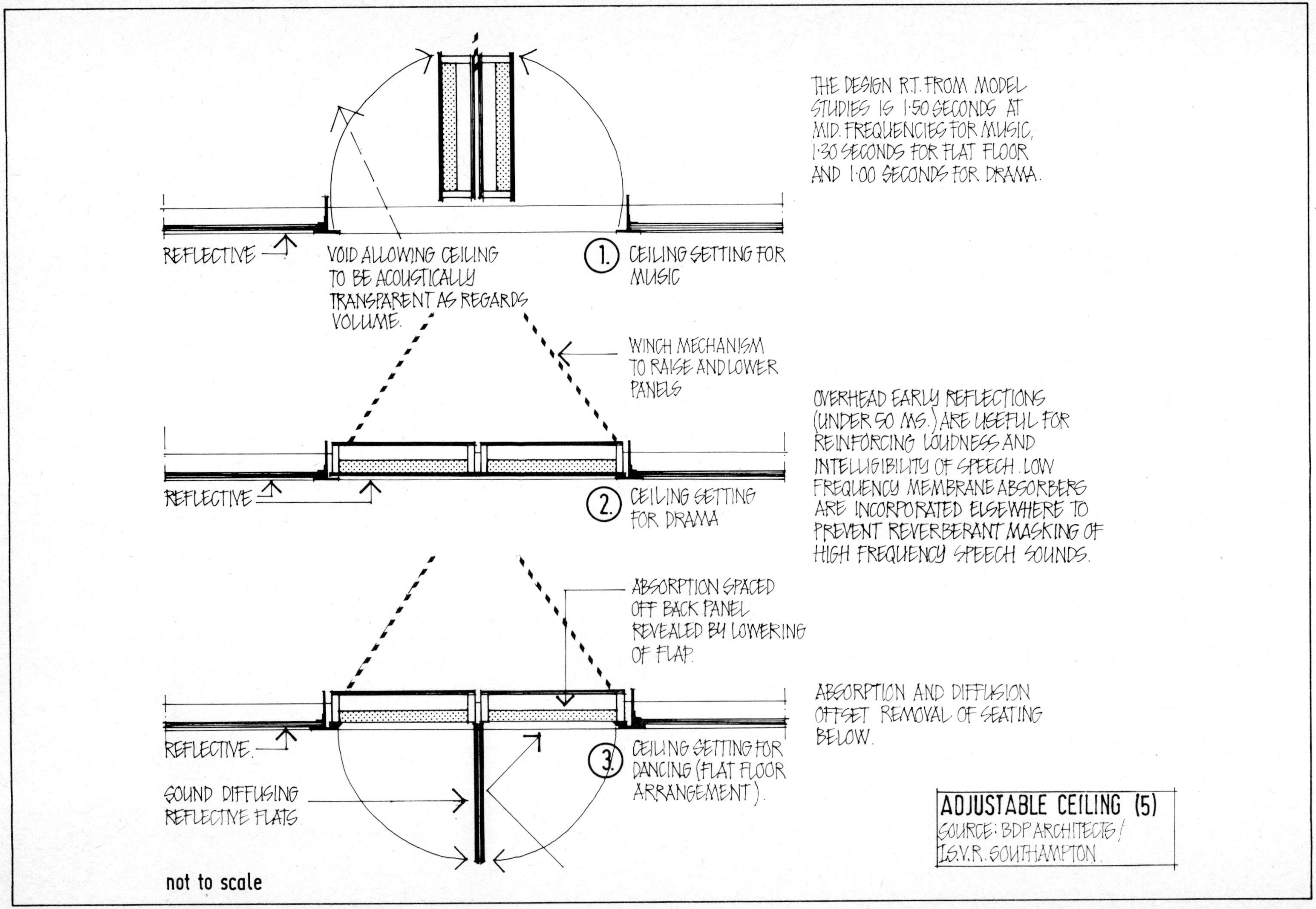
THE DESIGN R.T. FROM MODEL STUDIES IS 1·50 SECONDS AT MID. FREQUENCIES FOR MUSIC, 1·30 SECONDS FOR FLAT FLOOR AND 1·00 SECONDS FOR DRAMA.
REFLECTIVE
VOID ALLOWING CEILING TO BE ACOUSTICALLY TRANSPARENT AS REGARDS VOLUME.
1. CEILING SETTING FOR MUSIC
WINCH MECHANISM TO RAISE AND LOWER PANELS
OVERHEAD EARLY REFLECTIONS (UNDER 50 MS.) ARE USEFUL FOR REINFORCING LOUDNESS AND INTELLIGIBILITY OF SPEECH. LOW FREQUENCY MEMBRANE ABSORBERS ARE INCORPORATED ELSEWHERE TO PREVENT REVERBERANT MASKING OF HIGH FREQUENCY SPEECH SOUNDS.
REFLECTIVE
2. CEILING SETTING FOR DRAMA
ABSORPTION SPACED OFF BACK PANEL REVEALED BY LOWERING OF FLAP.
ABSORPTION AND DIFFUSION OFFSET REMOVAL OF SEATING BELOW.
REFLECTIVE.
3. CEILING SETTING FOR DANCING (FLAT FLOOR ARRANGEMENT).
SOUND DIFFUSING REFLECTIVE FLATS
ADJUSTABLE CEILING (5)
SOURCE: BDP ARCHITECTS/ I.S.V.R. SOUTHAMPTON.
not to scale

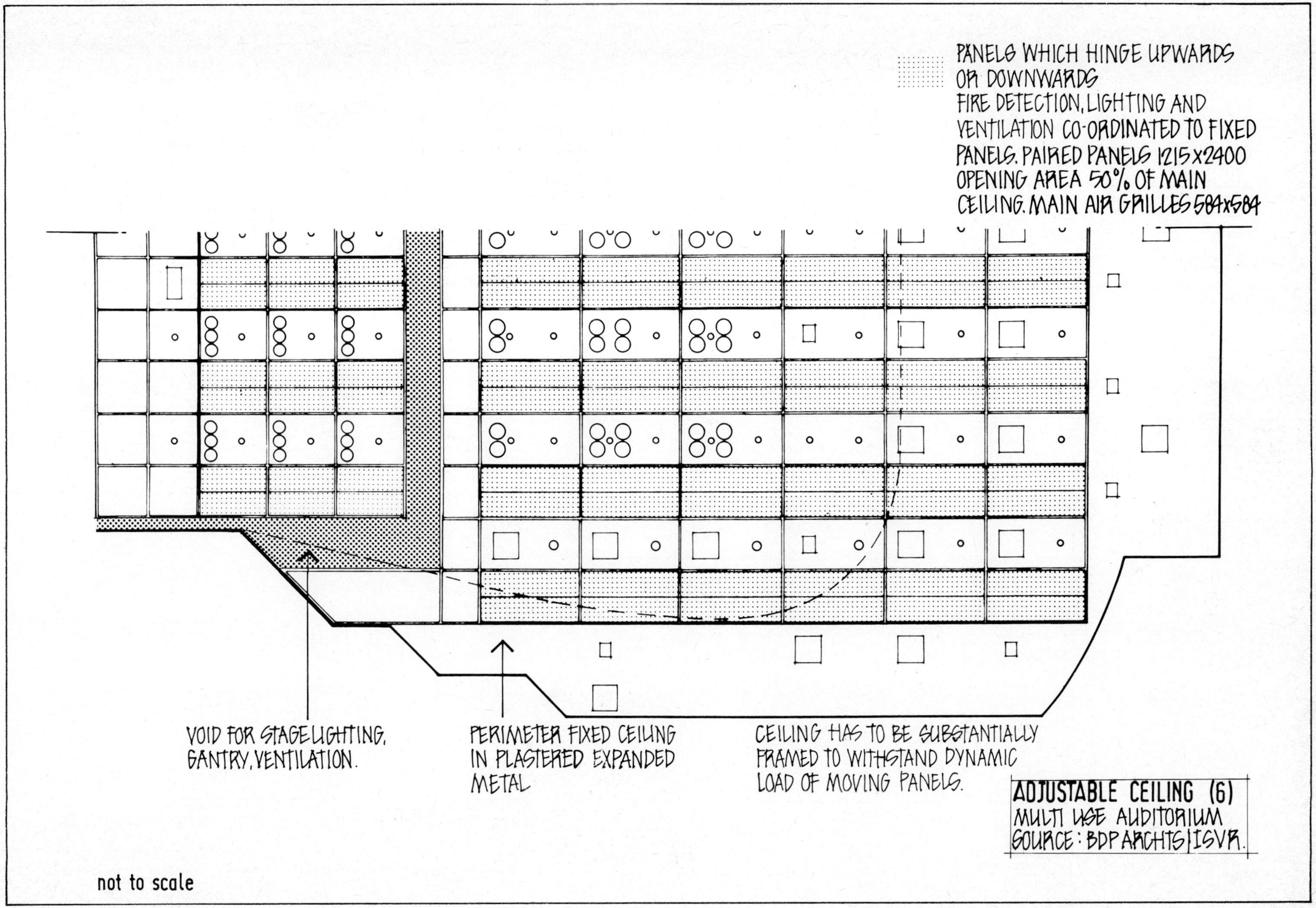
PANELS WHICH HINGE UPWARDS OR DOWNWARDS
FIRE DETECTION, LIGHTING AND VENTILATION CO-ORDINATED TO FIXED PANELS. PAIRED PANELS 1215x2400 OPENING AREA 50% OF MAIN CEILING. MAIN AIR GRILLES 584x584
VOID FOR STAGE LIGHTING, GANTRY, VENTILATION.
PERIMETER FIXED CEILING IN PLASTERED EXPANDED METAL
CEILING HAS TO BE SUBSTANTIALLY FRAMED TO WITHSTAND DYNAMIC LOAD OF MOVING PANELS.
ADJUSTABLE CEILING (6)
MULTI USE AUDITORIUM
SOURCE: BDP ARCHTS/ISVR.
not to scale

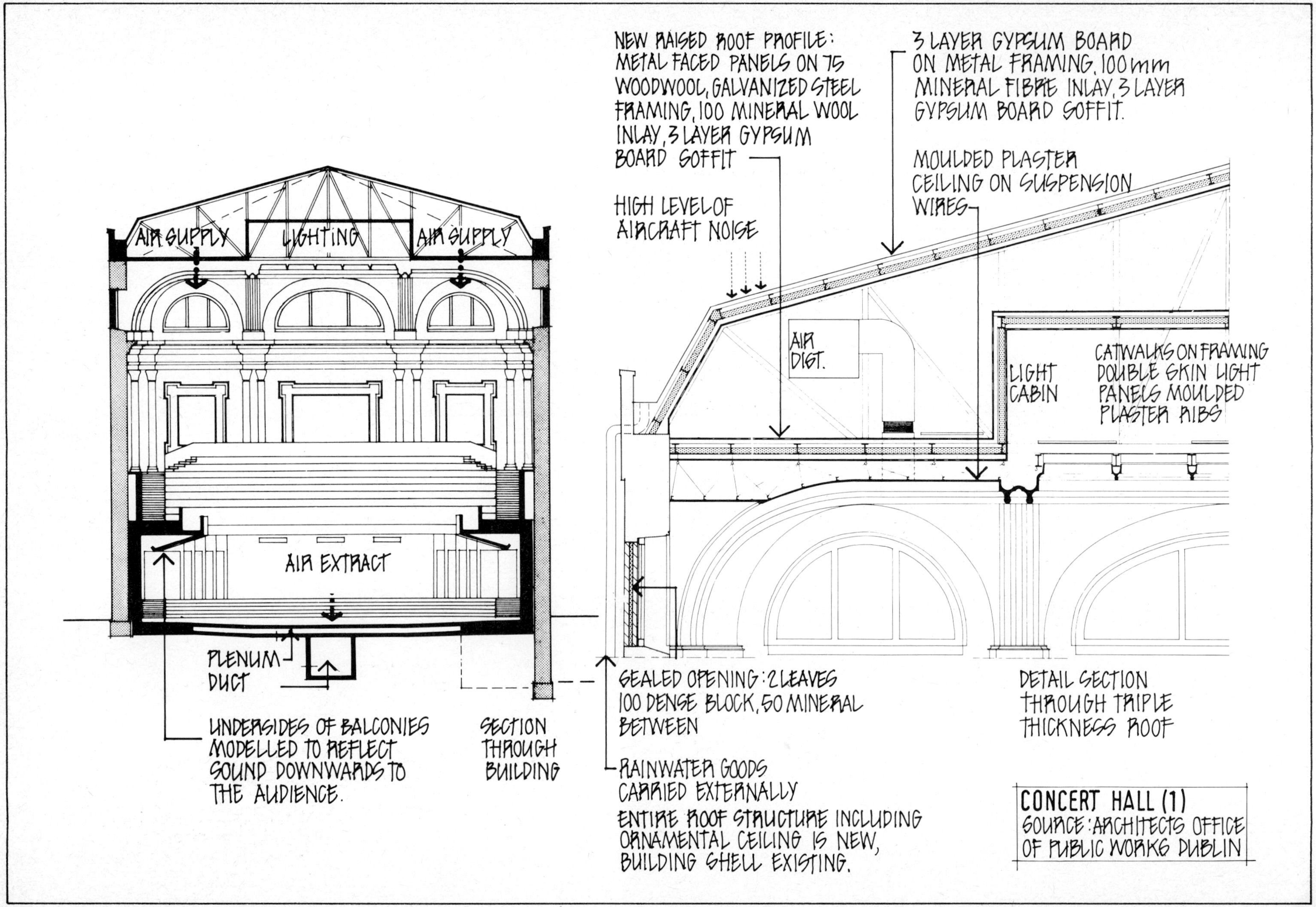

CONCERT HALL (1)
SOURCE: ARCHITECTS OFFICE OF PUBLIC WORKS DUBLIN

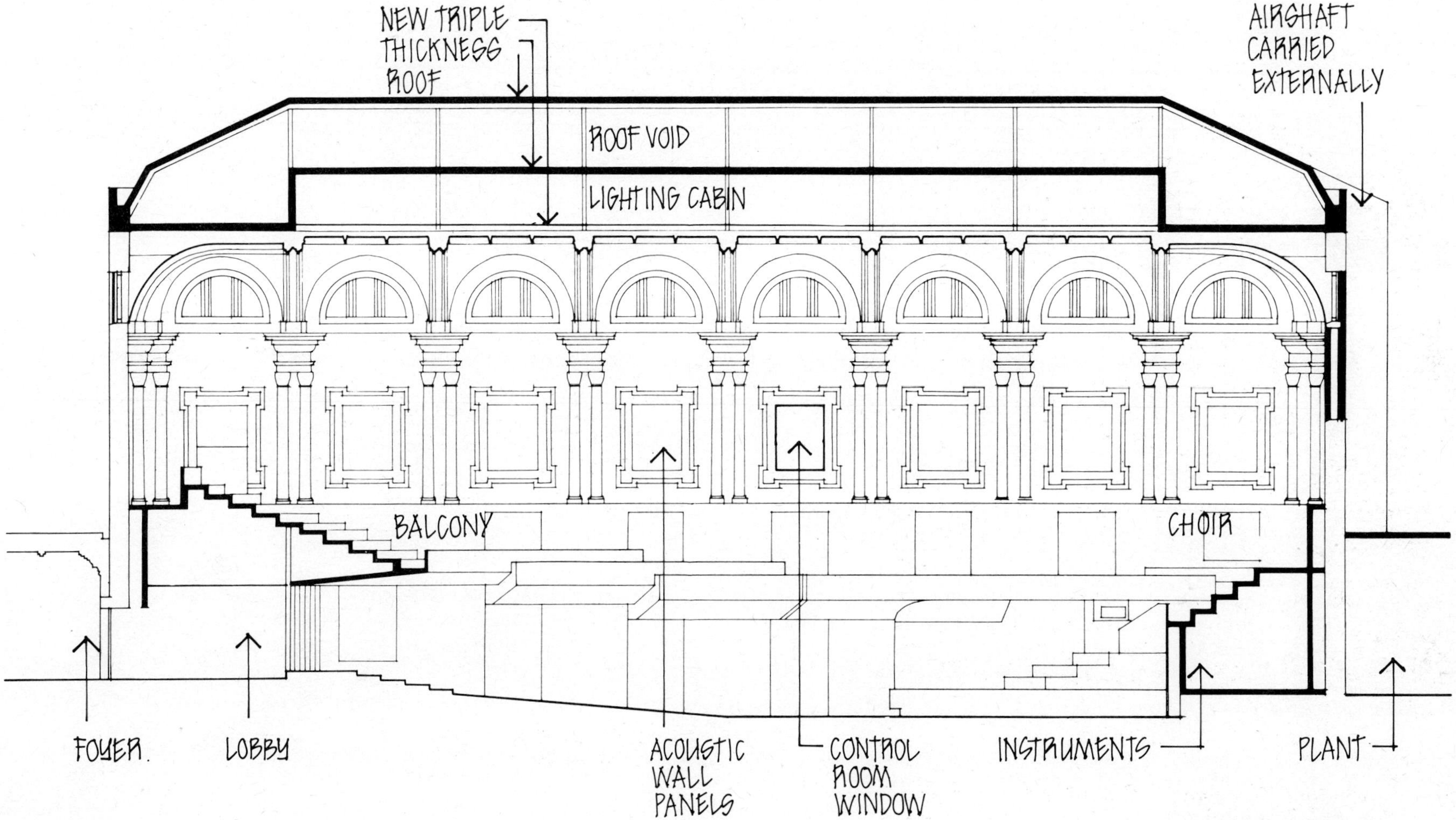

CONCERT HALL (2)
SOURCE: ARCHITECTS OFFICE OF PUBLIC WORKS DUBLIN

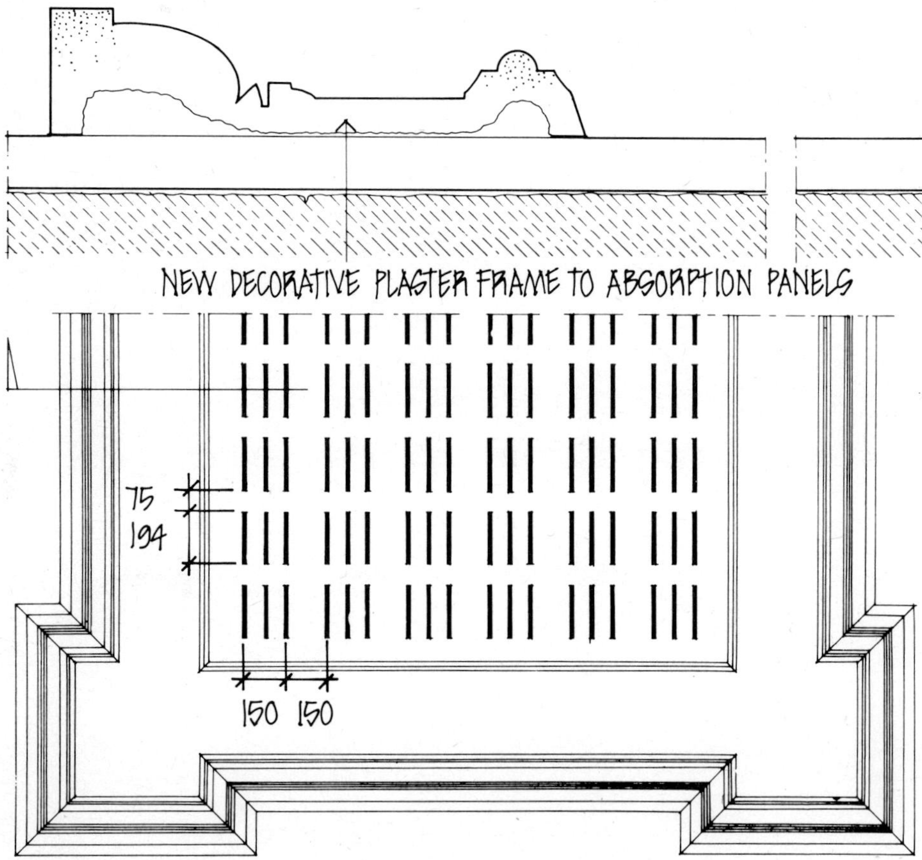

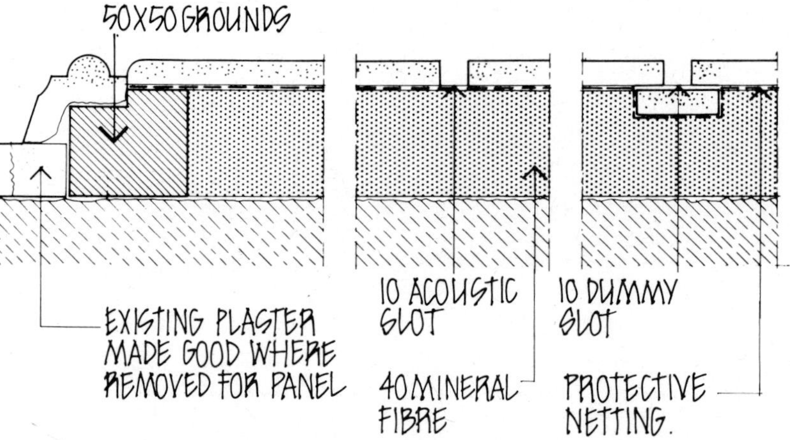

ADDITION OF PANELS ENABLED 'TUNING' OF AUDITORIUM DURING COMMISSIONING TO PROVIDE ABSORPTION AT A PARTICULAR MID-FREQUENCY RANGE TO ACHIEVE A FLAT REVERBERATION TIME CHARACTERISTIC THROUGHOUT THE FREQUENCY RANGE. PANELS ARE EVENLY DISTRIBUTED TO ACHIEVE EVEN SOUND QUALITY THROUGHOUT THE HALL. HELMHOLTZ RESONATOR EFFECT OF SLOTS RESTRICT PANELS USE TO NARROW BAND ABSORPTION.

ACOUSTIC WALL PANEL DETAILS

CONCERT HALL (3)
SOURCE: ARCHITECTS. OFFICE OF PUBLIC WORKS DUBLIN.

not to scale

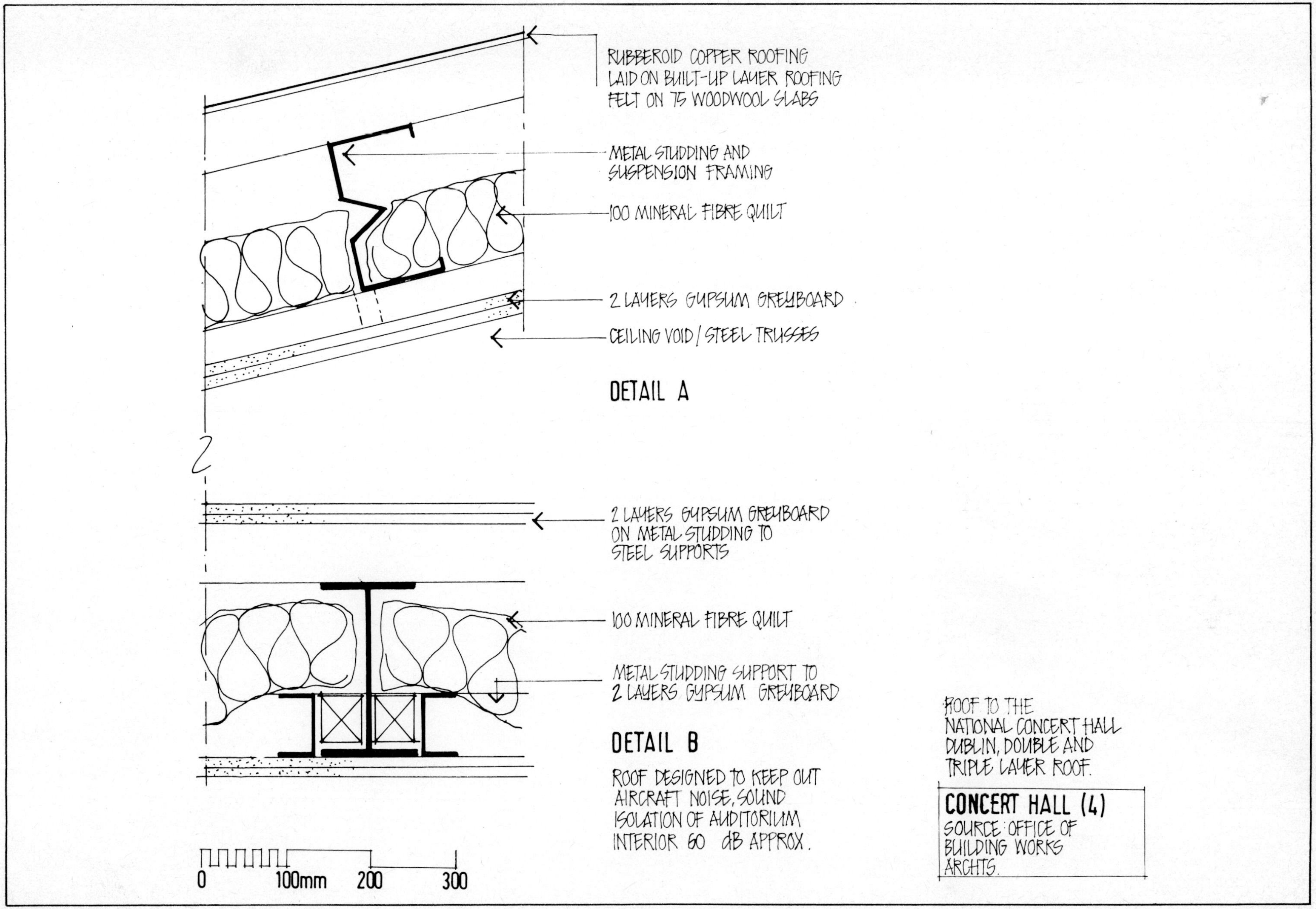
RUBBEROID COPPER ROOFING
LAID ON BUILT-UP LAYER ROOFING
FELT ON 75 WOODWOOL SLABS
METAL STUDDING AND
SUSPENSION FRAMING
100 MINERAL FIBRE QUILT
2 LAYERS GYPSUM GREYBOARD
CEILING VOID / STEEL TRUSSES
DETAIL A
2 LAYERS GYPSUM GREYBOARD
ON METAL STUDDING TO
STEEL SUPPORTS
100 MINERAL FIBRE QUILT
METAL STUDDING SUPPORT TO
2 LAYERS GYPSUM GREYBOARD
DETAIL B
ROOF DESIGNED TO KEEP OUT
AIRCRAFT NOISE, SOUND
ISOLATION OF AUDITORIUM
INTERIOR 60 dB APPROX.
0
100mm
200
300
ROOF TO THE
NATIONAL CONCERT HALL
DUBLIN, DOUBLE AND
TRIPLE LAYER ROOF.
CONCERT HALL (4)
SOURCE: OFFICE OF
BUILDING WORKS
ARCHTS.

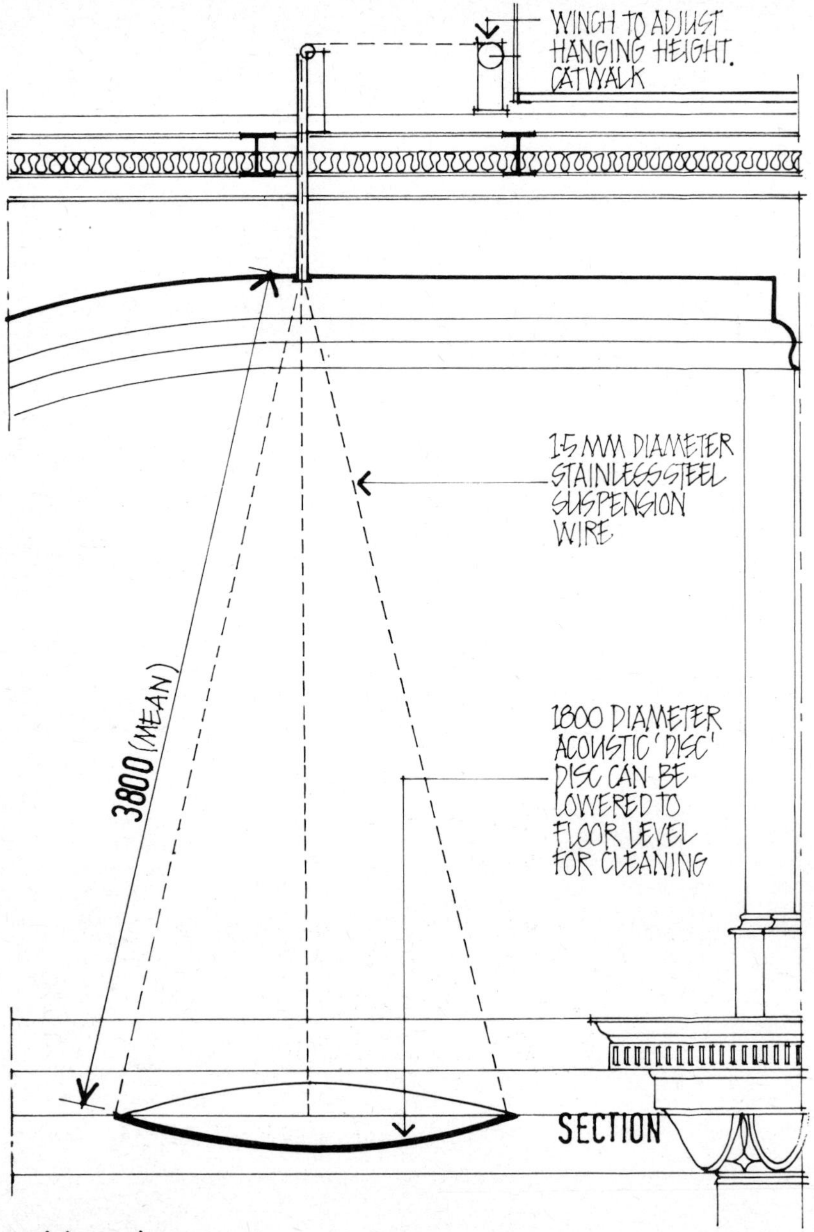

SECTION

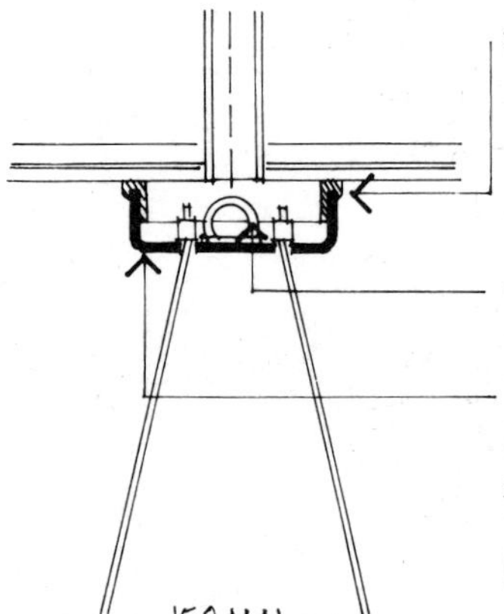

DETAIL SECTION

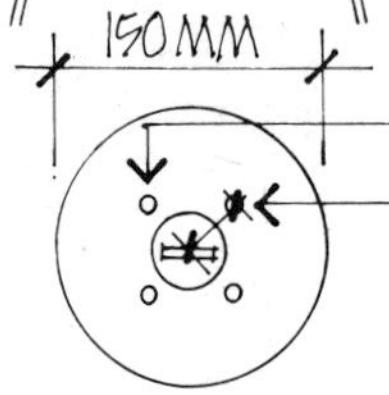

PLAN

SKETCH OF THE INTERIOR OF THE NATIONAL CONCERT HALL DUBLIN

IN THIS HALL 12 ASYMMETRICALLY-PLACED ACOUSTIC 'DISCS' OR 'CLOUDS' ARE USED TO GIVE LOCAL REFLECTION OVER THE STAGE TO ENABLE INDIVIDUAL PLAYERS TO DISTIN-GUISH THEIR OWN AND OTHER INSTRUMENTS DURING ORCHESTRAL PERFORMANCE. THEY CAN ALSO CORRECT FOR ACOUSTICAL DEFICIENCIES EG. ALBERT HALL LONDON 'SAUCERS.' REFLECTING SURFACES SHOULD BE SMOOTH, HIGH DENSITY, NON POROUS, AT LEAST 5 $KG.M^{-2}$ FOR SPEECH ONLY AND 25 $KG.M^{-2}$ FOR MUSIC. (IN ORDER TO MINIMIZE LOW FREQUENCY ABSORPTION) TO GIVE REFLECTION OF EQUAL CHARACTER OVER A WIDE FREQUENCY SPECTRUM, A LARGE SURFACE AREA IS NECESSARY. MATERIAL: ACRYLIC OR GRP.

CONCERT HALL (5)

SOURCE: ARCHITECTS OFFICE OF PUBLIC WORKS DUBLIN.

not to scale

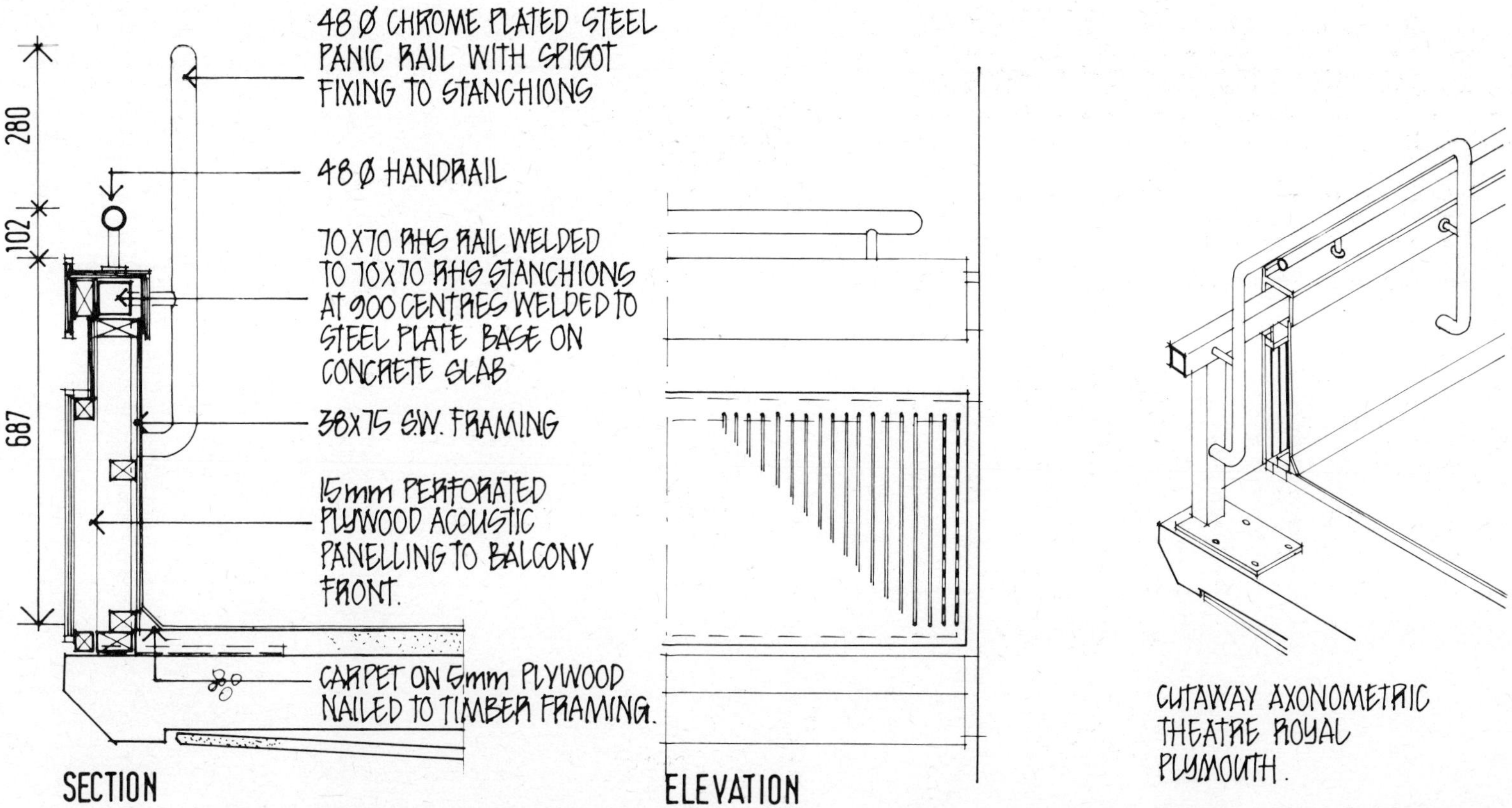

MANY AUDITORIA HAVE AS 'TUNING' TO THE NATURAL ACOUSTICS ARISING FROM FINISHES AND SEATING **SOUND ABSORBING** PANELLING OR WALL LININGS. THE PANELS ABSORB SOUND WELL AT A SELECTED FREQUENCY RANGE, BY A COMBINATION OF HELMHOLTZ AND RESONATOR ACTION. WHERE ACOUSTIC PANELLING IS REQUIRED IN SPACES IT MAKES ECONOMIC SENSE FOR IT TO BE BUILT IN AND FULFILL A USEFUL FUNCTION AS ILLUSTRATED RATHER THAN BEING HUNG ON A WALL AS AN AFTER THOUGHT.

THEATRE BALCONY
SOURCE: PETER MORO PTNRSHIP ARCHTS/AJ

not to scale

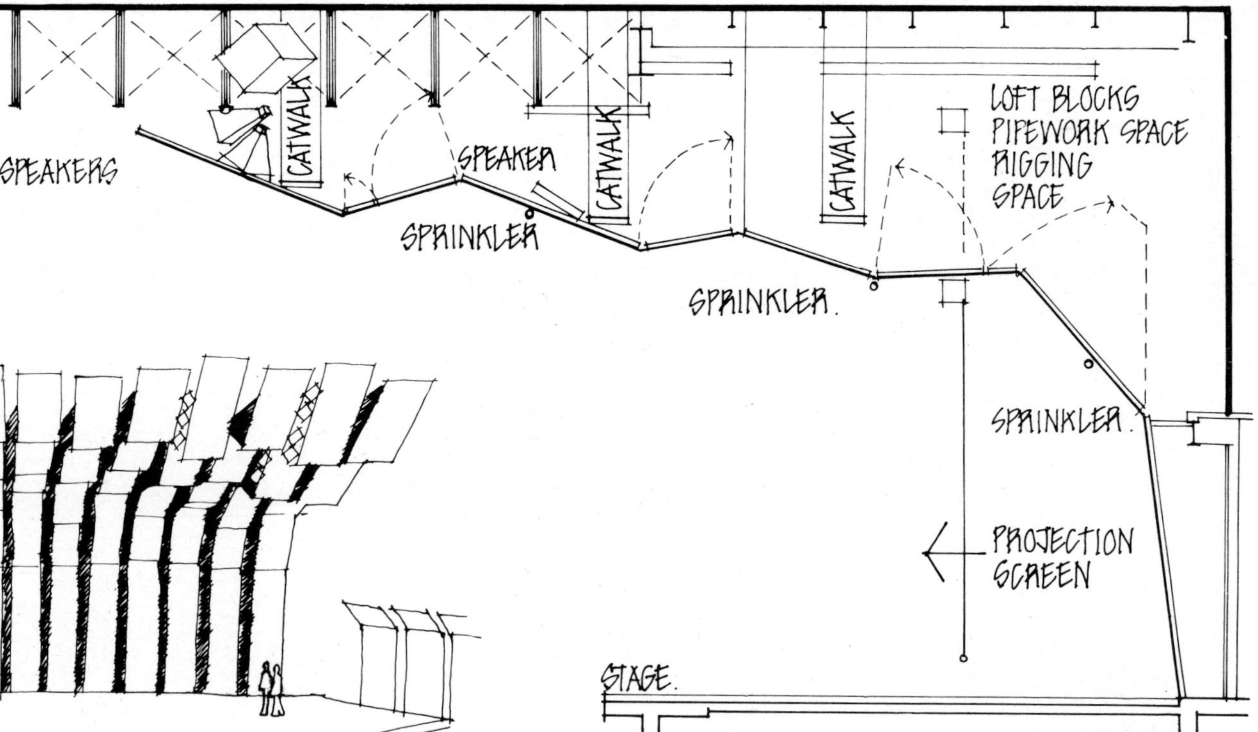

DETAIL SECTION

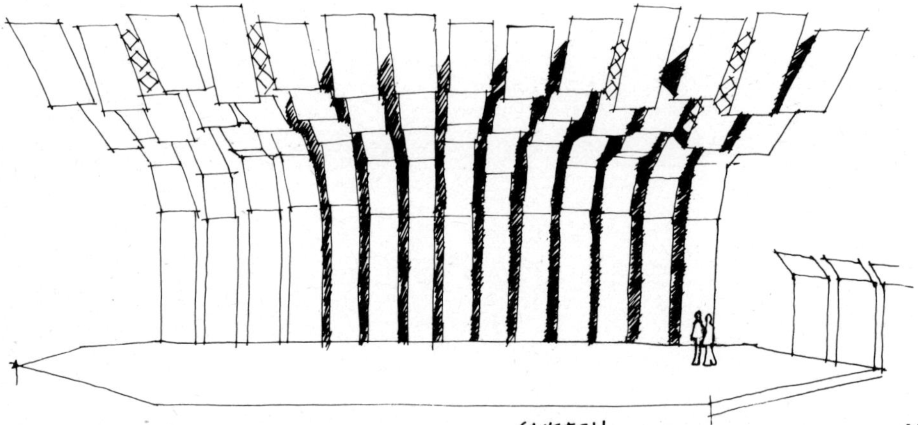

SKETCH

ADJUSTABLE PANELS, ASSEMBLED TOGETHER AS HINGED 'RIBBONS' WERE INCLUDED IN A REFURBISHMENT OF 2250-SEAT 58M DIAMETER CIRCULAR AUDITORIUM USED FOR ORCHESTRAL MUSIC, CINEMA, LECTURES, COMMUNITY AND UNIVERSITY EVENTS. ACOUSTICS IN THE HALL WERE PREVIOUSLY POOR BUT NOW REVIEWS COMMEND LACK OF DEAD SPOTS IN THE AUDITORIUM AND MUSICIANS REPORT EXCELLENT RECEPTION OF OTHER INSTRUMENTS. PANELS ARE REFLECTIVE GYPSUM BOARD ON METAL FRAMES

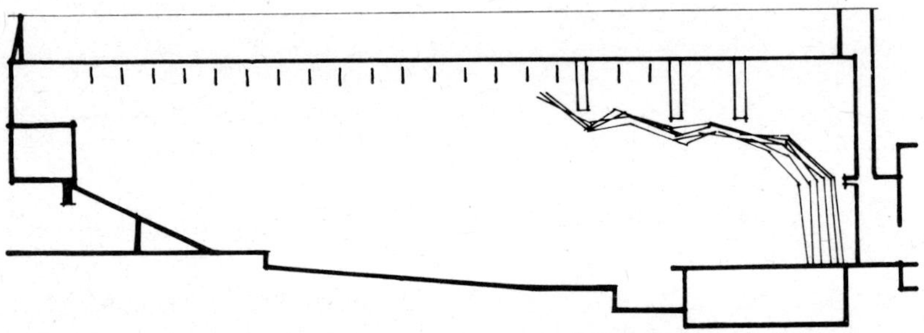

SECTION

STAGE CANOPY
ARCHITECTS: MITCHELL/GIURGOLA
NEW YORK.

not to scale

WALLS CONSTRUCTED TO SPECIFICATION PART 1. SCHEDULE 12 MUST BE EITHER:

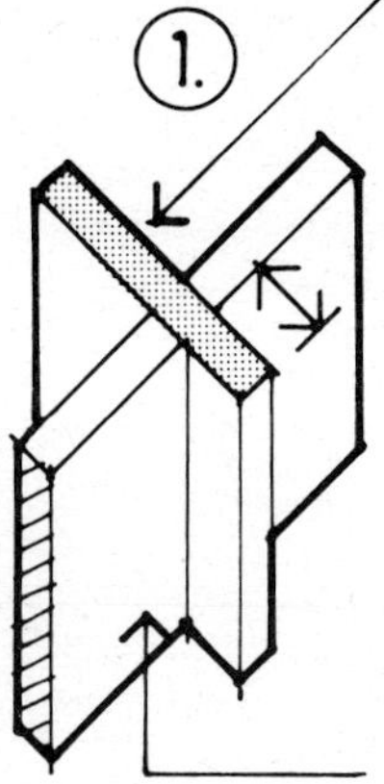

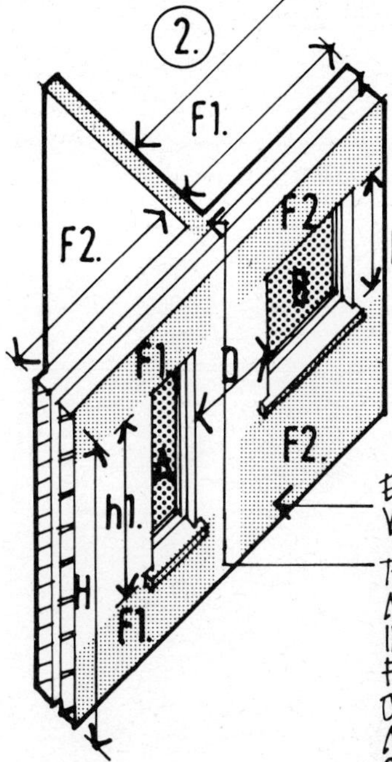

THE SEPARATING WALL MUST BE TIED OR BONDED INTO ONE LEAF OF EXTERNAL FLANKING WALL OF BRICKS OR CONCRETE WITH AVERAGE MASS OF 120 KG/M² MINIMUM OVER ANY PORTION OF THE LEAF 1 M SQUARE.

IF h_1 AND/OR h_2 ARE MORE THAN 2/3 H AND F_1 AND F_2 (ABOVE OR BELOW OPENINGS) ARE LESS THAN 3 M ON EITHER SIDE OF SEPARATING WALL THEN D MUST NOT BE LESS THAN 690 MM.

SEPARATING WALL MUST EXTEND TO OUTER FACE OF EXTERNAL FLANKING WALL OF TIMBER OR OTHER LIGHT CONSTRUCTION [BUT NOT IF TILE HUNG]

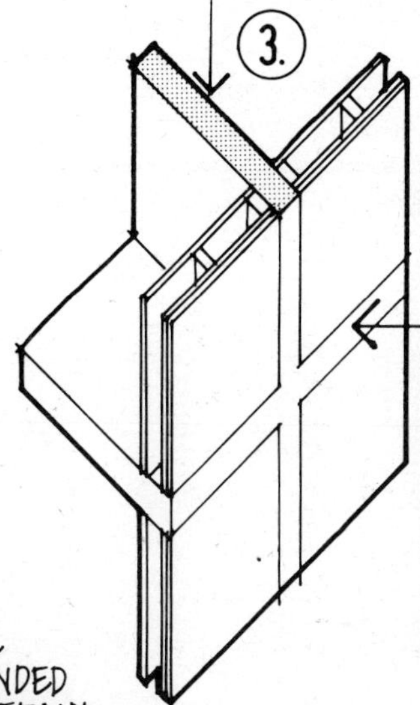

TOP AND BOTTOM OF EACH SEPARATING WALL MUST BE TIED OR BONDED TO SOLID GROUND FLOOR OR SUSPENDED CONCRETE FLOOR OF MIN. AVERAGE MASS 220 KG/M² MEASURED OVER ANY PORTION 1M² OR CONCRETE ROOF MIN. AVERAGE MASS 145 KG/M² MEASURED OVER ANY PORTION 1M SQUARE.

DEEMED-TO-SATISFY STANDARDS FOR ASSOCIATED STRUCTURE TO WALLS:
ENGLAND AND WALES
G2 (2)
BUILDING REGULATION REQUIREMENTS.

HOUSING (1)
SOURCE: NATIONAL BUILDING AGENCY.
BUILDING DESIGN MAG.

not to scale

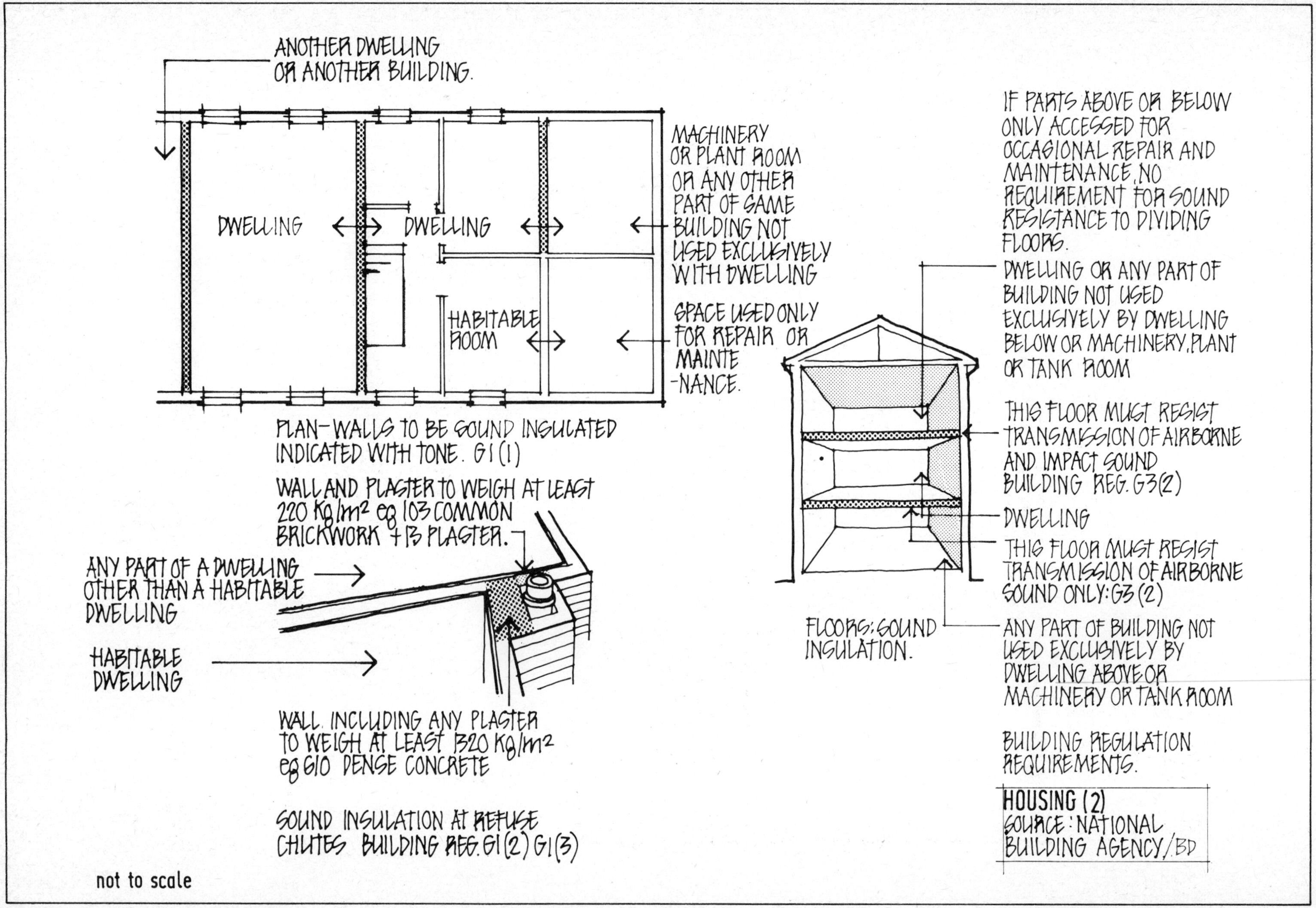
ANOTHER DWELLING OR ANOTHER BUILDING.
DWELLING
DWELLING
HABITABLE ROOM
MACHINERY OR PLANT ROOM OR ANY OTHER PART OF SAME BUILDING NOT USED EXCLUSIVELY WITH DWELLING
SPACE USED ONLY FOR REPAIR OR MAINTE-NANCE.
PLAN-WALLS TO BE SOUND INSULATED INDICATED WITH TONE. G1(1)
WALL AND PLASTER TO WEIGH AT LEAST 220 Kg/m2 eg 103 COMMON BRICKWORK + 13 PLASTER.
ANY PART OF A DWELLING OTHER THAN A HABITABLE DWELLING
HABITABLE DWELLING
WALL. INCLUDING ANY PLASTER TO WEIGH AT LEAST 1320 Kg/m2 eg 610 DENSE CONCRETE
SOUND INSULATION AT REFUSE CHUTES BUILDING REG. G1(2) G1(3)
not to scale
IF PARTS ABOVE OR BELOW ONLY ACCESSED FOR OCCASIONAL REPAIR AND MAINTENANCE, NO REQUIREMENT FOR SOUND RESISTANCE TO DIVIDING FLOORS.
DWELLING OR ANY PART OF BUILDING NOT USED EXCLUSIVELY BY DWELLING BELOW OR MACHINERY, PLANT OR TANK ROOM
THIS FLOOR MUST RESIST TRANSMISSION OF AIRBORNE AND IMPACT SOUND BUILDING REG. G3(2)
DWELLING
THIS FLOOR MUST RESIST TRANSMISSION OF AIRBORNE SOUND ONLY: G3(2)
FLOORS: SOUND INSULATION.
ANY PART OF BUILDING NOT USED EXCLUSIVELY BY DWELLING ABOVE OR MACHINERY OR TANK ROOM
BUILDING REGULATION REQUIREMENTS.
HOUSING (2)
SOURCE: NATIONAL BUILDING AGENCY/BD

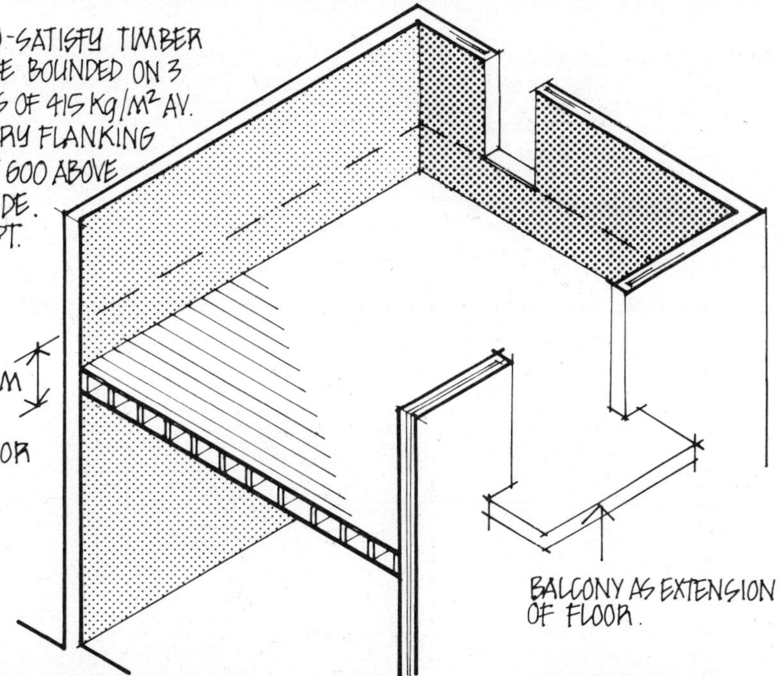

ENGLAND AND WALES BUILDING REGULATION G.4(2)

TIMBER FLOORS

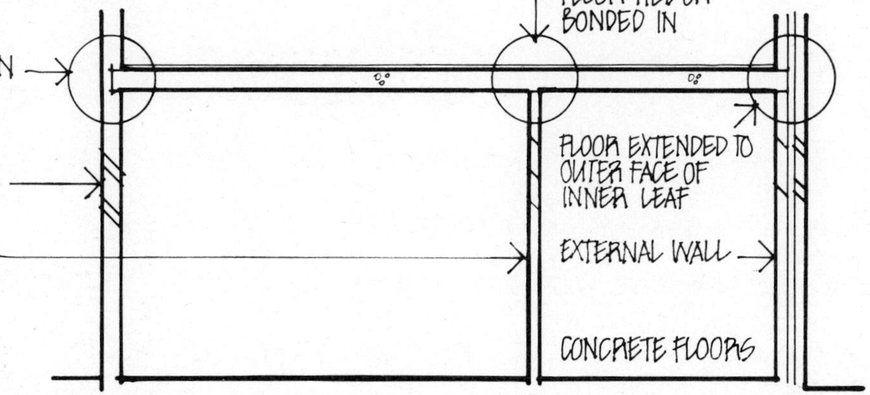

CONCRETE FLOORS

THERE ARE NO MANDATORY REQUIREMENTS FOR SOUND INSULATION WITHIN DWELLINGS, AS OCCUPANTS CAN CONTROL NOISE BY AGREEMENT WITHIN THE HOUSEHOLD. RECOMMENDATIONS HOWEVER SHOULD BE CONSIDERED

1. PARTITIONS BETWEEN WC AND LIVING ROOMS OR BEDROOM SHOULD REDUCE AIRBORNE SOUND TRANSMISSION BY 35 dB [CP3 CH III]
2. CASE OR MOUNT EXTERNALLY OR LOCATE IN CUPBOARDS SVP'S., RWP'S.

IF EXTERNAL NOISE IS A PROBLEM, CHECK DESIGN OF 'WEAK POINTS'—AIRBRICKS, FLUES PLUMBING WASTES. CHECK WHETHER NOISE INSULATION REGULATIONS 1975 APPLY. 18 HR. L_{10} LEVEL OF 68 dB IS LIMIT BEYOND WHICH DOUBLE GLAZING AND MECHANICAL VENTILATION MAY BE PROVIDED [OBLIGATORY POWER TO PROVIDE IF ROAD OPENED AFTER 16 OCTOBER 1972] OTHER LEVELS OR PERIODS MAY STILL ALLOW QUALIFICATION FOR GRANT. A GOOD DESIGN STANDARD IS CONSIDERED L_{10} = 40 dBA MAX. INTERNAL BACKGROUND LEVEL WINDOWS CLOSED DESIGN BULLETIN 26 : HMSO : 1974.

HOUSING (3)
SOURCE : NBA/BD

not to scale

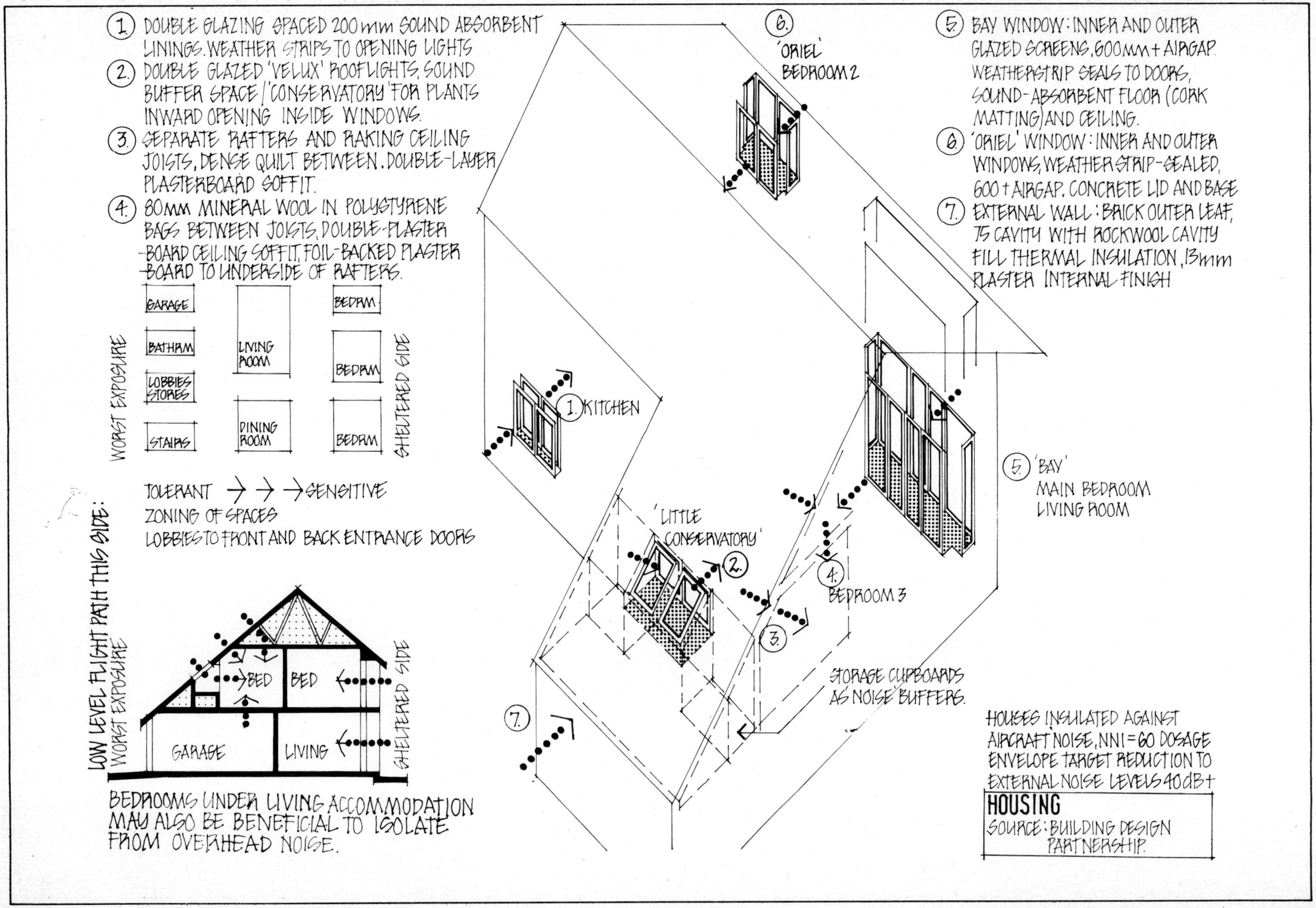

1. DOUBLE GLAZING SPACED 200mm SOUND ABSORBENT LININGS. WEATHER STRIPS TO OPENING LIGHTS
2. DOUBLE GLAZED 'VELUX' ROOFLIGHTS, SOUND BUFFER SPACE / 'CONSERVATORY' FOR PLANTS INWARD OPENING INSIDE WINDOWS.
3. SEPARATE RAFTERS AND RAKING CEILING JOISTS, DENSE QUILT BETWEEN. DOUBLE-LAYER PLASTERBOARD SOFFIT.
4. 80mm MINERAL WOOL IN POLYSTYRENE BAGS BETWEEN JOISTS, DOUBLE-PLASTER-BOARD CEILING SOFFIT, FOIL-BACKED PLASTER-BOARD TO UNDERSIDE OF RAFTERS.
GARAGE
BATHRM
LOBBIES STORES
STAIRS
LIVING ROOM
DINING ROOM
BEDRM
BEDRM
BEDRM
WORST EXPOSURE
SHELTERED SIDE
TOLERANT → → → SENSITIVE
ZONING OF SPACES
LOBBIES TO FRONT AND BACK ENTRANCE DOORS
LOW LEVEL FLIGHT PATH THIS SIDE:
WORST EXPOSURE
BED
BED
GARAGE
LIVING
SHELTERED SIDE
BEDROOMS UNDER LIVING ACCOMMODATION MAY ALSO BE BENEFICIAL TO ISOLATE FROM OVERHEAD NOISE.
6. 'ORIEL' BEDROOM 2
1. KITCHEN
'LITTLE CONSERVATORY'
2.
3.
4. BEDROOM 3
7.
STORAGE CUPBOARDS AS NOISE BUFFERS.
5. 'BAY' MAIN BEDROOM LIVING ROOM
5. BAY WINDOW: INNER AND OUTER GLAZED SCREENS, 600mm + AIRGAP. WEATHERSTRIP SEALS TO DOORS, SOUND-ABSORBENT FLOOR (CORK MATTING) AND CEILING.
6. 'ORIEL' WINDOW: INNER AND OUTER WINDOWS, WEATHER STRIP-SEALED, 600 + AIRGAP. CONCRETE LID AND BASE
7. EXTERNAL WALL: BRICK OUTER LEAF, 75 CAVITY WITH ROCKWOOL CAVITY FILL THERMAL INSULATION, 13mm PLASTER INTERNAL FINISH
HOUSES INSULATED AGAINST AIRCRAFT NOISE, NNI = 60 DOSAGE ENVELOPE TARGET REDUCTION TO EXTERNAL NOISE LEVELS 40dB+
HOUSING
SOURCE: BUILDING DESIGN PARTNERSHIP.

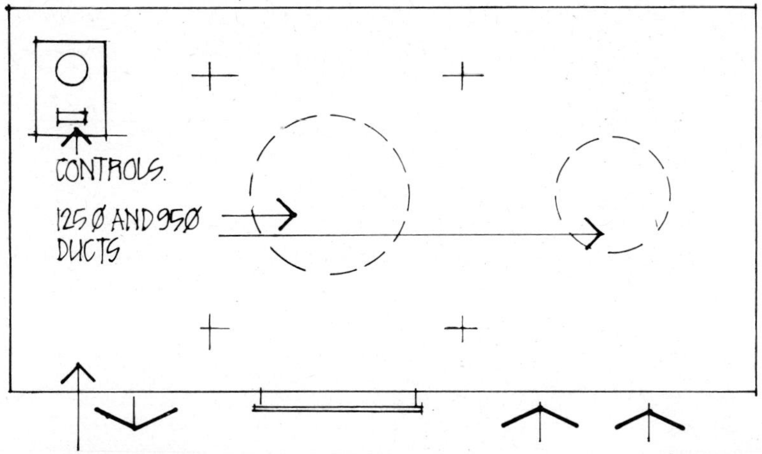

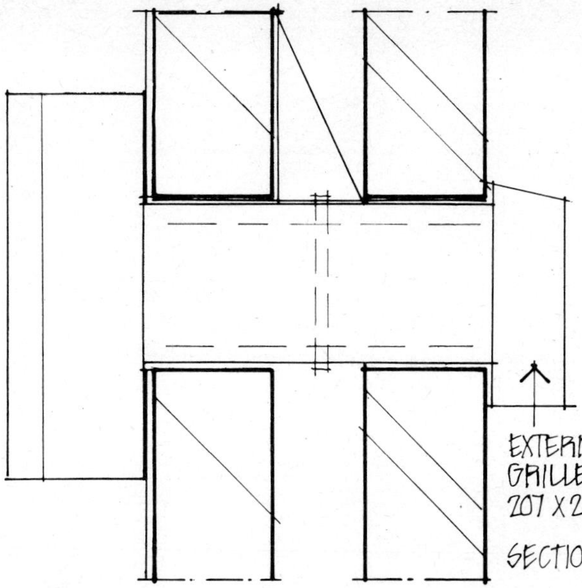

'ACCADIAIR AEROVENT' DAY 116A SLIMLINE VENTILATOR UNIT. PROVIDES NOT LESS THAN 3250 mm² NORMALIZED SELF-GENERATED SOUND PRESSURE LEVELS NOT EXCEEDING 35 dBA AT 31 LITRES PER SECOND OR 40 dBA AT MAX VOLUME.

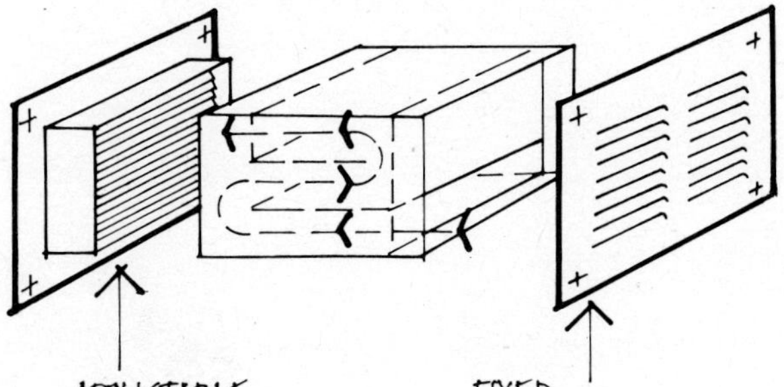

'AEROMUTE' TYPE B AMB 202 SOUND REDUCING AIRBRICK BUILT TO NOISE REDUCTION STANDARDS 129/1972 AND 1363/1973 PROVIDES 10500 mm² FREE AIR SPACE THROUGH UNIT. AV NOISE REDUCTION (100-3150 Hz) 48 dB TOTAL ADVERSE DEVIATION AT ALL 1/3 RD OCTAVE FREQUENCIES WITHIN 32 dB TOLERANCE ALLOWED.

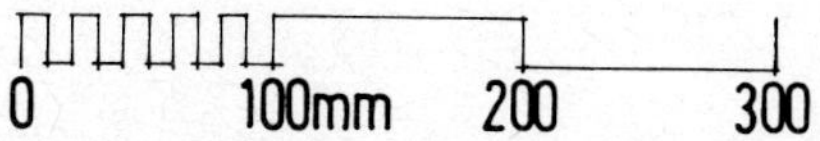

WALL VENTILATORS.

THESE TYPE OF VENTILATOR UNITS ARE WIDELY USED BY LOCAL AUTHORITIES IN HOUSES NEAR NOISY MOTORWAYS, RAILWAYS OR AIRPORTS TO ALLOW AIR TO BE TAKEN IN OR EXTRACTED.

HOUSING
SOURCE: SRL LTD / GLC.

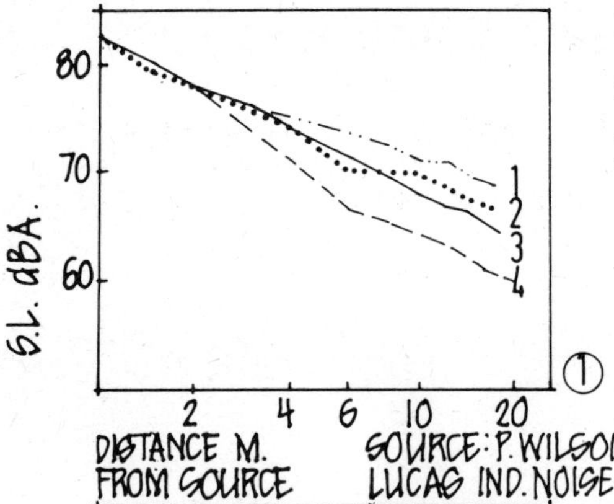

1. OPEN GANGWAY, UNTREATED FACTORY
2. ACROSS BENCHES, UNTREATED FACTORY
3. OPEN GANGWAY, ABSORBENT-LINED FACTORY
4. ACROSS MACHINES, ABSORBENT LINED MACHINES

① SOURCE: P. WILSON LUCAS IND. NOISE CENTRE

FACTORY TYPE	DECAY RATE: dB/DOUBLING DISTANCE
HARD EMPTY	−3
HARD WITH SCATTER ELEMENTS	−4
ABSORBENT LINED EMPTY	−4
ABSORBENT LINED WITH SCATTER ELEMENTS	−5

②

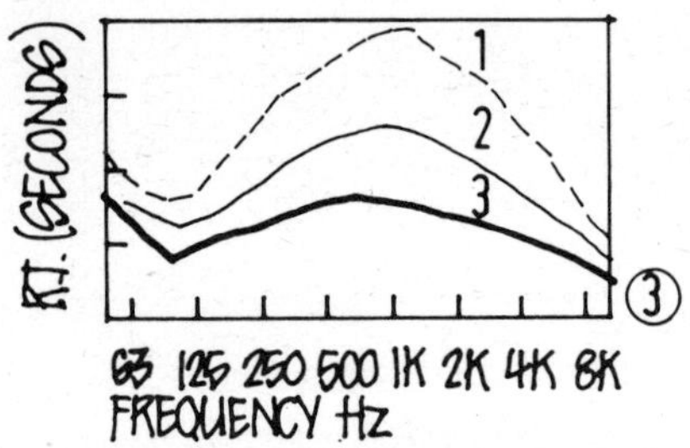

③

1. NO MACHINES
2. 25 MACHINES
3. 50 MACHINES

FACTORY 45X43X4-3m
HARD FINISHES
SOURCE: M. HODGSON
UNIVERSITY CAMBRIDGE

DESIGN GUIDE

1. WHERE EMPLOYEES EXPOSED TO ⩾ 8 HR LEQ. OF 90 dB (A),* NOISE REDUCTION MEASURES TO BE TAKEN AS FAR AS REASONABLY PRACTICABLE. EAR DEFENDERS TO BE WORN.
2. CLASSICAL DESCRIPTION OF SOUND IN ENCLOSURES (eg. SABINE) NOT APPLICABLE. TENDENCY FOR ABSENCE OF CONSTANT REVERBERANT SOUND LEVEL OUTSIDE DIRECT SOUND FIELD OF SOURCES. THIS MEANS ABSORBENT TREATMENTS DO NOT NECESSARILY PRODUCE MARKED SOUND LEVEL REDUCTIONS IF REMOTE.
3. LACK OF REVERBERANT SOUND FIELD MEANS ZONING OF NOISY FROM QUIET PROCESSES AND LOCAL SCREENING ARE EFFECTIVE MEASURES
4. OVERHEAD SUSPENDED ABSORBERS COUPLED WITH LOCAL SCREENING ARE OFTEN THE BEST FORM OF INCORPORATING SOUND ABSORPTION.
5. OPERATOR EXPERIENCE OF NOISE IS NOT IMPROVED BY TREATMENT IN FACTORY SPACE – NO VARIATION IN DECAY RATE > 2 M. (1).
6. FOR DECAY RATE FROM AN OMNIDIRECTIONAL SOURCE IN A FACTORY SPACE, GUIDE LINES (2.) MAY BE CONSIDERED
7. MACHINERY AND PLANT THEMSELVES PROVIDE SIGNIFICANT SCATTER AND ABSORPTION (3.).

* EEC PROPOSAL FOR COUNCIL DIRECTIVE (18 OCTOBER 1982): DAILY EXPOSURE MUST NOT EXCEED 85 dB(A) BUT TO TAKE ACCOUNT OF FEASIBILITY 5 YEARS ARE TO BE ALLOWED FOR IMPLEMENTATION

INDUSTRIAL BUILDINGS

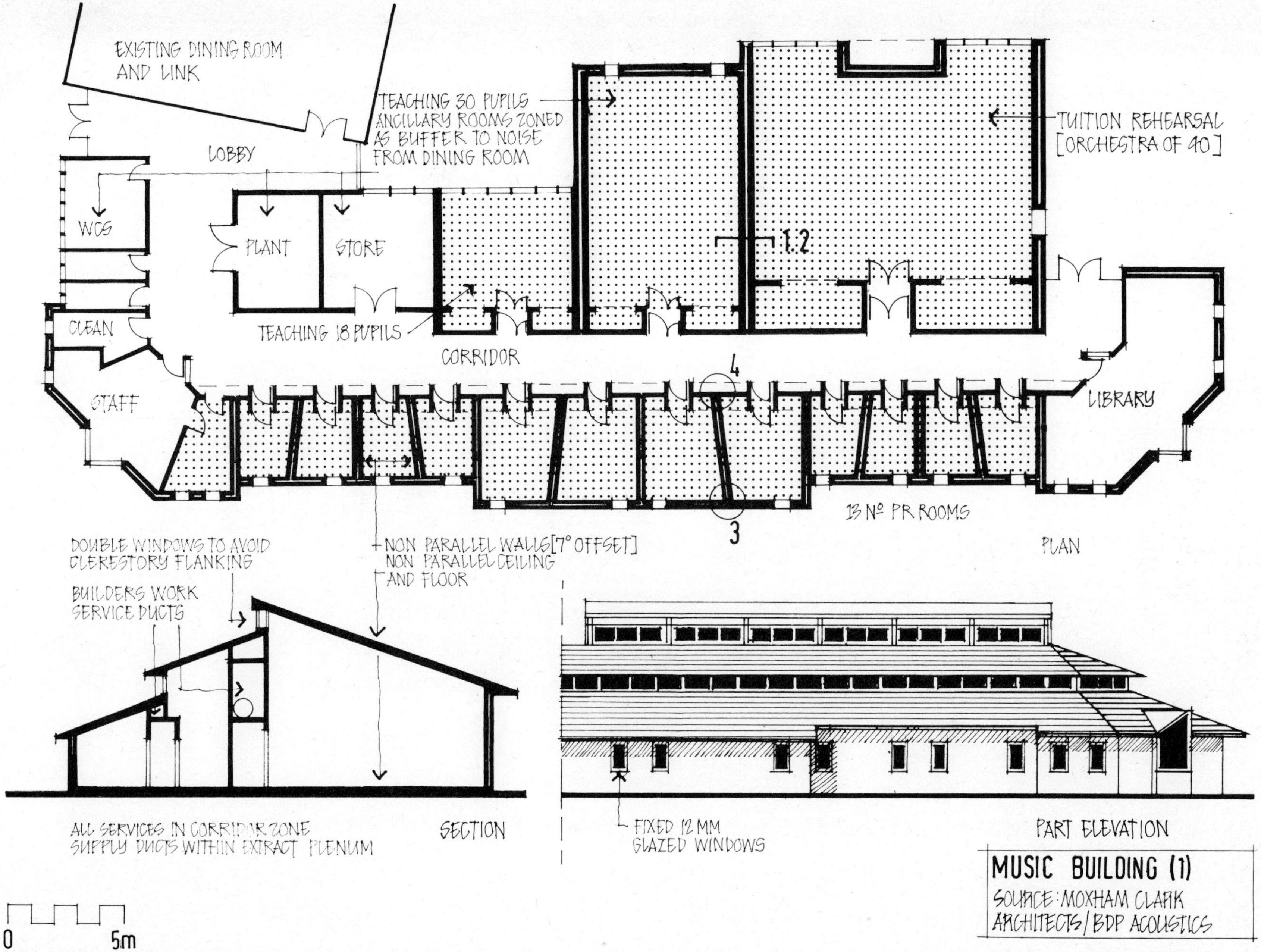

MUSIC BUILDING (1)

SOURCE: MOXHAM CLARK ARCHITECTS/BDP ACOUSTICS

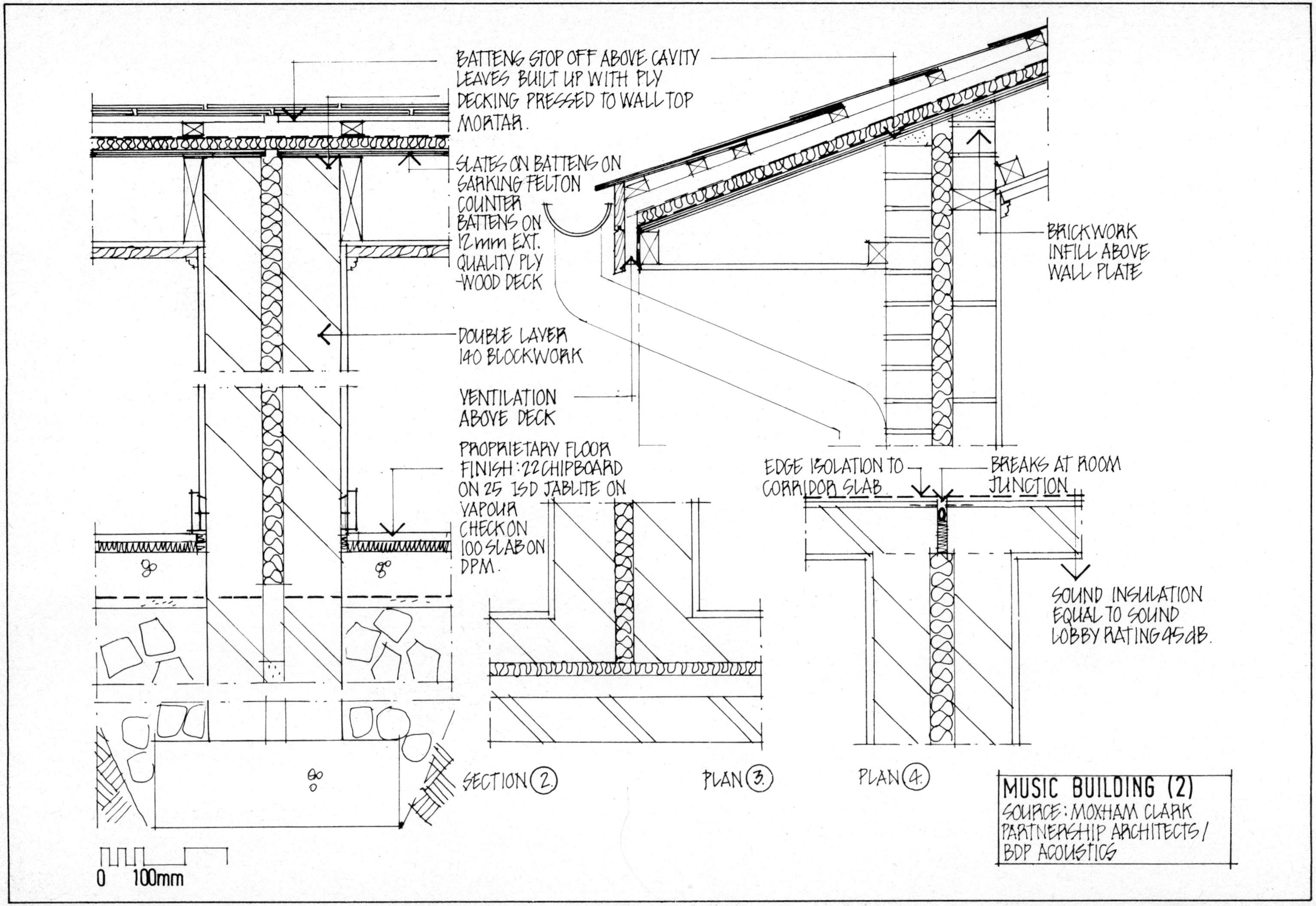
BATTENS STOP OFF ABOVE CAVITY LEAVES BUILT UP WITH PLY DECKING PRESSED TO WALL TOP MORTAR.
SLATES ON BATTENS ON SARKING FELT ON COUNTER BATTENS ON 12mm EXT. QUALITY PLY-WOOD DECK
BRICKWORK INFILL ABOVE WALL PLATE
DOUBLE LAYER 140 BLOCKWORK
VENTILATION ABOVE DECK
PROPRIETARY FLOOR FINISH: 22 CHIPBOARD ON 25 ISD JABLITE ON VAPOUR CHECK ON 100 SLAB ON DPM.
EDGE ISOLATION TO CORRIDOR SLAB
BREAKS AT ROOM JUNCTION
SOUND INSULATION EQUAL TO SOUND LOBBY RATING 45dB.
SECTION 2.
PLAN 3.
PLAN 4.
0 100mm
MUSIC BUILDING (2)
SOURCE: MOXHAM CLARK PARTNERSHIP ARCHITECTS / BDP ACOUSTICS

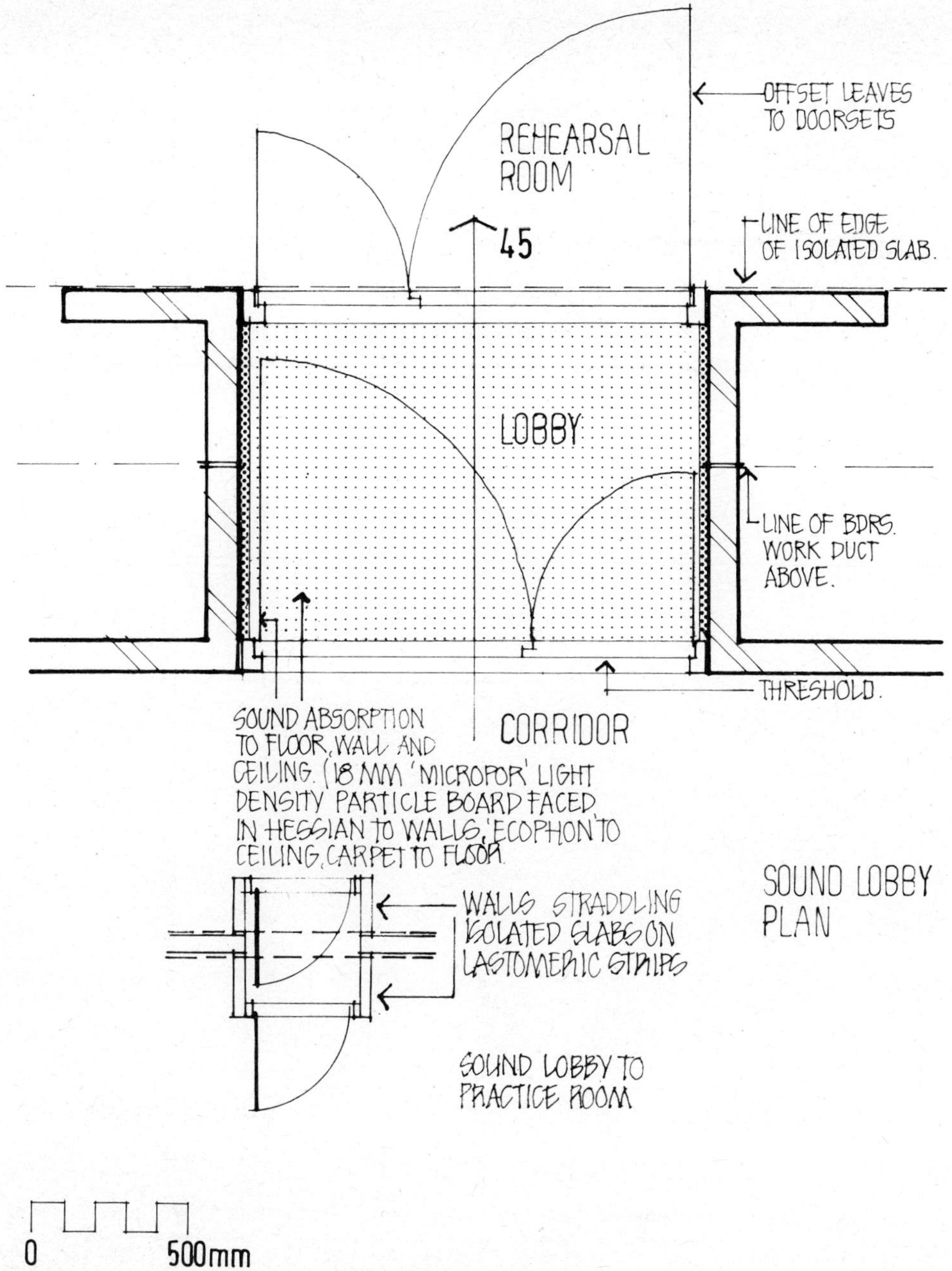

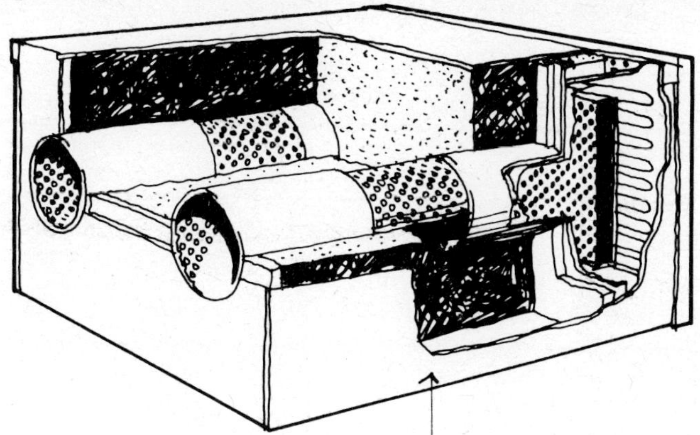

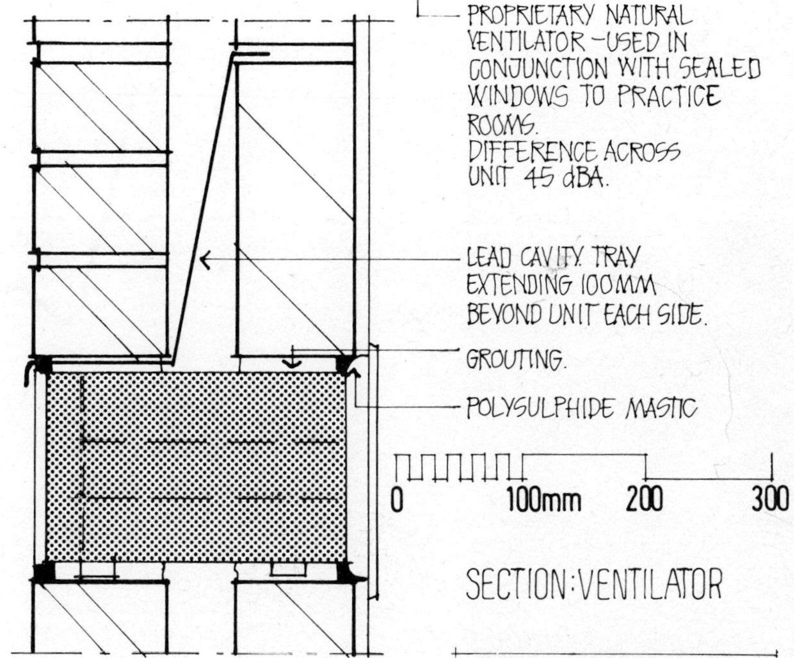

MUSIC BUILDING (3)

SOURCE: MOXHAM CLARK ARCHTS. / BDP ACOUSTICS

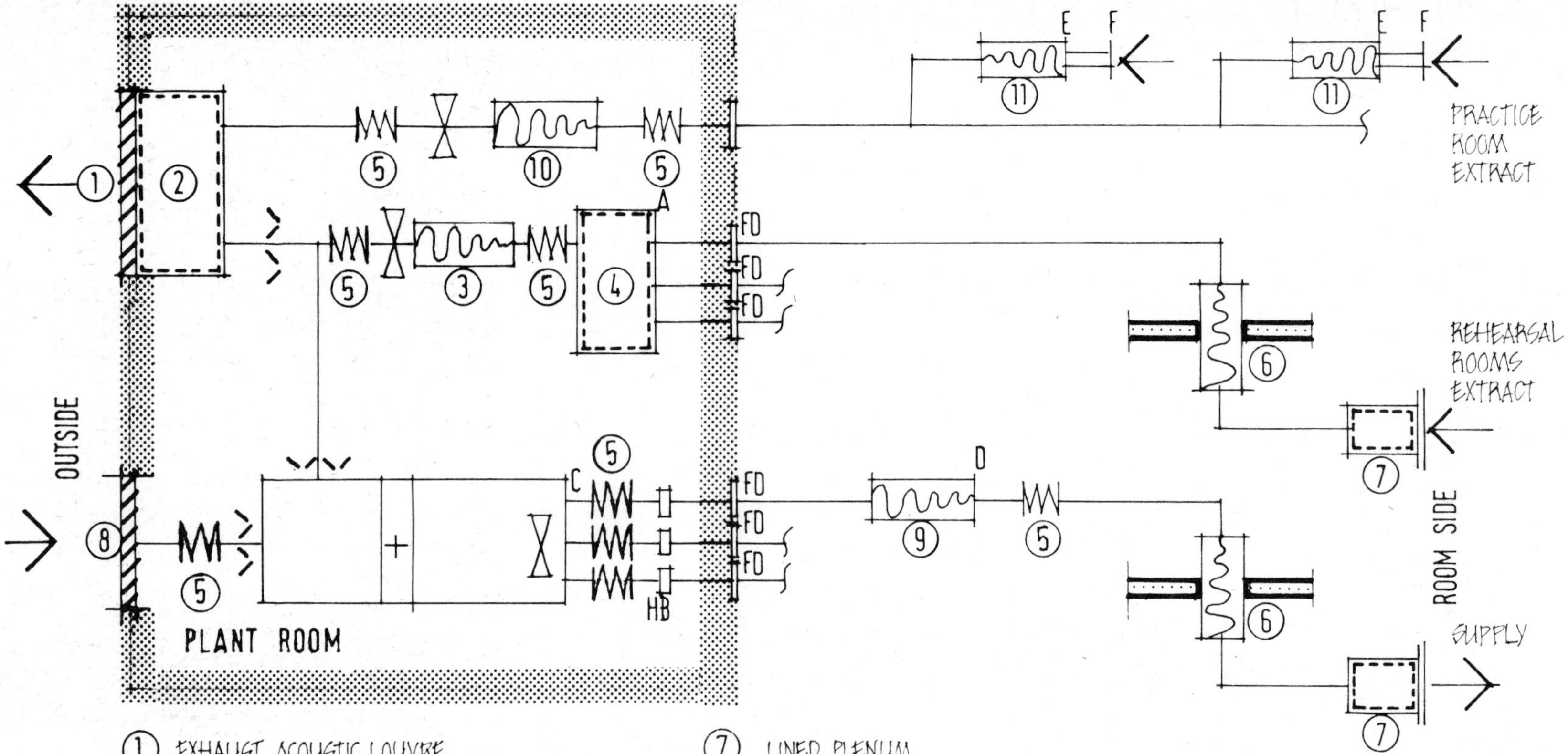

(1) EXHAUST ACOUSTIC LOUVRE

(2) EXHAUST CHAMBER (LINED)

(3) RETURN AIR ATTENUATOR

(4) RETURN AIR CHAMBER (LINED)

(5) FLEXIBLE CONNECTION

(6) CROSS TALK ATTENUATOR

(7) LINED PLENUM

(8) INLET ACOUSTIC LOUVRE

(9) SUPPLY AIR ATTENUATOR

(10) PR ROOMS RETURN AIR FAN ATTENUATOR

(11) EXTRACT CROSS TALK ATTENUATOR.

FD FIRE DAMPER.
HB HEATER BATTERY.

VENTILATION PLANT:
ACOUSTIC SCHEMATIC

MUSIC BUILDING (4)
SOURCE: MOXHAM CLARK ARCHITECTS / BDP ACOUSTICS.

3 APPENDIX A

SOUND ABSORPTION

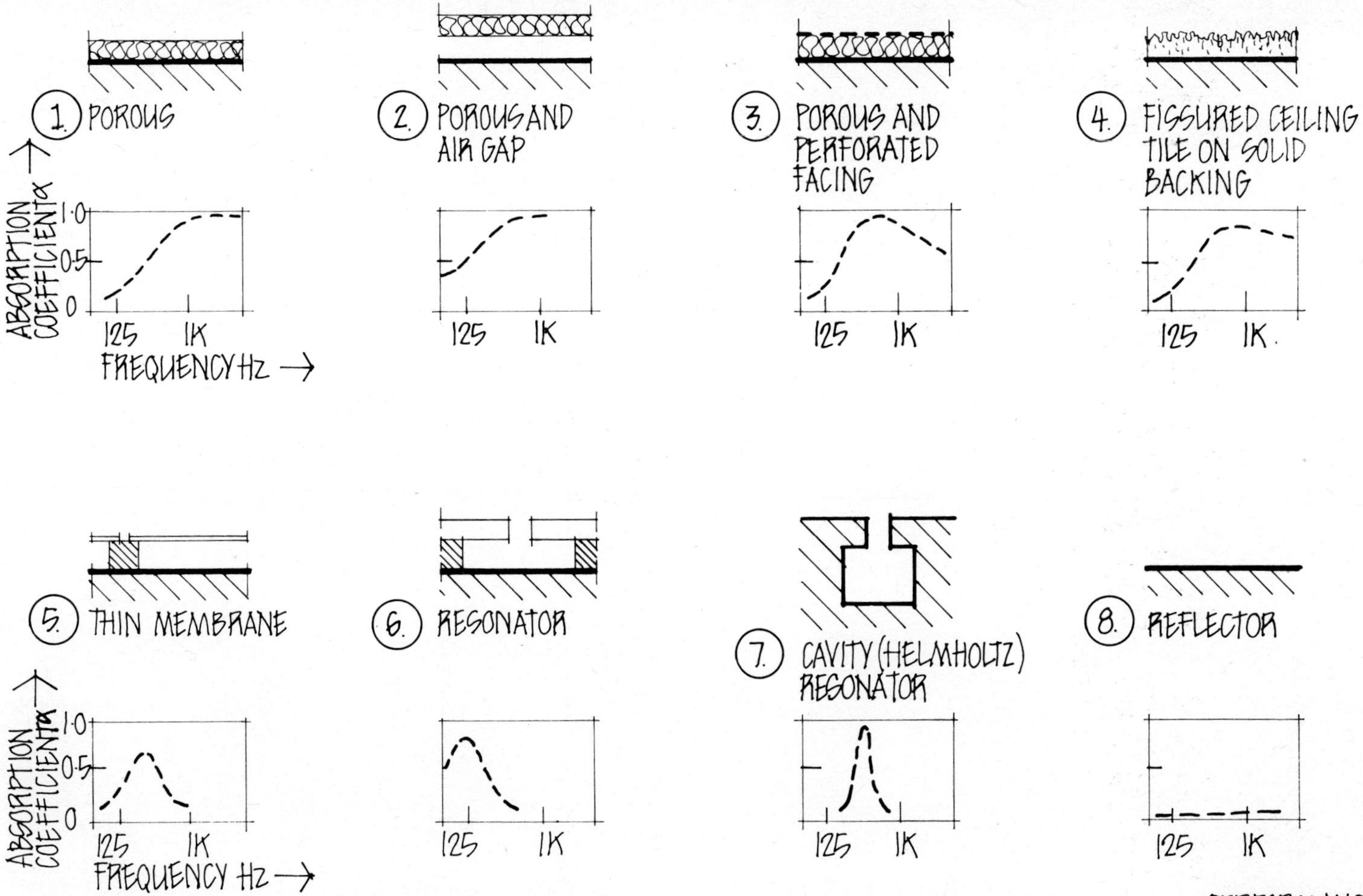

PERFORMANCE OF DIFFERENT WALL OR CEILING FINISHES

SOUND ABSORPTION

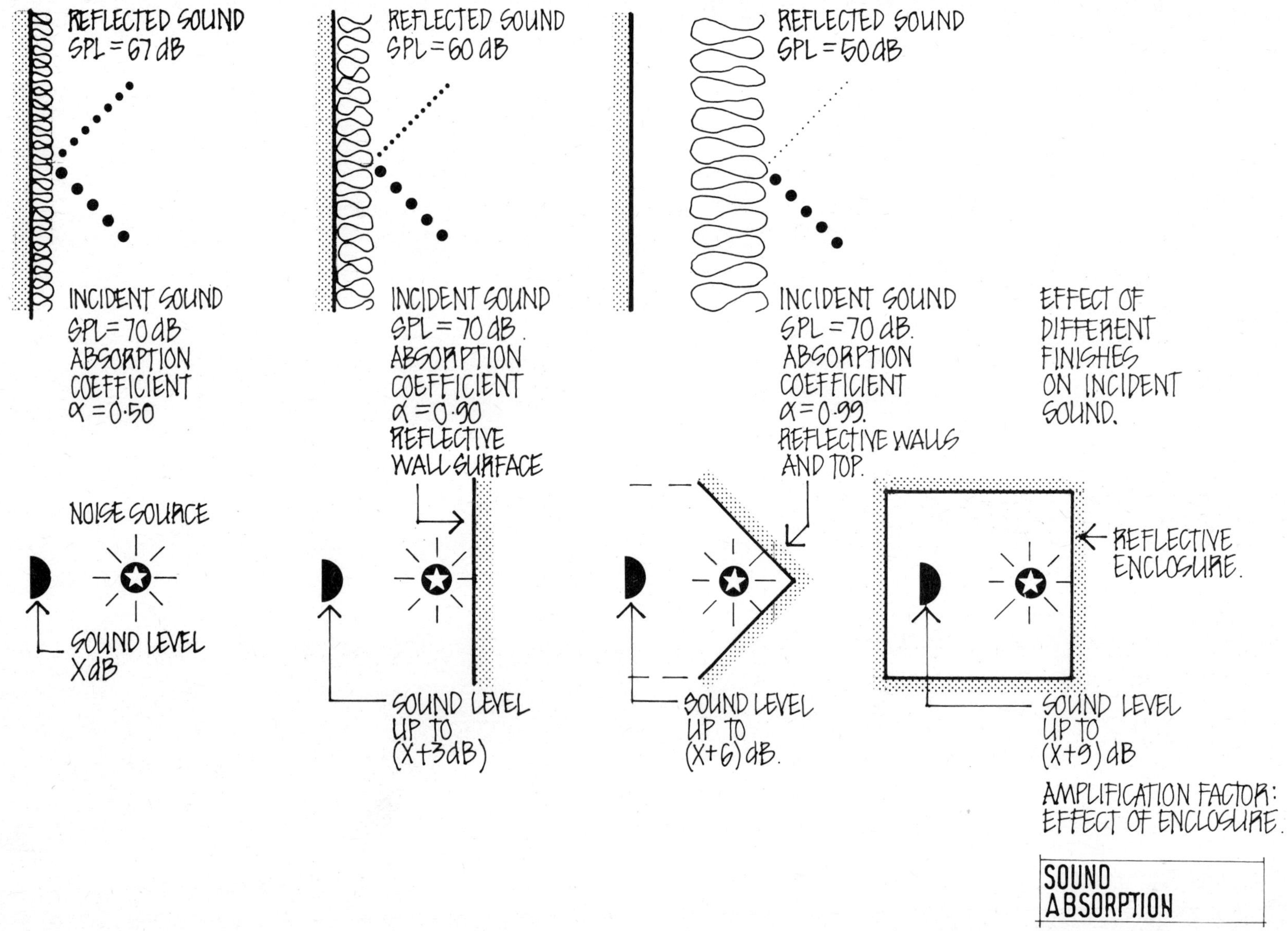
REFLECTED SOUND
SPL = 67 dB
INCIDENT SOUND
SPL = 70 dB
ABSORPTION
COEFFICIENT
α = 0·50
REFLECTED SOUND
SPL = 60 dB
INCIDENT SOUND
SPL = 70 dB.
ABSORPTION
COEFFICIENT
α = 0·90
REFLECTED SOUND
SPL = 50 dB
INCIDENT SOUND
SPL = 70 dB.
ABSORPTION
COEFFICIENT
α = 0·99.
EFFECT OF
DIFFERENT
FINISHES
ON INCIDENT
SOUND.
REFLECTIVE
WALL SURFACE
REFLECTIVE WALLS
AND TOP.
NOISE SOURCE
SOUND LEVEL
X dB
SOUND LEVEL
UP TO
(X+3dB)
SOUND LEVEL
UP TO
(X+6) dB.
REFLECTIVE
ENCLOSURE.
SOUND LEVEL
UP TO
(X+9) dB
AMPLIFICATION FACTOR:
EFFECT OF ENCLOSURE.
SOUND
ABSORPTION

SOUND INSULATION

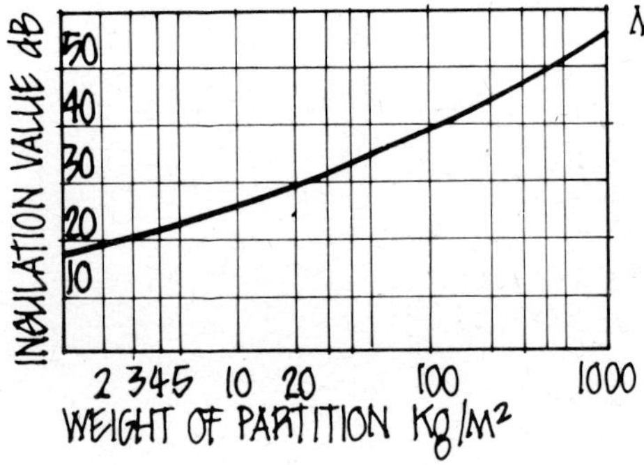

MASS LAW FOR SOUND INSULATION

FOR SINGLE WALLS THE WEIGHT PER UNIT AREA IS AN INDICATION OF INSULATION VALUE ONLY. RESULTS BELOW MASS EXPECTATION MAY OCCUR BECAUSE OF: RESONANCE AND COINCIDENCE EFFECT; FLANKING; EDGE FIXING. COMPOSITE CONSTRUCTION OF A SERIES OF LAYERS IMPROVES THE SOUND INSULATION ABOVE MASS LAW EXPECTATION AS DOES: RESILIENT MOUNTING OF PANELS; DISCONTINUITY OF CONSTRUCTION; DOUBLE OR TRIPLE LAYERS WITH WIDE AIR SPACE. A HEAVY DIVIDING ELEMENT IS PARTICULARY EFFECTIVE FOR LOW FREQUENCY SOUND.

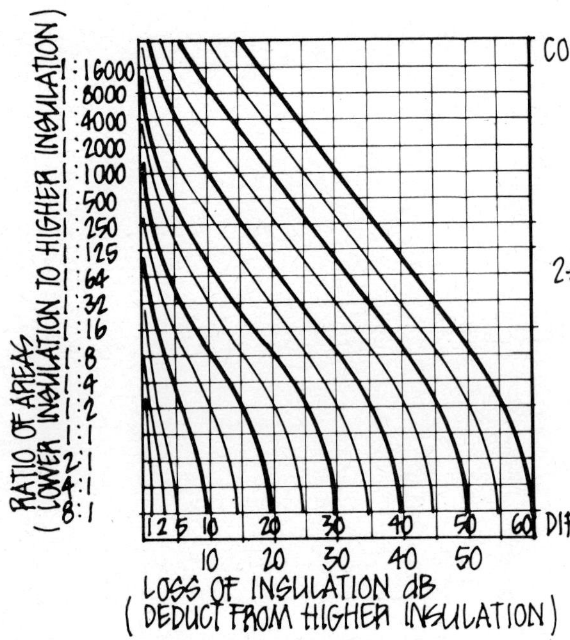

COMPOSITE ELEMENTS OF STRUCTURE

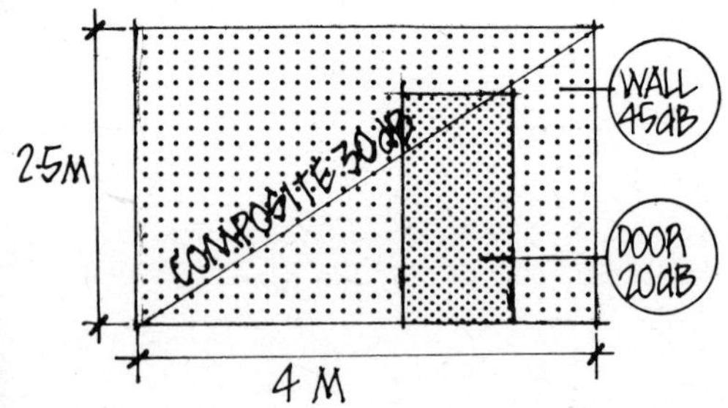

THE COMPOSITE SOUND REDUCTION INDEX (S.R.I.) OF PARTITION BETWEEN 2 ROOMS, OR OUTSIDE WALLS WITH WINDOWS IN A ROOM, CAN BE OBTAINED FROM THE FIGURE AT LHS. AN EXAMPLE OF THE EFFECT IS ALSO SHOWN AT LHS. A HOLE IN A 10 SQ.M. 45 dB-RATED WALL OF 10mm x 10mm DOWN RATES THE COMPOSITE RATING TO 30 dB

SOUND INSULATION
SOURCE: PARKIN HUMPREYS AND COWELL

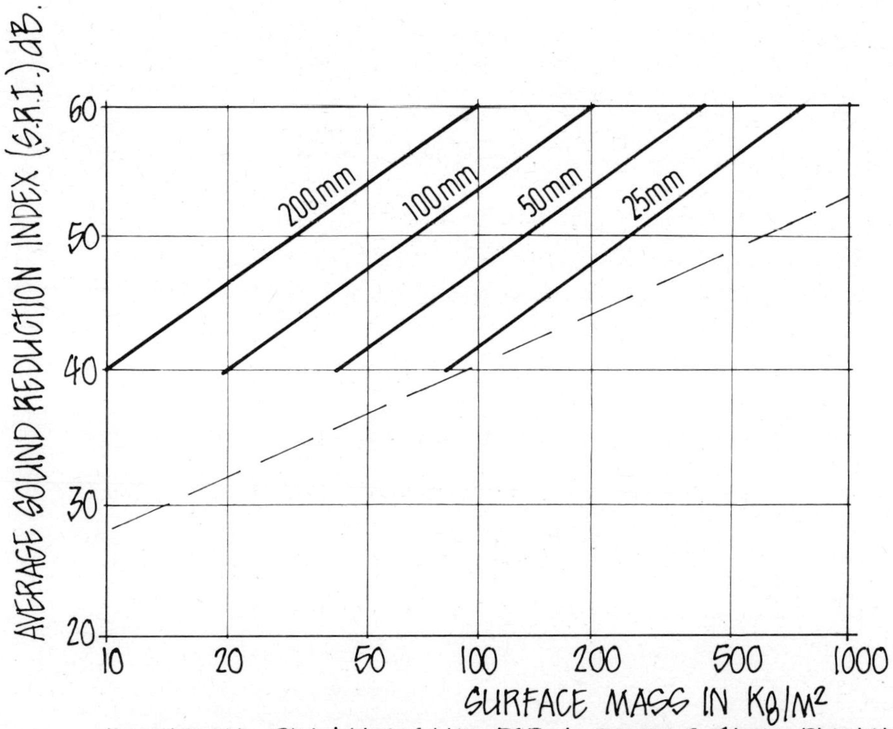

EMPIRICAL RELATIONSHIP FOR A DOUBLE SKIN PARTITION

— — — — SINGLE SKIN PARTITION

DIMENSIONS ARE FOR DIFFERENT CAVITY WIDTHS BETWEEN LEAVES.

THE 'MASS LAW' FOR SINGLE WALLS HAS A CORRESPONDING EMPIRICAL RELATIONSHIP BETWEEN S.R.I. AND WEIGHT PER UNIT AREA FOR DOUBLE LEAF PARTITIONS. AS WITH SINGLE WALLS, THESE CURVES ARE AN INDICATION ONLY BECAUSE THE S.R.I. WILL VARY WITH FREQUENCY.

SOUND INSULATION

EXTERNAL NOISE

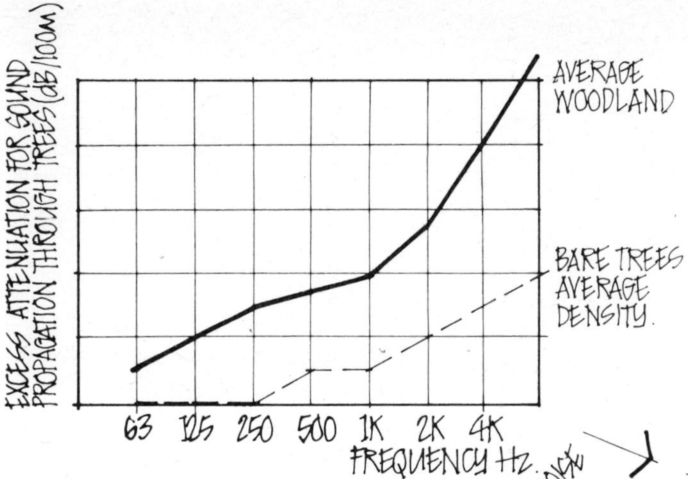

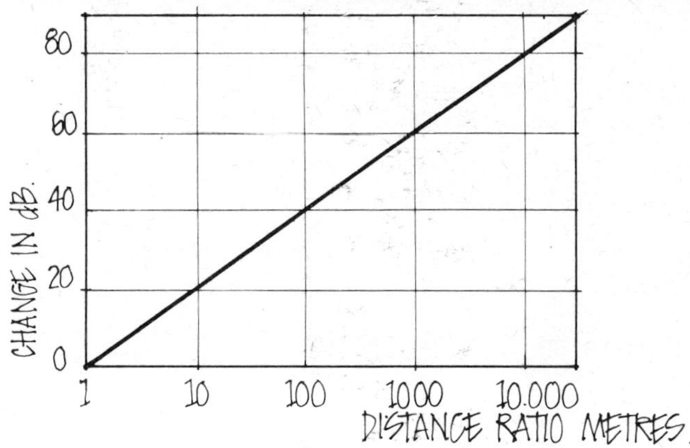

PLANTING.

PLANTING IS OF SOME BENEFIT FOR HIGH FREQUENCY ATTENUATION BUT OTHERWISE OF LIMITED USE IN SCREENING, SAY, TRAFFIC NOISE COMPARED WITH MOUNDING.

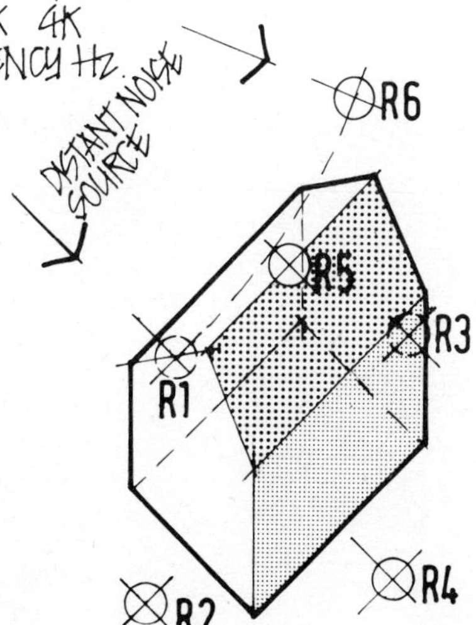

NOISE LEVEL AT FRONT	**R1** = X
NOISE LEVEL AT SIDES	**R2** = X-6dB
	R3 = X-6dB
NOISE LEVEL AT REAR	**R4** = X-10dB
NOISE LEVEL AT ROOF	**R5** = X-3dB
NOISE LEVEL WITH NO BUILD-UP RESULTING FROM WALL BEHIND	**R6** = X-2.5dB

DISTANCE

THE MAIN FACTOR AFFECTING REDUCTION OF NOISE LEVEL WITH DISTANCE IS THE SPREAD OF ENERGY. DOUBLING DISTANCE TENDS TO CAUSE A REDUCTION OF 6dB. OTHER FACTORS AFFECT THE REDUCTION WITH DISTANCE: GROUND EFFECTS, TEMPERATURE WIND, FOG, SNOW AND RAIN. FOR A LINEAR NOISE SOURCE LIKE TRAFFIC, THE REDUCTION WITH DISTANCE IS MUCH LESS - 3dB / DOUBLING.

BUILDINGS

BUILDINGS AND WALLS AFFECT SOUND LEVELS LOCALLY, SO LOCATING ROOMS AND WINDOWS TO MORE SHELTERED ELEVATIONS WILL GIVE USEFULLY LOWER INTRUSIVE NOISE LEVELS WITHIN.

EXTERNAL NOISE

4 APPENDIX B

Absorption
The ability of a material to absorb sound is measured by its 'absorption coefficient' α which is defined as the fraction of energy incident on its surface which is not reflected. The higher the decimal fraction value, the better the absorptive property of the material. This does not imply that some of the energy does not continue on through the material. The absorption A of a surface of area S is then given by:

$$A = S\,\alpha \qquad \ldots\ldots (1)$$

and if there are several absorbing surfaces of areas, S_1, S_2, S_3 etc. the total absorption A_T is obtained by:

$$A_T = \alpha_1 S_1 + \alpha_2 S_2 + \alpha_3 S_3 \qquad \ldots\ldots (2)$$

There are different types of absorber: porous absorbers where the sound energy falling on them is dissipated by viscous losses within the pores of the material, and resonant absorbers. Resonant absorbers themselves are grouped into those on the Helmholtz resonator principle, an example of which is a perforated panel separated from a solid backing containing some porous absorbent, and membrane resonators, where a solid panel is spaced off a solid backing but by virtue of panel flexibility and its thickness (less than 20 mm) vibrates on the trapped layer of air. For Helmholtz resonators, the frequency at which maximum absorption takes place is a function of the thickness of the panel, the area of the perforations and the spacing behind the panel. For membrane resonators, the frequency at which maximum absorption occurs depends on the spacing panel to backing, the superficial weight of the panel and whether the panel-backing cavity has any porous absorbent within. Both types of resonant absorber are used for low-frequency absorption, and combination of the two types is possible.

Acoustic, Acoustical
The adjectives are almost interchangeable in use. Acoustic tends to apply in describing the basic property of the sense of hearing. Acoustical is used to describe devices or personnel in the field of acoustics.

Acoustics
The science of sound (from the Greek, akouo, to hear).

Acoustical treatment
The application of design principles in architectural acoustics to isolate noise or vibration and to correct acoustical faults in spaces by addition of absorption, reflectors or other devices.

Ambient noise
Background or general noise level characteristic of an area, often used in comparison with a specific noise source overlay. Sound pressure level in dBA exceeded for 90% of the time (L_{90}). Ambient noise control may be by use of rating curves—see text.

Assisted resonance
An electronic method of lengthening the reverberation time of a hall by means of a series of channels each consisting of a microphone mounted in a tuned resonator in the ceiling feeding on an amplifier connected to a loudspeaker also in the ceiling.

Clarity (Deutlichkeit)
Defined as 10 log (energy over first 80 milliseconds expressed as a fraction of the remaining energy). The degree of clarity in sound reception depends on the strong receipt of direct rather than reverberant sound.

Coincidence
This occurs when the wavelength of the incident sound wave projected onto a partition or panel matches the bending wavelength of the partition or panel.

Critical frequency
Lowest frequency when coincidence occurs. Critical frequency is raised for thinner and less stiff surfaces in the sound path.

Decibel
Sound pressure level is the usual characteristic expressed in decibels. The scale is based on 10 times $\log_{10}$ of the relative intensity of a sound and a reference pressure. 'dBA' is a decibel unit with an 'A' weighting applied. This weighting has reduced response to extremes of frequency in order to simulate the response of the human ear.

Definition
Ratio of useful energy arriving within the first 50 milliseconds to total energy comprising signal; high figure means good definition.

Diffraction
Ability of a sound to pass round a screen or barrier. Lower frequency sounds can diffract around obstacles more easily because of their longer wavelength.

Diffusion
Complexity of reflecting surfaces causing an even dispersion of sound in a room, with no directionality of sound waves.

Dryness
A characteristic of an auditorium which has a short reverberation time; to some extent the opposite of 'richness'.

Echo
Reflected sound discernible as separate from the initial sound, by virtue of the long reflected sound path.

Flanking
Ability of acoustic energy to by-pass a sound barrier at the edges. Good airborne sound insulation through a floor construction, for example, may be flanked by sound transmission down the walls or through ducts.

Flutter
Rapid echo pattern between parallel walls which can be discerned.

Focusing
Acoustic energy, like light energy, can be reflected from concave surfaces into a concentrated focus. This may leave 'dead spots' elsewhere.

Frequency
The number of cycles per second that a vibrating system completes. Units of frequency are cps. or more usually Hertz; both have identical value. The audibility of a sound depends on its level and also on its frequency. The human ear can detect sounds with frequencies ranging from 20 to 20,000 Hz although increasing age reduces the upper limit. The frequency, f, velocity, C, and the wavelength, λ, of the sound waves are related by the expression:

$$C = f\lambda \qquad \ldots\ldots (3)$$

Intimacy
Impression in sound quality, sometimes called 'presence', that the source is near. In an auditorium this can occur in surprisingly distant seat locations if they benefit from close reflecting surfaces.

L_{10}
Noise level in dBA which is exceeded for 10% of the time.

L_{90}
Noise level in dBA which is exceeded for 90% of the time (background or lower limit noise condition).

L_{eq} (equivalent continuous sound level)
A-weighted energy mean averaged over the measurement period, ie the continuous steady noise level which would have the same total A-weighted acoustic energy as the real fluctuating noise measured over the same time.

Loudness
The subjective judgement of sound intensity by an individual which tends to be influenced by sound pressure and frequency. A typical response is for a three-fold change of sound pressure level to be considered a doubling of loudness.

Loudness level
The loudness level in phons of a noise is defined as the sound pressure level in decibels of a 1000 Hz tone which sounds equal in loudness to the sound which is being rated.

Masking
The effect whereby the threshold of audibility of a sound is raised by the presence of another sound. Masking is most effective when the masking sound is of lower frequency than the sound to be masked.

Noise
Sound unwanted by the recipient.

Noise reduction coefficient (NRC)
A single value to express absorption coefficients averaged over the octave bands centred on 250, 500, 1K and 2K Hz.

Normalized level difference
Difference in dB between energy levels in rooms on opposite sides of a partition corrected for a standard amount of absorption representative of normal furnished conditions in the receiving room.

Normalized level difference, $Dn = L_1 - L_2 + 10 \text{ Log } RT - 10 \text{ Log } .5dB$ where RT = measured reverberation time in the receiving room during tests.

Reflection
Sound energy returned after impact on a surface, rather than being absorbed as heat energy within the surface.

Resonance
The natural vibration of an area of material at a particular frequency as a result of excitation by a sound at that frequency.

Reverberation
The effect whereby a sound builds up in a space or at a point in a space because of multiple reflections from surrounding enclosing walls, floors and ceiling. This may enhance the sound if constant and the sound will gradually die away after the sound source ceases.

Reverberation time
The reverberation time of a room is the time taken for a steady sound when switched off to die away to inaudibility. More exactly, it is the time taken for the sound pressure level to fall by 60 dB. The reverberation time, T, is related to the volume of the room and the total absorption of the room by:

$$T = \frac{0.16V}{A} \text{ (seconds)} \qquad \ldots\ldots (4)$$

where V = volume in m^3, A = absorption in m^2 units.

Richness
A property of sound in an auditorium where there are many repetitions and reflections within a short period. Said to occur higher up in an auditorium because the sound is arriving up from more surfaces.

Room-to-room transmission
When a sound is transmitted from a reverberant room through a partition with sound reduction index (SRI) and Area S, into a room of absorption A, then the sound pressure level in the second room is:

$$SPL_2 = SPL_1 - SRI + 10 \text{ Log} \frac{S}{A} \text{ (dB)} \qquad \ldots\ldots (5)$$

SPL_1 in this expression is the average sound pressure level in the first room. If the sound power level is known for the first room, then SPL_1 can be found from the equation:

$$SPL_1 = SWL - 10 \text{ Log } A + 6 \text{ (dB)} \qquad \ldots\ldots (6)$$

If instead of a second room there is just open air, then the difference in sound pressure levels between one side of the partition and the other is:

$$SPL_1 - SPL_2 = SRI + 6 \text{ (dB) (dB)} \qquad \ldots\ldots (7)$$

If SPL_1 near the inside of the wall of a building is known, then SPL_2 at some distance from the building is found from:

$$SPL_2 = SPL_1 - SRI + 10 \text{ Log } S - 20 \text{ Log } r - 14 \text{ (dB)} \qquad \ldots\ldots (8)$$

Sensitivity
The ear's sensitivity varies with frequency. It is much less sensitive at low rather than high frequencies, reaching a maximum at or around 1000 Hz. The sound pressure levels of sounds which are perceived as of equal loudness are referred to as equal loudness contours.

Simple sound source
A sound source radiating sound equally in all directions is called a simple source. The sound pressure level at a distance r from the source is given by:

$$SPL = SWL + 10 \text{ Log} \left(\frac{1}{4 \pi r^2}\right) \qquad \ldots\ldots (9)$$

$$\text{or } SPL = SWL - 20 \text{ Log } r - 11 \text{ (dB)} \qquad \ldots\ldots (10)$$

This is modified by different constraints which depend on the directivity of the source. As an example, if the sound source is close to the ground the sound radiates into a hemisphere and then 3 dB must be added to the SPL:

$$SPL = SWL - 20 \text{ Log } r - 8 \text{ (dB)} \qquad \ldots\ldots (11)$$

6 dB would be added if the source were situated in the angle between two walls. If the sound source is in a reverberant room rather than outdoors, the average reverberant level will be:

$$SPL = SWL - 10 \text{ Log } A + 6 \text{ (dB)} \qquad \ldots\ldots (12)$$

The total sound pressure level resulting from the combination of reverberant and direct sound fields is given by:

$$SPL = SWL + 10 \text{ Log} \left(\frac{1}{4 \pi r^2} + \frac{4}{A}\right) \qquad \ldots\ldots (13)$$

Sound
Sensation of hearing wave motion in an elastic medium.

Sound insulation
The fraction of incident sound energy falling per second on a partition which is transmitted through it is called the transmission coefficient τ and depends on mechanical and material properties of that partition. The performance of a partition is given in terms of an index known as the sound reduction index (SRI):

$$SRI = 10 \text{ Log} \left(\frac{1}{\tau}\right) \text{ (dB)} \qquad \ldots\ldots (14)$$

The SRI varies with frequency and is measured in a laboratory in either octave or one-third octave bands.

Sound power level
When a source of sound radiates sound waves in all directions, the total energy radiated in one second is called the sound power of the source and is measured in watts. The decibel range is again used to define sound power level (SWL) as follows:

$$SWL = 10 \text{ Log} \left(\frac{W}{Wo}\right) \text{ (dB)} \qquad \ldots\ldots (15)$$

W is the sound power of the source and Wo is the internationally recognized reference power 10^{-12} watts.

Sound pressure level
The ear responds to the average value of pressure in the wave or root-mean-square pressure. The smallest detectable pressure is 2×10^{-5} N m^{-2} or 2×10^{-5} Pa (Pascals) and the largest tolerable pressure is 20 Pa, referred to as the threshold of pain. The ear is approximately a logarithmic detector of pressure and so it is convenient to express sound pressure in decibels, and the sound pressure level (SPL) is defined as:

$$SPL = 20 \text{ Log} \left(\frac{P}{Po}\right) \text{ (dB)} \qquad \ldots\ldots (16)$$

Po is 2×10^{-5} Nm^{-2} and P is the sound pressure of interest.

Sound reduction index
Difference in dB measured between the amount of energy flowing towards the wall in the source room and the total amount of energy entering the receiving room (usual range 100–3150 Hz).

$SRI = L_1 - L_2 + 10 \text{ Log } S - 10 \text{ Log } A$ (dB) (for negligible flanking)

Where S = common area of partition in m^2, or more conveniently:

$$SRI = L_1 - L_2 + 10 \text{ Log} \frac{ST}{0.16V} \text{ (dB)}$$

Sound spectrum
Sounds can be analysed to reveal their frequency content. This can be achieved by dividing the frequencies into octave or one-third octave bands and the sound pressure levels measured in those bands.

Sound waves
Sound waves are longitudinal pressure waves usually propagated through the air but also through solid material. In air they travel with a speed of 343 ms^{-1} at 20°C.

Structure-borne sound
Sound energy that has passed through the solid elements of the building structure.

5 APPENDIX C

TABLES

Typical sound reduction index values for elements of structure (20–50m^2 size) **(dB)**

	Octave 125	Band 250	Centre 500	Frequency (Hz) 1K	2K	Average (100–3150 Hz)
Sheet materials						
3 mm lead sheet	30	31	27	38	44	
20 g galvanized steel sheet	8	14	20	26	32	
18 g corrugated steel sheets	30	20	22	30	28	
Plasterboard on studding	15	20	25	29	32	
4 mm single glazing	20	21	24	27	26	26
6 mm single glazing	23	25	28	32	25	27
12 mm single glazing	28	30	33	29	32	31
12 mm laminated	30	32	36	36	37	36
25 mm single glazing	27	31	30	33	43	
6–12–6 glass/airgap/glass	21	25	29	34	29	28
6–150–4 lined reveals	22	28	36	43	47	35
4–200–4 lined reveals	29	35	43	46	45	39
6–188–6 lined reveals, separate frame	33	39	42	48	50	
Walls						
110 brick, 13 plaster each side	33	36	41	51	57	45
230 brick, 13 plaster each side	41	45	49	56	62	50
280 cavity wall, plastered brick with ties	38	42	51	59	63	52
150 dense concrete	33	39	44	52	60	47
100 lightweight blockwork	27	31	39	45	53	
12 plasterboard either side of 50 × 100 studding	16	22	28	38	50	
9.5 plasterboard either side of 50 × 100 studding	25	28	34	47	39	
Floors						
T & G boards or chipboard deck to joists, plasterboard and skim soffit	18	29	37	40	44	35
As above with glass-wool mat under deck, 50 mm sand to lath and plaster ceiling	37	42	47	52	60	49
125 RC concrete plus 50 floor screed	35	37	42	49	58	45
As above, screed reinforced and on glass-wool mat	38	43	48	54	60	49
200 RC concrete + screed	38	45	47	52	60	48
Doors						
Flush, hollow-core doors, normal edge cracks	12	13	14	16	18	
Solid, normal edge cracks	17	21	26	29	31	
Metal skinned acoustic door in heavy metal frame, seals and special furniture	36	39	44	49	54	

Sources Parkin Humphreys and Cowell/Woods

Co-efficients of sound absorption (m^2 units)

	125 Hz	500 Hz	2000 Hz
Slate or tiled roof with boarded soffit	.15	.1	.1
Match-boarding 19 mm over air space to solid wall	.3	.1	.1
Brickwork, open texture unpainted	.02	.03	.04
dense painted	.01	.02	.03
painted 2 coats	.01	.02	.03
Lightweight concrete block unpainted	.2	.6	.5
Carpet, thin, on thin felt to solid floor	.1	.5	.6
thin on thin felt on boards and joists	.2	.3	.3
Concrete, rough finish	.02	.02	.05
smooth	.01	.02	.02
Cork tiles (thin), PVC, linoleum, woodblock floor finishes	.05	.05	.1
Curtains, medium-weave density, close to walls	.05	.25	.3
medium-weave density, hung in folds 300 mm off	.1	.4	.5
wall dividing spaces	.03	.1	.2
Fibreboard, 12, on battens, unpainted	.3	.3	.3
Hard floor tiles, 'Compo' floor	.03	.03	.05
Glass, windows with 4 mm glass	.3	.1	.05
windows with 6 mm plate glass	.1	.04	.02
Polished marble wall finish	.01	.01	.02
Glass fibre, 25 mm, resin bonded	.1	.55	.75
50 mm, resin bonded	.2	.7	.75
Glass fibre or mineral wool 25 mm over air space	.4	.8	.9
Granolithic cement screed	.01	.02	.02
Metal 3% perforated panels, 25 mm fibreglass in air space over	.2	.9	.25
10% perforated 150 fibreglass in air space	.2	.75	.7

	125 Hz	500 Hz	2000 Hz
Openings, ventilation grilles 50% void	.15	.35	.4
Plaster, lime or gypsum on solid backing	.03	.02	.04
lime or gypsum; large air space	.2	.1	.04
13 acoustic plaster on solid backing	.13	.35	.45
Plasterboard, on joists, large air space	.2	.1	.04
Plywood panels on studs to solid backing, mineral wool in void	.4	.15	.1
Water, as in swimming pool	.01	.01	.02
Wood boarding, 25 boards on joists	.1	.1	.05
Wood match-boarding 19 on battens to solid wall	.3	.1	.1
Wood panels 15% perforated panels on battens, 50 fibreglass, air space	.33	.8	.4
Wood strips, 38 strips spaced to give 5 slots, 25 mineral wool	.25	.65	.65
75 woodwool slabs	.2	.8	.8
Sound absorption factors			
Air, average temperature and humidity, per m^3	—	—	.0066
Audience, close spacing, per person	.21	.46	.51
Seating, wood, per chair	.01	.02	.02
upholstered, perforated under seat, per seat	.19	.28	.28

Note: Occupation of seats partially discounts seating absorption. Similarly, seating and audience partially mask floor finishes under.
Sound absorption factors are approximate and should be used with extreme caution.

6 GUIDE TO FURTHER READING

American Society for Testing and Materials
Annual Book of ASTM Standards, Part 18
US 1982

BBC Engineering
Guide to Acoustic Practice
1980

Brick Development Association
Low Rise Domestic Construction in Brick: Part 3 Thermal and Sound Insulation
1979

BRE/BRS Digests
103 **Sound Insulation of Traditional Buildings**
128, 129 **Insulation Against External Noise**
143 **Sound Insulation: Basic Principles**
184 **Demolition and Construction Noise**
226 **Thermal, Visual and Acoustic Requirements in Buildings**
266 **Sound Insulation of Party Floors**
5/81 **Field Measurements of the Sound Insulation of Timber Joist Party Floors**
8/81 **Field Measurement of the Sound Insulation of Floors with Floating Screeds and Hollow Precast Bases**

British Gypsum
White Book
4th Edition 1981

Construction Industry Research and Information Association (CIRIA)
Report 1988: Sound Insulation of House Separating and External Walls
1981
(R H Clough and R D Jones)

Cremer, Muller, Schultz
Principles and Applications of Room Acoustics
vols 1 and 2
Applied Science Publishers Ltd
1982

Department of the Environment
Planning and Noise
Circular 16/73, 1973

Department of the Environment
Sound Insulation in Buildings: Cost and Performance of Walls and Floors
1971
(H R Humphrey and D J Melluish)

Fader, Bruce
Industrial Noise Control
Wiley-Interscience US 1981

Furrer and Lauber
Raum und Bauakustik Lärm Abwehr
Birkhauser-Verlag
Basel and Stuttgart 1972

K B Ginn
Architectural Acoustics
Brüel and Kjaer
1978

Herman Miller Research Corporation
The Action Office Acoustic Handbook

Heinrich Kuttruff
Room Acoustics
Applied Science Publishers Ltd
2nd Edition 1979

Lord, Gatley and Evenson
Noise Control for Engineers
McGraw-Hill, US 1980

J E Moore
Design for Good Acoustics and Noise Control
The MacMillan Press Limited
1978

K A Mulholland and K Attenborough
Noise Assessment and Control
Construction Press
1981

P H Parkin, H R Humphreys and J R Cowell
Acoustics, Noise and Buildings
Faber and Faber
4th Edition 1979

Ian Sharland
Woods Practical Guide to Noise Control
Woods of Colchester Limited
1972

Theodore J Schultz
Community Noise Rating
Applied Science Publishers Ltd
2nd Edition 1982

Sound Research Laboratories
Practical Building Acoustics
SRL Limited
1976

Lyle F Yerges
Sound, Noise and Vibration Control
2nd Edition 1978

7 INDEX